P9-APG-491

29.95
754

BEYOND SUCCESS

THE RUFFIN SERIES IN BUSINESS ETHICS
R. Edward Freeman, *Editor*

THOMAS DONALDSON
The Ethics of International Business

JAMES W. KUHN AND DONALD W. SHRIVER, JR.
Beyond Success: Corporations and Their Critics in the 1990s

R. EDWARD FREEMAN, EDITOR
Business Ethics: The State of the Art

FORTHCOMING TITLES TO BE ANNOUNCED

Beyond Success

CORPORATIONS
AND THEIR CRITICS
IN THE 1990s

JAMES W. KUHN
Columbia University
Graduate School of Business

DONALD W. SHRIVER, JR.
Union Theological Seminary
in the City of New York

HF
5387
.K83
1991
Arab

New York   Oxford
OXFORD UNIVERSITY PRESS
1991

The United Library
Garrett-Evangelical/Seabury-Western Seminaries
2121 Sheridan Road
Evanston, IL 60201

Oxford University Press

Oxford   New York   Toronto
Delhi   Bombay   Calcutta   Madras   Karachi
Petaling Jaya   Singapore   Hong Kong   Tokyo
Nairobi   Dar es Salaam   Cape Town
Melbourne   Auckland

and associated companies in
Berlin   Ibadan

Copyright © 1991 by Oxford University Press, Inc.

Published by Oxford University Press, Inc.,
200 Madison Avenue, New York, New York 10016

Oxford is a registered trademark of Oxford University Press

All rights reserved. No part of this publication may be reproduced,
stored in a retrieval system, or transmitted, in any form or by any means,
electronic, mechanical, photocopying, recording, or otherwise,
without the prior permission of Oxford University Press.

Library of Congress Cataloging-in-Publication Data
Kuhn, James W.
Beyond success : corporations and their critics in the 1990s /
James W. Kuhn and Donald W. Shriver.
p.   cm.   — (The Ruffin series in business ethics)
ISBN 0-19-506433-X
1. Business ethics.   2. Industry—Social aspects.
I. Shriver, Donald W.   II. Title.   III. Series.
HF5387.K84   1991
174'.4—dc20   90-7084   CIP

9 8 7 6 5 4 3 2 1

Printed in the United States of America
on acid-free paper

Dedicated to
Our Children
and Theirs

PREFACE

This study developed out of the two authors' eight-year experience as co-teachers in an academic course designed to promote dialogue between future business managers and future theologians. We know from this experience that those two groups are, initially, more likely to see each other as antagonists than as partners. The theme of our study is that there are partnerships to be formed in the modern world of global economics, partnerships that our professional training may have poorly prepared us to consider, let alone explore. This book will serve its purpose if it helps any reader to undertake both a careful consideration and an exploration of the worth of partnership with other groups, institutions, and organizations, either inside or outside the traditionally defined business firm.

In a broad sense the study concerns the relation of economics and ethics: but here we have examined and written about the ideas with a critically important awareness that the business manager, like any organizational leader, does not have the luxury of dealing with ethics, economics, organizational principles, laws, customs, and religions as so many separate boxes. Human action requires an integration of these and other realms of awareness. Increasingly, in business, as in the rest of modern society, "all things considered" means *many* things indeed. This study focuses on one major contemporary source of concerns clamoring loudly for the attention of business managers: the "new constituencies" of the corporate responsibility movement.

Reduced to its simplest statement, the thesis of the study is that both managers and the new constituencies have much to learn from each other. As scholars we have tried to overhear the conversation already begun between the two, and we want to offer our interpretations of what they are saying. Our aim is to contribute to and encourage their dialogue by underscoring its importance and delineating its economic and ethical content, and also by pointing out some directions in which it might yet go.

A study written at the intersection of disciplines, professions, and institutions of our society may have too much philosophy to suit managers, too much economics to suit philosophers, and too little specific

problem-solving advice for people who meet payrolls, engage in boycotts, and fight the injustices of our new global economic system. It will
probably annoy capitalists, who chafe at the constraints that hem in
"free" competition in the market, and irritate socialists, who have yet
to see anything democratic in the market-idea. We are particularly interested in how a market economy can become—slowly is becoming—
an instrument for the expression of certain democratic and ethical values. We believe that the new constituencies, who vigorously press their
criticism of business policy, decisions, and activities, are often new actors in the market whose voices often and authentically express those
values. We believe they are worth listening to, partly because, as colleagues, we have found each other worth listening to, but most of all
because all human beings are equally worthy.

We have discussed with many of our colleagues the notions and ideas
presented here. In particular, we are grateful for the many contributions of the late Paul McNulty, of the Graduate School of Business,
Columbia University. We thank Roger Shinn, of Union Theological
Seminary, and Robert Jackall, an early member of our joint seminar,
who has given us encouragement, insight, and helpful criticism. Most
of all, we thank the many students who, over the years, have taught us
much more than we were ever able to contribute to them.

New York   J.W.K.
April, 1990   D.W.S.

CONTENTS

BEYOND SUCCESS

1

Managers and New Corporate Constituencies: Ethics in Business Tomorrow

Over the past decade, and with increasing frequency, public commentators have accused business managers of unethical doings and an alarming lack of morality in the conduct of their firms' affairs. The business press, regular newspapers, and television news programs regularly report the latest accounts of low standards of behavior in high corporate offices. Managerial response has been varied, but usually has been of three kinds.

First, a few managers have maintained that business behavior is little different from what it has long been, except that now it is publicized to an unprecedented degree. Some have argued that, although in fact ethical business practices are more prevalent now than in the past, the public has been raising its expectations faster than practices have changed, thereby producing the illusion of a worsening ethical climate. Since there are almost no data on either past or present unethical behavior among businesses, there is little that can be offered as evidence for or against this argument. "White-collar crime," as illegal (and unethical) business practices are called, has not been a subject of much scholarly study,[1] nor is it an aspect of business on which government collects much information.[2]

A second managerial response to business's more highly advertised misdeeds is the argument that little can be done, except to prosecute lawbreakers and enforce government regulations. If those in business have not learned by the age of ten the differences between right and wrong, they are not apt to learn it as employees and managers.[3] Implicit in such a response is the notion that ethical behavior is largely a matter of individual choices in a business setting and furthermore, that those choices are relatively simple and clear-cut. This argument can be

3

dismissed as entirely simpleminded, grossly underrating the complexities of modern business activities. Such complexities involve the problems and opportunities of a global economy. They confront business managers with dilemmas, difficult trade-offs, and paradoxical situations. Both business managers and the rest of us need far more systematic information, more meticulous expert study and painstaking research than has yet been undertaken if the problems of ethical behavior are to be more fully understood.[4]

The third response is that business managers had better assume that they can influence the ethical behavior of those within their organizations. The adoption of codes of behavior and standards of ethics by most large business firms suggests that they believe such action is at least necessary as a public relations ploy. In companies like IBM and Johnson & Johnson, to name two of many, there is good reason to think that the company codes serve more substantial purposes. The codes help both shape and reflect a lively, continuing concern by management that ethical behavior be explicitly considered in all aspects of the business. More recent efforts through training sessions, role playing, and frequent discussions between both top and lower level employees of emerging ethical problems indicate that more and more managers have decided ethics can be learned at the place of work just as in the family and home.

Recent public and managerial interest in ethics does not mean that business is only now involving itself in ethical behavior and showing concern for the values by which it is judged. All significant activities in which human beings are involved carry a moral—an ethical—charge. If economists and managers insist that the only purpose of a business firm is to maximize shareholder wealth, then they declare an ethical standard. Thus, teachers and students in business school, in all their various courses—finance, marketing, production operations, accounting, planning, or human resource management—implicitly discuss ethical issues, and accept or reject them as guides to action. Daily, business managers are involved in ethical or unethical behavior as they make decisions, enforce policy, negotiate, bargain, plan, scheme, and monitor. Those affected by their activities will certainly judge managers as good or bad, in varying degrees, depending upon the standard by which they measure them. Their standards may not be the same as those followed by the managers, of course, and therein lies the subject of this study.

Usually both business managers and the scholars, particularly the economists, who follow their doings carefully treat the ethical dimensions of business, the market, and production implicitly rather than explicitly. In a pluralistic society infused with secular professionalism, where positivist philosophy reigns, people of affairs have long been wary about discussing ethics openly. The result is that both business managers and economists often lack not only a vocabulary of ethics but also any theory to guide them in analyzing the ethical side of their professional lives.[5] They have preferred to hide from themselves the values that inform their business judgments, economic theories, and the tacit assumptions on which they rest their analyses.

BUSINESS'S AMBIGUOUS ETHICAL POSITION IN SOCIETY

Nonmanagerial working people and their religious prophets, philosopher guides, and political governors treat ethics more explicitly than business managers. Those engaged in commerce, trade, finance, and business are viewed with suspicion and are perceived to stand in a morally ambiguous position. Consider the ambivalence with which our forebears endowed commerce and trade. An ancient symbol of both was the Hermes Four, used by the great trading companies and banking houses of Italy during medieval times as the mark of their trade. The sign of Hermes (or Mercury) was appropriate, for the Greeks considered him the god of commerce. But he was also the god of thieves! The Hebrew tradition, quite as old, endowed commerce with a similar ambiguity. The founder of cities and the first trader was Cain—also the first murderer!

True, traders, financiers, and business people helped provide useful fabricated goods and provided needed, sophisticated services, but they also appeared to enrich themselves with amazing ease and speed, especially when compared to the efforts required by most of humankind, through most of history—and unto this day—in winning their livelihood through farming or herding.

Farmers and pastoralists have always had to wrestle with Nature or with nature's God to win their living, and by the sweat of their brows sow, tend, and harvest their produce. They have had to contend with variable soils, irregular rain, uncertain sun, and unexplained plagues. In contrast, those in commerce, with little apparent physical labor and some insulation from the vagaries of nature, won for themselves, on average, more comfortable lives, and sometimes fabulous wealth.

Farmers and herders wrested their living with their own hands and labor. They served their neighbors by selling their surplus when nature was bounteous; if they did not enjoy surpluses they gained no profit. Those involved in trade, commerce, and business employed others in their service, and all too easily could reap rewards and profits by exploiting their employees; and while they could and often did serve their neighbors well, they could also profit by *not* serving. In the view of ancient societies, withholding grain from the market to raise prices was extortion. So too was charging high prices for grain during a famine, and adulterating olive oil to increase profits was hardly serving one's neighbors. It is not surprising that over millennia whole cultures and societies came to view business, trading, finance, and commerce as being imbued with a certain invidiousness. That view persists, for the chances of exploitation and profit by *not* serving still offer themselves regularly to those in business.[6] It is a widely held public view that business managers have to accept as a given: that view will be changed, gradually, only as business leaders convince those whose lives they affect, that they seldom, if ever, exploit and that they almost always earn their reward through service.

MANAGERS' GREAT FUNCTIONS

In an address at the inaugural of Harvard's Graduate School of Business Administration in 1929, Alfred North Whitehead asserted that "A great society is a society in which its men of business think greatly of their functions."[7] He did not precisely explain what those functions might be, but it seems unlikely that among them were maximization of profits or increasing the bottom line. While producing profits and "making money" may be a necessary, even vital, managerial function, few people would describe either as a "great" function. The great functions performed by business managers, more likely, are those of *service*—serving the varied needs and diverse demands of people, helping them to sustain their lives and increase their enjoyment by providing goods and services. These are great, even noble, functions.

But who is to be served, why, and how? Adam Smith assumed that *who* was consumers. They were to be provided with those things they could effectively demand:

> Consumption is the sole end and purpose of all production; and the interest of the producer ought to be attended to, only so far as it may be necessary for promoting that of the consumer. The maxim is so perfectly self-evident, that it would be absurd to attempt to prove it.[8]

It is an assumption that mainline modern economic theory also makes; indeed, economic efficiency is defined as using resources in such a way as to provide optimum consumer satisfaction, as expressed by individual choices, within the limits of spending budgets. A market in which consumers (as all or the largest portion of the community) are served, allows division—and thus specialization—of labor and encourages productive cooperation.

Even two centuries ago, it was obvious to Smith that individual effort was not sufficient to provide a viable society or an abundant, productive economy; human beings were special creatures who needed cooperative relationships to survive.

> In civilized society [man]* . . . stands at all times in need of the co-operation and assistance of great multitudes, while his whole life is scarce sufficient to gain the friendship of a few persons. In almost every other race of animals each individual, when it is grown up to maturity, is entirely independent, and in its natural state has occasion for the assistance of no other living creature. But man has almost constant occasion for the help of his brethren . . .[9]

How was the necessary cooperation to be achieved? Smith offered an answer arising from a realistic, if pessimistic and perhaps Presbyterian, understanding of human shortcomings. It is in vain for people to expect help out of the benevolence of others:

> He will be more *likely* to prevail if he can interest their self-love in his favour, and shew them that it is for their own advantage to do for him what he requires of them. Whoever offers to another a bargain of any kind, proposes to do this. Give me that which I want, and you shall have this which you want, is the meaning of every such offer; and it is in this manner that we obtain from one another the far greater part of those good offices which we stand in need of.

What he first proposes as a likelihood, Smith then offers as an absolute declaration about human nature, which is widely quoted among economists and accepted by many business managers as an apt description of economic motivation:

> It is not from the benevolence of the butcher, the brewer, or the baker, that we expect our dinner, but from their regard to their own interest. We address ourselves, not to their humanity but to their self-love, and *never* talk to them of our own necessities but of their advantages.[10]

Modern economists and business leaders, in their fondness for this quotation, are guilty of a notorious failure to note the moral theory that peeps out in Smith's reference to humanity. Seventeen years before the publication of *The Wealth of Nations,* he had spelled out his notion of normative humanity in *The Theory of Moral Sentiments.* There he made clear what sort of self a human being should be interested in, a self quite different from the narrow, economic self that Karl Marx and many a modern market economist thought they discovered in the later classic.

Human selves, for Smith, were multidimensional. Economic striving

---

*In this book we endeavor throughout to use gender-neutral language. In quotations, however, we reproduce the text as originally written.

was an expression of the virtue of prudence, or knowledge of those activities that are necessary to a person's very survival. But alongside prudence were the equally human virtues of justice and benevolence. Justice was rooted in the equal worth that a good person and a good society accords to every person. Benevolence was one's active entry into the feelings, the needs, and the situations of other persons. In the practice of all three virtues, people demonstrate that the only "self" is social. Like Aristotle, Smith saw people as social animals. The active practice of the virtues results in their mutual reinforcement. Our interest in physical survival impels us to cooperate with one another. Our sense of justice persuades us that the survival of others is as justified as our own, and our capacity for benevolence enables us to "enter into or identify ourselves with the joys and sorrows of others." Balance between these three makes one truly human. Such balance can only be achieved "by imagination and genuine feeling for others."[11]

> And hence it is, that to feel much for others, and little for ourselves, that to restrain our selfish, and to indulge our benevolent, affections, constitutes the perfection of human nature; and can alone produce among mankind that harmony of sentiments and passions in which consists their whole grace and propriety. As to love our neighbor as we love ourselves is the great law of Christianity, so it is the great precept of nature to love ourselves only as we love our neighbor, or, what comes to the same thing, as our neighbor is capable of loving us.[12]

One wonders what would have been the course of Adam Smith's reputation in modern economic theory if this quotation were regularly put alongside the famous one about "the benevolence of the butcher." Obviously Smith's human being is a far more complex moral creature than the "salesman" that Karl Marx found portrayed in *The Wealth of Nations,* just as Marx's human being is far more morally complex than the "materialist" that his capitalist critics have found in him. For Smith, no aspect of human nature was simple. He was well aware, for example, that human wants were many, unpredictable, and changeable over time. We "want new wants which will only reveal themselves in exploratory activity," that is, in time and history. In sum, for Smith, "answers to economic questions thus depend on judgments and insights into human nature."[13]

We will not hesitate in this book to revert to this general rule. One might simply characterize the book as an attempt to redress the imbalance between prudence, justice, and love as "moral facts" relevant to human behavior in the economic world. Along with Smith, we are interested most in the interchanges, the conflicts, and the complementarities among these virtues—especially in an era when economic questions, these 200 years after *The Wealth of Nations,* involve a whole Earth and the whole of humankind. Late in *The Theory of Moral Sentiments* Smith writes that

> . . . the wise and virtuous man is at all times willing that his own private interest should be sacrificed to the public interest of his own particular order or society.

The just laws of a particular society may require it, and the voluntary benevolence of individuals as well. The widening circle of larger interests encompassing lesser has no foreseeable boundary; indeed, if we knew "all the connections and dependencies of things," Smith speculates, we would promote the "prosperity of the universe" at the expense of many a costly "misfortune" for oneself, one's friends, and one's country. Every human life and society may have at some point to take the part of a "good soldier," willing to die for the interests of others.[14]

Since he wrote these words, many a scholar has risen up to mock the notion that justice or benevolence has anything to do with how an economy works or ought to work. The view of this book is that neither Adam Smith nor economic history since his time justifies this mockery.

MANAGERIAL USE OF BENEVOLENCE
AND ECONOMIC SELF-INTEREST

In 1954 Professor Dennis Robertson, great English economist and contemporary (and rival) of John Maynard Keynes, explored the implication of Smith's conclusion that market participants address *only* economic self-interest when dealing with each other. In a witty essay delivered on the occasion of Columbia University's bicentennial, he posed the question, "What Does The Economist Economize?"[15] His answer, *love*, may surprise the reader or it may confirm a suspicion that economists have slight regard for any but selfish motivations. He used the word "love" as, we believe, a dramatic, summary term encompassing altruism, loyalty, trustworthiness, civic spirit, and additional other-regarding ethical behaviors.

In approving the economist's attempt to economize love, Robertson presumed what is probably obvious, that love (and the other values he subsumed under it) is a scarce resource. As an economic good, managers and others in the market do not want to waste it, or rely too much upon its presence. Further, since it is even scarcer than economic self-interest, love should not be used where self-interest will accomplish the same efficient end. By not discussing ways in which the supply of love might be increased, he implicitly suggests that it is not only limited but perhaps also fixed and therefore as a matter of practicality not available to market participants.

In support of his notion that economy theory details the ways in which a competitive market economizes love, Robertson refers to the declaration of his respected teacher and mentor, Alfred Marshall. Marshall, as influential an economist in his day as Paul Samuelson has been in his, wrote, "progress chiefly depends on the extent to which the *strongest* and not merely the *highest* forces of human nature can be utilized for the increase of social good."[16] Robertson may have read this statement too broadly. "Chiefly" does not mean "exclusively," and since Marshall admitted the possible use of the "highest forces of human nature," we may presume that he believed they played some role in market motivations. Robertson, like Smith, skipped too easily from "likelihood" that economic self-interest will prevail to the conclusion that in a market economy one can appeal *only* to a certain narrow definition of interest.

Business managers, learning economic theory and the language of the market from the texts that repeat the abstracted and simplified conclusions of such men as Smith, Marshall, and Samuelson, explain their own values and organizational activities in economic terminology. In effect, managers presume themselves to be economizers of "love." They are "realists" relying on economic self-interest to motivate those with whom they work. They pursue profits for themselves and their shareholders only as a selfish endeavor. In using the economists' language, though, managers ignore much of what they do, and even lose many of their own ways of thinking and their strategies for enlivening their organizations. These are problems that we will be exploring in later chapters. As we do so, we will move from certain descriptions of what business managers actually do, to prescribe more complex ethical standards for them than they actually verbalize.

Experience teaches almost everyone that we expect, use, and rely on varying degrees of benevolence or love—trust, altruism, duty, and ethics—in our various economic encounters, both in and outside the firm. Professor Amartya Sen, a philosopher-economist at Harvard points out that:

> . . . maximizing behavior does not entail maximization specifically of self-interest. The objective function can include other goals and commitments. I believe it is true that self-interest must be a major motivation among the various motives that we have. [But if prudential pursuit of self-interest] . . . excludes all other motivation and all other modes of behavior, we get a modelling of human beings that can scarcely accommodate rationality in general. . . . If people have other goals and motivations, why should they be compelled by economic theory to pursue self-interest?[17]

Most managers, we believe, find it obvious as a matter of experience that trust and loyalty among colleagues often have "bottom-line," efficiency consequences. Anyone who has dealt with a trustworthy business partner and concluded a deal with a handshake rather than a legal contract drawn with detail by lawyers, recognizes the savings involved. John Shad, former chair of the Securities and Exchange Commission, pointed out that every year some $40 trillion worth of securities are traded by brokers who make their deals over telephones, with very little fraud or cheating. If brokers could not rely upon trust, the efficiencies of the sales and trades would be much less than they are.[18]

Other possibilities come readily to mind. If one may presume the honesty and integrity among subordinates with whom one works, then monitoring, measuring, and auditing costs are lessened significantly. Being able to rely upon truth telling among one's colleagues, even when it entails self-revelation of faults or shortcomings, permits more organizational flexibility, easier identification of problems and their causes, and faster remedies than would otherwise be the case.

This experience may be more obvious and compelling than in Smith's day because the world of business has changed markedly. Would Smith today conclude that in the large business firms, managers *never* talk to those with whom they deal—customers, employees, suppliers, franchisers, community leaders—of the firm's necessities but only of the participant's advantages? We think not. True, employees will not work for little or no wage, suppliers expect at least the market price for their raw materials, and investors pursue optimum returns, but there are other ties that bind the parties together, and values other than economic self-interest that each pursues. In 1776, when Smith published *The Wealth of Nations,* the largest firms employed at the most hundreds of workers, and were managed by individual owner/operators. Relationships within the firm as well as with those outside were direct, immediate, and, by today's standards, simple. The elements of love—trust, honesty, loyalty, truth telling—that probably modulated the self-interested economic relationship of investors, suppliers, producers, employees, and consumers may easily have been assumed to be merely normal parts of personal encounters. In reaching generalized conclusions about the operation of the market and the motivations of those within it, the practiced values involved in personal, face-to-face encounters that went beyond mere economic self-interest were simply assumed.[19] Today, managers are not apt to presume such values in their relationships, nor to overlook them. They have the responsibility of overseeing complex organizations that employ thousands, tens of thousands, and sometimes hundreds of thousands of people. They operate plants and facilities in dozens or even hundreds of locations around the world. They directly affect the activities and lives of millions of people and indirectly influence multiples of those millions.

To restrict managers' functions *only* to the manipulation of people's Smithian economic self-interest does not serve modern managers very well. The quantitative differences between the world of the Smithian producer and that of today's managers are vast. Modern economic environments have created qualitative—and ethical—differences large enough to encourage all of us to reexamine the economy's provisions of the quality and ethics of economic life.

A MANAGERIAL ETHIC BEYOND
ECONOMIC SELF-INTEREST?

Will managers reexamine their customary explanation of necessary and exclusive reliance upon economic self-interest in the conduct of business? Many observers of the business scene, including economists and business-school teachers, are doubtful. David Vogel, an astute student of business, for example, maintains that managers are not apt substantially to include nonpecuniary values in their decisions and policies. "The most important decisions made by any firm are out of the control of those who govern it; they are dictated by the imperatives of a market economy."[20] He appears to agree with Walter A. Weisskopf that insofar as managers are imbued with the economists' version of the market creed, morality for them has become a matter of formal rationalism and subjective hedonism. Any subjective goal pursued in a rational, efficient, systematic fashion is assumed to be right.[21] Such views are those of market determinists. Managers do not create markets, but merely respond to them.

Other economists of the past did not perceive the market and the economic system to be as compelling as Vogel supposes. Joseph Schumpeter believed that managers, particularly entrepreneurial ones, displayed much more initiative than Vogel allows. Although traditional market pressures are often urgent, they do not require any particular response. Business managers are much like the rest of us who live within the strictures of the market in that they cope with, but also innovate within economic constraints. Coping strategies—many and varied—provide continuing options. They, like life itself, are more open and contingent than market apologists presume.

As detailed later in this study, business managers often do not enrich themselves at the expense of all the other groups with which they deal; they do not sacrifice all other interests to maximize profits. Herbert Simon and his colleague, James March, long ago called attention to managers' regular practice of "satisficing."[22] It is not uncommon for managers to take advantage of oligopolistic situations to reward themselves, but in the United States managers also commonly share many of the reaped rewards with their employees, suppliers, retailers, and local communities as well as shareholders. To label such sharing as simply the pursuit of economic self-interest obscures the meaning of "selfhood" more than it illuminates it. One may more satisfactorily admit that managers respond to a number of values, of which their own economic interest is but one. Identifying economic activities where managers in fact respond only to economic self-interest—what economists perceive as the only market value—will reveal pathologies. An historical example is the selling of human beings into slavery. A contemporary example might be dealing in illicit drugs. In such "businesses" violence replaces persuasion, trust is minimal, and position is maintained by killing.

Henry Clark, a professor of social ethics, asserts that not only does society articulate values different from those usually perceived in the market but that celebration of those values helps change managers and those with whom they are involved.[23] We argue in this study that both business managers and those with whom they deal through social and political, as well as market, arrangements respond to values other than those of economic self-interest. They follow "habits of the heart" common among Americans.[24] Without being able to articulate their thoughts well, managers may comprehend the need for a sea of ethical praxis in which the bark of competitive markets may float. Experience, in some unexamined and unanalyzed way, has taught many senior managers that trust, morality, loyalty, and altruism—presumably nonmarket values—provide the rules and boundaries that permit market competition, to be a contest and not an excuse for murder—a playing field, not a killing field![25]

Economists and business professors may too readily dismiss Reinhold Niebuhr's observation that justice will gradually drift into disorder unless it is constantly "pulled upward" by values other than economic self-interest. A rather-too-cynical "realism" will intensify the evils it claims to understand if it is not tempered by idealism. History, wrote Niebuhr, has

> . . . revealed that political encounters and debates in a free society involved not only contests of interest and power, but the rational engagement and enlargement of a native sympathy, a sense of justice, a residual moral integrity, and a sense of the common good in all classes of society.[26]

He was critically aware that in their collective sense of "common good" people can be *wrong* and murderously opposed to another group's sense of *its* common good. Hence arise the ironic wars of ideals against ideals. He was not alone in understanding that ordinary human society requires a certain ethical minimum for its ordinary transactions.

In a sweeping reference to the nature of economies, Fred Hirsch declared that "Elemental personal values of honesty, truthfulness, trust, restraint and obligation are all necessary inputs to an efficient (as well as pleasant) contractual society. . . ."[27] Albert Hirschman made the

same observation as Hirsch, pointing out that "if all these needed personal values are added up, the amounts of benevolence and morality
required for the functioning of the market turn out to be quite impressive!"[28]

Vogel and others, then, may take too seriously the public declarations of business managers that they pursue economic self-interested
profits and nothing else. Again, ironically, the profit-criterion of "success" in business becomes an ideal in these declarations, with all the
force of an ethical norm. "I'm in business to make money and put my
competitors out of business!" is one plucky way of announcing this norm.
But behind such an announcement usually hides some shy array of
other ideals. Only sheer genius allows economists, in their ordinary,
workaday opinions and writings, to ignore the ways in which Americans, including business managers, do *not* always pursue economic self-
interest; and only their faith in nineteenth century scientism and ready
acceptance of correlation as proof of cause (as in astrology) allow them
to accept neoclassical theory as reality rather than a simulacrum.[29]

INTERDEPENDENCE AS A BUSINESS REALITY

Interdependencies pervade business organizations. Organizations sustain relationships with a great variety of groups and interests formerly
defined as "outside" their boundaries. While the interest of a collective,
corporate "self" is present, active, and powerful, most of those involved
in any organization recognize that one is not apt to negotiate a satisfactory outcome unless one takes into account the needs and desires of
others. Professor Sen suggests that daily business life makes clear that
other-serving as well as economic self-serving may contribute to productive and allocative efficiency.

> We are not only competitive with each other; much of the time we can be
> primarily cooperative with each other. If workers in a factory, for example,
> were to pursue their narrowly perceived interests or goals, I don't think that
> you would get a very high achievement of productivity. Many systems flour-
> ish precisely because people have codes of conduct; there are certain things
> to be done . . . [they] may be guided also by the recognition of the strategic
> interdependences—the social instruments in the form of demands of fit and
> proper behavior. This involves practical recognition of the goals of others,
> of the enterprise, of their colleagues, and the fact of living in situations of
> social interdependence.[30]

Such interdependence means that modern business activity and market transactions hardly resemble the elementary illustrations used in
beginning economic courses. Economic self-interest may dominate when

two people negotiate a one-time sale, for example, of a sack of grain, when the seller is seeking the highest possible price and the buyer the lowest possible price. More often, those engaged in business find themselves buying and selling with each other, again and again, in a variety of different settings and situations. Each needs the other, being interdependent, and immediate economic self-interest hardly comprehends the intricacies of their relationship over time. In their interdependency, and given the many plays of the "business game," they find themselves in a prisoners' dilemma: If they cooperate and trust each other, they can, over time, improve the benefits and outcomes for both; if they distrust each other and neither cooperates, they both may end with low payoffs. If one party trusts, and therefore cooperates, while the other distrusts and "defects," the trusting party loses heavily, while the other party gains greatly. Once trust has been broken, however, it is not easily restored, and the benefits of mutual action and cooperation will not easily be realized.[31] Customers, once "cheated," take their business elsewhere if they can.

Leading economic theoreticians like Kenneth Arrow recognize how distantly abstracted from life are economic models that presume only individual, economic self-interested choice and decision.

> . . . all significant actions involve joint participation of many individuals. Even the apparently simplest act of individual decision involves the participation of a whole society. . . . Formally, a social action is then taken to be the resultant of all individual actions . . . as being factored into a sequence of individual action. I certainly do not wish to deny that such factoring takes place, but I do wish to emphasize that the partition of a social action into individual components, and the corresponding assignment of individual responsibility, is *not* a datum. Rather, the particular factor in any given context is itself the result of a social policy and therefore already the outcome of earlier and logically more primitive social values.[32]

The enduring truth of Marxism as a theory of society is that there are social structures, organizational power and cultural norms by which people shape human behavior. The enduring truth of democratic capitalism is that, given a chance, individuals will act and interact to change their society. Between these two abstract truths lies the more ordinary truth that social life arises from both agent-initiators and agent-responders. A regard for this truth will occupy later discussion about and analysis of "responsibility."

Insofar as economists assume individuality in taste, purpose, and behavior, when in fact people often respond, as well, to group norms, social obligations, and associational values, their policy recommendations and their analyses may often miss the managerial or professional mark. Attempting to explain all behavioral change through individual

response to price and income differences, not only ignores the influence of the social context, but as Albert Hirschman pointed out, ne-
glects one important source of change—autonomous, reflective change
in values.[33] Since managers continually strive to influence the behavior
of those both inside and outside their firms, they may well find it both
efficient and profitable to pay close attention to more than prices, wages,
and individual demands. As investors, producers, and marketers, they
are also community members, share many general social values, and
may find the autonomous, reflective changes in the values of those with
whom they deal at once both obstacles and advantages. They may experience, on occasion, a genuine change in their own values, which root
mysteriously in something profoundly personal and profoundly social
at the same time.

Managers may find in some, perhaps even many, situations that it is
easier, more efficient, and less expensive to attempt an increase in the
supply of "love" than to depend upon market processes driven by economic self-interest alone. Developing and sustaining ethical behavior is
apt to resemble encouraging the acquisition of job skills. Policies and
programs that generate loyalty, trust, and self-policing ethical behavior
may effectively reinforce, supplement, or, in some cases, even replace
unreliable economic incentives.[34]

So prestigious have economists become that managers tend to deny
their own strategies of appealing to a complex of human loyalties and
the various senses of "self." They feel obligated to use the only publicly
accepted language of economists, that of economic self-interest, and
proclaim that it is the only motivator worth considering.

The reality is quite obviously different and more perplexing. In different situations managers implement strategies both novel and diverse.
To study those strategies and consider the obstacles to their development and the opportunities they may offer, we invite the reader to
examine the following chapters. The perspective is different from that
of mainline economics, and in a sense it is more optimistic than the
conclusions of many business critics. Albert Hirschman submits that:

> the paradigm about self-interest leading to a workable and perhaps even
> optimal social order without any admixture of "benevolence" has now been
> around so long that it has become intellectually challenging to rediscover the
> need for morality. To affirm this need has today almost the same surprise
> value and air of paradox which the Smithian farewell to benevolence had in
> its own time. Second, and more important, it has become increasingly clear
> that, in a number of important areas, the economy is in fact liable to perform
> poorly without a minimum of "benevolence."[35]

In short, the study of managers' use of different strategies, in practice,
will be far richer and more complex—and certainly more compelling—

than studies that assume that managers do no more than fulfill the narrow assumptions of the neoclassical model.

Recent business history suggests an even more compelling perspective: Those who do not expect much change in business's ethical behavior and social responsibility ignore the influence of groups and communities learning to mobilize the market as a device for forwarding their values. The pessimists may take too seriously the legal fiction that corporations are persons, and thus may be treated as unitary decision makers and self-interested economic maximizers. Even in pessimists' analyses of individuals, they assume choices as given, and seldom inquire into the source of the choices or the values to which they may be responses.[36] Whatever neoclassical assumptions about the market may be, business managers and professionals operate in real-world situations. Neither individualism nor economic self-interest accounts for their organizational successes, failures, and processes. Managers contend with market forces practically, and although economic theory provides insights about the nature of that practice, managers seldom find it adequate to guide them in their grappling with daily predicaments and unexpected operating problems.

BUSINESS ETHICS AND
NEW CORPORATE CONSTITUENCIES

Increasingly, grappling with the unexpected involves one or another group that has declared its interests to be affected by corporate decisions and its values to be ignored by corporate policies. From traditional managerial perspectives, such groups and their interests or concerns may be beyond the limits of the business, extraneous to it and any possible responsibility. As groups use their abilities artfully, employing modern means of political persuasion to secure government regularity restraints, or mobilizing the market through modern communication technology to create new opportunities or to inflict extra costs upon firms, managers have come to realize that who and what is a part of the firm is not just a matter of either legal niceties or managerial discretion.

Groups can and do decide for themselves whether they are a part of a business corporation or industry—they declare themselves to be constituencies, parts of the whole, that must be taken into account in decisions, policies, and activities. An illustration offered itself in late 1989. The National Toxics Campaign (NTC), a Boston-based environmental group, became impatient with slow-moving Environmental Protection Agency (EPA) efforts to ban pesticides that members considered hazardous to human health. EPA's leisurely pace had been dictated in part by the earlier Reagan officials who did not like the agency's mandate, and partly by the pesticide industry's fight to keep the EPA from acting at all. Rather than wait upon the government body, NTC began negotiating with a number of supermarket chains in the United States and Canada and a food distributor in California.[37] It persuaded them to agree to provide consumers with information about the kinds of pesticides used on the fruits and vegetables offered for sale, and also secured promises from the managers of the firms to encourage growers to phase out by 1995 the use of 64 pesticides considered potential carcinogens by EPA.

NTC had declared itself a part of the food-distributing industry and an intimate in the policy making of the grocery corporations with whom it dealt. It participated, not as an outsider, but as one expert in its knowledge about food markets, consumers' changing tastes, and how to move with the times. The potential power it wielded was derived from a skillful understanding of how to exploit the market—and serve consumers. The business firms recognized that the endorsement by NTC might increase sales to those as concerned about pesticides as NTC members. Any advocate of the free market could find little objectionable in either the agreement or its terms. Providing information to consumers and serving their demands can hardly violate market norms, and if grocers believe consumers will prefer pesticide-free food, it is in the interest of both themselves and the consumers to provide it.

Ironically, both government bureaucrats and representatives of the affected trade associations, including the Produce Marketing Association and the National Agricultural Chemicals Association (representing the $4 billion a year domestic pesticide industry) criticized the private action. An EPA spokesperson paternalistically declared:

> The coalition [of grocers and NTC] puts the onus of pesticide regulation on grocery retailers, who are not and should not be responsible for such decisions. E.P.A. is committed to moving faster to stop the use of any pesticides that present unreasonable risk. It is important for the Government to base its decision on top quality science. In addition, as required by law, E.P.A. must balance both the value to society and the risk which may be presented by these pesticides.[38]

A spokesperson for a food trade association denounced NTC as a "narrowly focused special-interest group," "activists . . . casting public doubt on fresh fruits and vegetables—some of the most healthful foods people can eat."[39] The denunciation is peculiar, for only a month before, Ray A. Goldberg, a professor at Harvard's Graduate School of Business Administration, indicated that consumers interested in environmentally safe and clean products are a new and growing market. "This is not just a small market niche of people who believe in the 'greening' of America. It is becoming a major sentiment of the consuming public."[40]

A defender of the market might have wondered why NTC was not praised for helping create additional choices for consumers and pursuing solutions that avoid the stultifying hand of government. The government and business representatives, denouncing a voluntary agreement in line with market—and even broader social—values, projected their own turf-prerogatives. They perceived NTC neither as a supermarket constituency nor as one of their own. They could categorize it only as an interfering "outsider."

MANAGERS AND NEW CONSTITUENCIES: HOW TO INCLUDE EACH OTHER IN A MUTUAL STORY

Rapidly changing social conditions and the swift integration of the world economy are pressing both business managers *and* constituency leaders to learn from each other, as never before. In the past each conveniently tended to characterize the other stereotypically, assigning roles, motives, and values that were seldom checked against actual performance and deeds in practice. Each party accepted the conventional stories of the other—more often than not, stories that each told about itself, even when they were quite different from the stories acted out. Note how the lobbyists of the Food Trade Association, above, excluded NTC from *its* story, though its very success in striking a bargain with a number of large grocers, was proof that it had become now very much a part of the food business story.

Throughout much of the nineteenth and well into the first half of the twentieth centuries, American managers accepted as undisputed fact that unionists were violent outsiders—anti-business, anti-capitalist, third-party adversaries whose recognition would seriously impair production and lessen productivity gains, to the detriment of owners, consumers, and the public generally. Given such "facts," managers presumed that they had nothing to learn from unions, union leaders, or union members—their criticisms were already discounted, their demands immediately suspect, and their perspectives were assumed to be distorted. Unionists responded with their own caricaturing stories of managers as greedy profit-seekers, heartlessly unconcerned about worker welfare. Each found enough validating facts in experience to assure themselves that their respective stories required no modifications in response to the stories of the other. (The pertinence of story-telling and story-changing to human relations we explore extensively later, in Chapter 11. For the moment we can observe simply that human conflict consists not only in a clash of interests, principles, and goals but also in ways of answering the questions: What is the story of your life until now? Where have you come from, where do you think you are going, and what is the role you are trying to play in relation to other people with other roles in the story?).

As they interacted with each other, first through strikes, walkouts, lockouts, boycotts, picketlines, and demonstrations, and later through negotiations, bargaining, and joint administration of agreements, many on both sides came to realize that the oft-repeated old stories no longer described reality well, though American managers generally have maintained a marked distaste for unionism.[41] By the late 1980s, a number of leaders on both sides were beginning to recognize that they possessed many more common interests than they had understood in earlier decades, and the possibilities for cooperation could be as large as those achieved through conflict.

Business managers and the newer corporate constituencies have generally dealt with each other much as employers and unionists have long interacted. They, too, have tended to view each other as adversaries, with whom they could deal only at arm's length, and from whom they could learn nothing. Corporate managers generally have defended such wariness based on their experience with the newer constituencies. They viewed the constituencies as they had the unionists, as hostile outsiders at worst, and as uninformed, irrelevant critics at best. Constituencies have often responded much as unionists did in characterizing business managers as grasping exploiters at worst, and as bumbling, careless, or irresponsible power wielders at best.

As examples in succeeding chapters will show, neither managers nor constituencies always, or even usually, enact the roles that each has assigned to the other. In the quarter-century after World War II, for example, however much General Motors managers sought large profits, neglected the quality of their products, and failed to keep up with changing consumer demand, they also displayed a remarkable sense of responsibility to a variety of other constituencies. They shared their gains with both employees and suppliers, long protecting from the harsh forces of competition the communities in which they lived and worked. Few of the management critics—including conservationists, consumerists, and dissident shareholders—ever explicitly recognized the gains and benefits the company provided its various, conventional constituencies, such as its employees, steel suppliers, parts vendors, and the communities in which they were situated. To be sure, General Motors managers granted company favor unequally and with an element of arbitrariness, but they also acted with an appreciation for values of fairness and social cohesion.

In their turn, General Motors' managers, like many another American manager, preferred caricatured critics to critics-in-fact. When called to account through the market, beginning in the late 1960s and over the next decades, for ignoring large numbers of their consumers, the managers rushed to conclusions about the aims, purposes, and values of such groups as civil right advocates, consumerists (particularly Ralph Nader and his activist followers), and environmentalists. Rather than taking consumerist criticisms seriously, as indicators that something was wrong with the way they were producing and selling automobiles, managers perceived a "socialist" threat to the American way, an impairment of private property and an attack upon profits. Such threats were always more rhetorical than substantial, and in fact they never materialized. Had the managers focused on the faults identified and the shortcomings revealed by their critics, rather than on their own fears of the critics, the automobile industry in the United States might be far stronger and sturdier than it is today.

Such a focus would have required them to listen carefully to their critics with a realization that they, as managers, just might not be doing everything right. They might have more wisely responded to their critics by seeking a dialogue rather than hiring detectives to dig up "dirt" about their critics or by taking a confrontational stand against them. They would have had to understand their critics' experience rather than assume they knew it already. They would have had to be ready to accept the then-strange and awkward notion that their critics could actually be considered some kind of members of the firm, a part of the company in a more inclusive sense than conventional business explanations allow and a more encompassing way than lagging legal definitions permit. There was a simple synonym for most of the people not on the payroll: *customers,* actual or potential. How did automakers ever learn to talk about a "market" in abstraction from the living human reality of flesh and blood customers?

To recognize those in the community at large who criticize, complain, and protest as possible constituents, to comprehend their story as they play it out, takes more than merely an act of will, or a directive from the CEO to create a new department to foster corporate goodwill. We believe that it requires the insight that the world continually surprises us. To manage in such a world requires an artistic transcending of the accepted, tried, and routine relationships and an ability *to redefine the managerial story so that it may include, but not be sacrificed to, the demands and offers of others.* In dealing with such potential constituents, each party will have to be clearly aware of its own values, those practiced as well as those professed; each will also have to be able to search perceptively for overlapping, congruent, or cooperative values, even when they are buried in angry rhetoric and acrimonious threats.

AN OVERVIEW OF THE STUDY

The following chapters explore the opportunities and problems confronting business managers and the leaders of both old and new corporate constituents as they are learning to deal with each other. Both are discovering that they must devise new, untested strategies for meeting the demands of a rapidly changing world economy. They are also having to adapt traditional values to meet the requirements imposed by a swiftly changing social scene at home. The parties often take adversarial positions, each fighting the other to maintain long-accepted roles and purposes that underlie their conception of a productive and livable society. We argue that their disputes frequently revolve not so much around what is the right thing to do—upon which there is remarkable agreement—as around the role and position of those who are to do it.

Managers usually insist that they have the necessary right to command the kind, pace, and quality of production and distribution, while constituencies insist that they be allowed to participate in setting policy that promotes justice and equity. If allowing such participation subordinates managerial claims, managers all too often respond defensively, arguing that efficiency and business necessities are ignored. Though there may be decided and important differences in the formal emphasis managers put on values—productive efficiency over equity or justice, for example—the reader will soon examine situations among large manufacturing firms and in whole industries where the parties' emphases are reversed, as in the case of NTC and the trade associations.

Though their interests and values overlap in some areas and coincide at some times, they also clash, particularly in the emphases each seeks to provide. In negotiating the trade-offs and in their general dealings with each other, both have much to learn from the other. We believe that business managers need to listen to and respond to the needs of their corporate constituencies, and that constituency leaders and their members need to pay close attention to and consider carefully the concerns of managers. In a poor world the expertise in producing goods and services efficiently should not—dare not—be taken for granted; but production (and distribution) carried out with little regard for justice and fairness provides no sure or stable foundations for business firms. Managers and constituencies need each other for essential social reasons. Each helps correct partial views and understandings of a complex, interdependent society.

This study begins by exploring the emergence of self-declared corporate constituents, the first of which were employees organizing unions. They experimented with a variety of tactics to exert their influence upon both managerial policies and practices, such as picketing, public boycotts, strikes, legal action, and adverse publicity. The Civil Rights movement of the 1950s and 1960s borrowed many of those tactics, and in the process of winning legislative and legal rights for minorities, helped train a cadre of activists, early representatives of the well-schooled, idealistic "Immense Generation" born in the decade after World War II, upon whose talents and skills the modern corporate constituency movement has been built. Both unionists and civil righters appealed to old, well-established, and honored American rights of free speech and association in the pursuit of justice. In fact, they were the same rights upon which business corporations themselves have been legally established.

The study then considers the widely argued notion that the new constituencies have created much, if not most, of the problems with which business managers have had to cope over the last twenty years. According to this notion, the "political" attacks by constituencies upon business have led to intrusive, expansive government regulation, besmirched business reputations, ongoing hostility to necessary business practices, spread of an anti-business, anti-capitalist ideology, national decline in productivity, and increased public dissatisfaction with business performance. We consider the possibility that these problems have arisen not out of alleged or perceived managerial failures so much as out of the past remarkable successes of managers, their businesses, and the dynamic of capitalism.

Public disillusionment with business and most constituency criticism probably arise, paradoxically, more from managerial delivery of certain implicit economic promises than from any shortcomings. Managers too easily assumed that ever-increasing production and a rising standard of living would solve almost all social problems and increasingly relieve economic tension. Experience suggests that our amazing production of the last four decades has created as many, and perhaps more, problems than it solved; new, daunting issues have been uncovered, even while the economy's enrichment has opened up possibilities for more Americans than ever before. Business leaders have not prepared themselves well for recognizing, let alone dealing with that opening.

Today's managers who look longingly back to the quarter-century after the end of World War II, when American business enjoyed enormous prestige and its production was the envy of the world, do not always appreciate the large numbers of the disadvantaged, dispossessed, and discriminated-against, who were not able to participate fully in the economy or to wield effective political influence. Leaders and members of constituencies, on the other hand, who routinely condemn managers of large, multinational corporations as heartless profit seekers, fail to note the remarkable degree of corporate social responsibility those same managers exhibited over that same period. True, it usually took the form of paternalistic concern for favored constituencies while ignoring others; and the responsibility displayed manifested economic self-serving aspects. With very few exceptions the leaders of these corporations were white, college-schooled males, but they were hardly models of "economic men," who made all their decisions according to a bottom line.

In fact, business managers have always been Americans and economic actors; they believe in, respond to, and value traditions and follow norms common to people who live in the United States. Like all Americans, managers continually confront social tensions arising out of their conflicting notions of how life should be and is lived. They proclaim the values of individualism and competitive markets but live their lives and accomplish most of their work in cooperative corporate endeavors. Managers' narrow interpretations of social and economic traditions and their particular business perspectives served their interests, to be sure, but they have always been ready to abandon the free, competitive market in practice, if they found reason to do so. They preferred a world of business both more controlled and more human. They then could be accused of hypocrisy; but more accurately they might have been accused of betraying singlehearted loyalty to economic determinism.

Reinterpreting the nature of competition as specified by economists, they more often sought to shelter themselves and their organizations from its harsh effects. They preferred social responsiveness, as they defined it, over market response, and when through luck and the fortunes of history, they discovered they could manipulate the market, they insisted upon organizational trappings and claimed a social position, particularly inside their firms, which were almost sure eventually to undermine not only their autonomy but also much of their economic power. Both their insistence on their own importance and their claims for social preference could not be sustained as both world and domestic competition pressed them harder and harder. From today's vantage point the glory days of the large American manufacturing corporation, from the turn of the century to past mid-century (excepting the dismal Great Depression of the 1930s) now appear to be more a parenthesis in our history rather than the climax of an extraordinarily successful business form.

New constituencies found no ready acceptance and have had to generate political power to create economic power for themselves; thus their effects upon managers were often marginal, even if innovative. It was the well-established, legal, and property-based power of stockholders who forced managers in recent years to come quickly to terms with the changing economic system. Stockholders found new ways of exercising the power that had long been inherent in their votes, but never before effectively mobilized. Learning how to appeal to constituencies, who often have substantial stockholdings, and amassing other shares through innovative financial arrangements, stockholders have been able to exert great economic power upon managers, requiring the quickest, most massive restructuring of business in our history. Other constituencies' leaders have taken note of how shareholders can mobilize financial power and are experimenting with ways in which they may apply it themselves, or in alliance with other institutional investors. The take-over phenomenon has been the nightmare of many a manager in the 1980s, subjecting them to the experience of being judged strictly (and sometimes unjustly) by the narrow rule of current stock performance.

In the third part, the study scrutinizes the values that managers' both practice and profess. Business managers usually feel constrained pub-licly to defend their policies in terms of competitive market theory, though their practices hardly conform to its tenets. As governors and sources of authority in large, cooperative, and voluntary organizations, managers have long understood the necessity of offering their col-leagues, associates, and fellow employees more rewards than those con-templated by conventional economics. Neither the market's promise of freedom nor its offer of efficiency is sufficient to weld together a firm's productive community of workers. Market freedom is too conditionally dependent upon one's income to be fully convincing—if you don't have much of an income, you don't enjoy much freedom. Efficiency can be specified fully only if there is prior agreement on the values of both inputs and outputs. Efficiency in the building of the Egyptian pyramids was one thing; efficiency in a Japanese auto factory is quite another.

Managers simply cannot escape the necessity of treating the issue of fairness-equity-justice. When pressed, they show no reluctance in deal-ing with the issue, though their definitions of either equity or justice seldom coincide with that of constituency members. Managers, like other Americans, have always highly rated fairness in their relationships with other firms, government, and community members. Fair trade is not a strange or unheard-of term among business managers. And under-neath most notions of fairness is the more comprehensive notion of justice.

In the last part, the study considers new styles of managing, first examining the various ways in which philosophers suggest managers can best define their responsibilities and understand community values or community principles. In a changing environment that is transform-ing the context within which managers must operate, both managers and we, the authors of this book, may need to be more experimental and tentative in our conclusions than has been thought sensible in the past. For both managers and the constituencies, perplexity may be wiser than certainty. Promising process rather than authoritative logic may be the best contribution to the discovery of business ethics in the nine-ties. Philosophers have usually focused on economic justice as primarily a problem of distribution, but managers are apt to be more impressed with justice as a matter of production. The ability, the need, the oppor-tunity, even the right to contribute to productive effort—enhancing one's own welfare and that of the community at large—all loom large for many constituencies and for managers as well. How productive and distributive processes can be more *just* constitutes the major economic-political agenda for world society in the coming decades.

Production and its constellations of values and institutions are woven into the very fabric of society: An industrial economy succeeds because it mobilizes its people into a vast cooperative and voluntary effort that goes far beyond our imagined boundaries of individual activity, values, and virtues. It is time, we think, that Americans recognize and explore what Alasdair MacIntyre calls the moral facts of human life.[42] The phrase sounds odd, but the cases, stories, and situations explored in this study indicate that both managers and constituency members regularly consider the moral facts of their common experiences and shared lives. They have at times constructed for themselves, and for others who join them, storylines of who they are, what they are about, and the values that inform their activities. Responsibility for both managers and constituency members, we believe, means taking seriously the moral facts upon which our various storylines are based. Few can tell a story about their lives without assuming that the meaning, direction, and purpose of the story are real in fact, because they are the best descriptions of why some events occurred and others did not.

Responsible managers and responsible constituencies increasingly are going to have to learn how to recognize, to listen to, and to include each other's stories about themselves—particularly those moral facts that give special, human meaning to the world in which we live together. Stories and facts can sometimes be erroneous. They will always be partial. Correction and addition are usually required. Insofar as managers and constituencies weave a web of responsiveness to each other, sometimes conflicting with each other and at other times jointly acting, they will be calling each other to account, promoting social solidarity and justice with efficiency and productiveness remaining as their major responsibility. We do not see much likelihood that disagreements will disappear nor will all differences be finally resolved; responsiveness does not mean an absence of conflict, but rather a recognition that it can transcend even our diversities and contradictions. Our storylines will be left incomplete, ready for adjustment and the inclusion of new, unknown, and previously unrecognized moral facts and moral actors.

In the last chapter, by way of defining the frontiers of the next decade, we present several business-constituency encounters of the past decade. Some are histories of missed opportunities and familiar, hostile, unresponsive, battles; others are stories of trials and tribulations on both sides, through which both parties learned moral facts about which they were ignorant earlier. Some of the encounters are recent experiments, too new for the authors or the readers to see likely outcomes. All of the encounters, however, point to a newly emerging style of management that will become more and more common in the coming decade.

NOTES

1. See Edwin H. Sutherland, *White Collar Crime* (New York: Holt, 1940, and Marshall B. Clinard and Peter C. Yaeger, *Corporate Crime,* (New York: The Free Press, 1980).

2. Between 1975 and 1986 civil cases commenced in U.S. District Courts increased in almost all categories, except antitrust. Actions taken under statutes; civil rights, environmental, and labor laws; securities, commodities, and exchanges laws; and protected property rights at least doubled. However, there are no data available that indicate the number of cases successfully prosecuted, the amounts involved, or the kinds of companies affected. See U.S. Department of Commerce, Bureau of Census, and *Statistical Abstract of the United States, 1988,* 108th Edition, (Washington: Government Printing Office, 1988), Table No. 295, p. 172.

3. See Felix G. Rohatyn, "Ethics in America's Money Culture," *New York Times,* June 3, 1987. "I no more believe that ethics can be taught past the age of 10 than I believe in the teaching of so-called creative writing. There are some things that you are born with, or they are taught by your parents, your priest or your grade-school teacher."

4. Improving ethical behavior is not likely to be possible if ethics is treated as an add-on or decoration of the "basic" purposes of business. Ethics are necessarily an integral part of any activity, and must be examined and enacted as a part of the whole. Albert O. Hirschman warns that "it seems to me impractical and possibly even counterproductive to issue guidelines to social scientists on how to incorporate morality into their scientific pursuits and how to be on guard against immoral 'side effects' of their work. Morality is not something like pollution abatement that can be secured by slightly modifying the design of a policy proposal. Rather, it belongs into the center of our work; and it can get there only if the social scientists are morally alive and make themselves vulnerable to moral concerns—then they will produce morally significant works, consciously or otherwise." *Essays in Trespassing: Economics to Politics and Beyond* (Cambridge: Cambridge University Press, 1981), p. 305. [Reprinted with permission.]

5. Albert O. Hirschman noted, *"there is today an increased concern with moral values, even in economics*—that rock of positivist solidity." In "Morality and the Social Sciences: A Durable Tension," Ibid., p. 294. [Italics added.]

6. Miriam Beard, *A History of Business: from Babylon to the Monopolists* (Ann Arbor: The University of Michigan Press, 1962), pp. 1–7. Despite business managers' bad reputation, she wittily points out they hardly compare to the tyrants and military masters so often praised by historians.

7. *Adventure of Ideas* (Harmondsworth, Middlesex: Penguin Books, 1948), p. 120. Dr. Courtney C. Brown, long the dean of the Graduate School of Business, Columbia University, had this quotation inscribed on the plinth of Uris

Hall, the school's main building. The fuller quotation is: "The behavior of the community is largely dominated by the business mind. A great society is a society in which its men of business think greatly of their functions. Low thoughts mean low behavior, and after a brief orgy of exploitation, low behavior means a descending standard of life. The general greatness of the community, qualitatively as well as quantitatively, is the first condition for steady prosperity, buoyant, self-sustained, and commanding credit."

8. Adam Smith, *An Inquiry Into the Nature and Cause of the Wealth of Nations* (New York: The Modern Library, 1937), Book IV, p. 625. The reader should note that Smith did not know managers of the kind now found in business firms. Most firms were small and thus owner operated; he was interested in their profits, as owners, but not in their function of "inspection and direction." These activities were merely a kind of labor, for which they should receive a wage. He referred to one who often carried out these activities as "some principal clerk." See Book I, p. 49.

9. Ibid., p. 14.

10. Ibid. Italics added.

11. E. G. West, in Adam Smith, *The Theory of Moral Sentiments,* (New Rochelle, New York: Arlington House, 1969, p. xx.

12. Ibid., pp. 27–28.

13. Ibid., Introduction, p. xxx.

14. Ibid., p. 346.

15. Dennis Robertson, "What Does the Economist Economize?" *Economic Commentaries* (London: Staples Press, 1957), pp. 147–155.

16. Ibid., p. 148.

17. Arjo Klamer, "A Conversation with Amartya Sen," *Journal of Economic Perspectives* 3 (Winter 1989):142. [Reprinted with permission.]

18. Answers to questions on the *Dreyfus Roundtable*, New York: CBS, April 30, 1989. Television program.

19. There is good reason to think that Smith did, in fact, presume such moral ties among people in both the market and society. See E. G. West, in Adam Smith, *A Theory of Moral Sentiments,* pp. xxxi–xxxii.

20. David Vogel, *Lobbying the Corporation* (New York: Basic Books, 1978), p. 225.

21. Walter A. Weisskopf, *Alienation and Economics* (New York: E. P. Dutton, 1971), pp. 116–139.

22. See Herbert A. Simon, *Administrative Behavior,* 2nd Edition (New York: Macmillan, 1957), and James G. March and Herbert A. Simon, *Organizations* (New York: John Wiley & Sons, 1958). In the latter book (p. 65), the authors note that "Humans, in contrast to machines, evaluate their own positions in relation to the value of others and come to accept others' goals as their own." John Kenneth Galbraith also noted that those who join organizations are motivated by more than pecuniary gain or compulsion. Though he did not expressly mention altruistic interests in motivation, his discussion of motivation allowed for it as well as self-interests. See John Kenneth Galbraith, *The New Industrial State* (Boston: Houghton Mifflin Company, 1967), pp. 128–58.

23. Henry Clark, "Editorial: The Dilemma of the Business Ethicist," *Quarterly Review*, vol 7, no. 4 (Winter 1987):3–14.

24. Alexis de Tocqueville, *Democracy in America*, trans. George Lawrence, ed. J. P. Mayer (New York: Doubleday, Anchor books, 1969), p. 287. Robert Bellah and his coauthors used the term as the title of their book. Robert N. Bellah, Richard Madsen, William M. Sullivan, Ann Swidler, and Steven M. Tipton, *Habits of the Heart: Individual and Commitment in American Life* (Berkeley: University of California), 1985.

25. Both Adam Smith and Thomas Malthus qualified the pursuit of self-interest. Smith qualified it in *The Wealth of Nations* once. On page 651 of the Modern Library edition, he added to a reference to self-interest the phrase "as long as . . . [one] does not violate the laws of the justice." Malthus systematically added a reservation to the self-interest rule, that one should also adhere to the rules of justice. See Thomas Malthus, *Principles of Political Economy* (London: John Murray, 1820) pp. 3 and 518. See Albert O. Hirschman's discussion of this point in *Essays in Trespassing: Economics to Politics and Beyond*, pp. 35–136.

26. Henry Clark, "Editorial: The Dilemma of the Business Ethicist," *Quarterly Review*, vol. 7, no. 4, (Winter, 1987), p. 6 (Nashville: United Methodist Publishing House), quoting from Reinhold Niebuhr, *Man's Nature and His Communities: Essays on the Dynamics and Enigmas of Man's Personal and Social Existence* (New York: Scribner & Sons, 1965), p. 68.

27. Fred Hirsch, "The Ideological Underlay of Inflation," in Fred Hirsch and John H. Goldthorpe, eds., *The Political Economy of Inflation* (Cambridge: Harvard University Press, 1978), p. 274.

28. Hirschman, *Essays in Trespassing*, p. 301.

29. Donald N. McCloskey, *The Rhetoric of Economics* (Madison: The University of Wisconsin Press, 1985). See also Donald N. McCloskey, "The Rhetoric of Economics," *The Journal of Economic Literature* xxi (June 1983):481–517.

30. Arjo Klamer, "A Conversation with Amartya Sen," *Journal of Economic Perspective*, vol. 3 (Winter, 1989), p. 145.

31. Robert Axelrod, *The Evolution of Cooperation* (New York: Basic Books, 1984).

32. Kenneth J. Arrow, "Values and Collective Decision-Making," in P. Laslett and W. G. Runciman, eds., *Philosophy, Politics and Society*, vol. 3 (New York: Basil Blackwell, 1987), pp. 121–22. [Reprinted with permission.]

33. Albert O. Hisrchman, "Against Parsimony: Three Easy Ways of Complicating Some Categories of Economic Discourse," *Papers and Proceedings*, American Economic Association, vol. 74, 1984, p. 90.

34. See Amitai Etzioni, *The Moral Dimension: Toward A New Economics* (New York: The Free Press, 1988). He refers to a number of economists who have pointed out the way in which economic transaction costs may be significantly lowered—and thus efficiency enhanced—by ethical behavior. See also Kenneth Arrow, *The Limits of Organization* (New York: W. W. Norton, 1974) and Albert O. Hirschman, *The Strategy of Economic Development* (New Haven: Yale University Press, 1958).

35. A. Hirschman, *Essays in Trespassing*, p. 299.

36. See William S. Vickery, "An Exchange of Questions between Economics and Philosophy," in Edmund S. Phelps, ed., *Economic Justice* (London: Penguin Books, 1973), p. 46.

37. The firms were (1) Raley's Supermarkets, a $1 billion a year chain of fifty-seven markets based in Sacramento, California. In 1987 it became the first firm to test produce for pesticide residues and to advertise the results, a program adopted, up through 1989, by at least ten other major chains; (2) ABCO Markets Inc., a Phoenix-based chain of seventy-five stores in Arizona with annual sales of $800 million; (3) Provigo Inc., a Montreal-based chain of more than 1,000 stores in Quebec and owner of thirty Petrini Supermarkets in the Middle West; (4) Bread & Circus Wholefood Supermarkets, which owned five stores in Massachusetts; and (5) American Brothers Produce, a medium-sized distributor in San Jose, California.

38. Keith Schneider, "5 Supermarket Chains Open Effort Against Pesticide Use," *New York Times,* September 12, 1989. [Copyright © 1989 by The New York Times Company. Reprinted by permission.]

39. Ibid.

40. Michael Freitag, "Luring 'Green' Consumers," *New York Times,* August 6, 1989.

41. See Richard B. Freeman, "Contraction and Expansion: The Divergence of Private Sector and Public Sector Unionism in the United States," and Melvin W. Reder, "The Rise and Fall of Unions: The Public Sector and the Private," in *The Journal of Economic Perspectives* 2 (Spring 1988):63–110. Both scholars argue that hostile, anti-union activities have been a significant cause of the three-decades-long drop in union membership in the United States.

42. See Alasdair MacIntyre, *After Virtue: A Study in Moral Theory* (Second Edition; Notre Dame, Ind.: University of Notre Dame Press, 1984), especially pp. 57–59, 83–84. We give major attention to MacIntyre's theory in Chapter 11. The notion of "moral fact" flies in the face of much Western moral philosophy over the past three centuries. MacIntyre summarizes this theory: ". . . there are no facts about what is valuable. 'Fact' becomes value-free, 'is' becomes a stranger to 'ought,' an explanation, as well as evaluation, changes its character as a result of this divorce between 'is' and 'ought.' " (Ibid., p. 84). A contrary theory, which philosophers like MacIntyre want to recover, claims that there is purpose, meaning, and value *in* objects, in our very perception and definition of them. . . . [T]he concept of a watch cannot be defined independently of the concept of a good watch nor the concept of a farmer independently of that of a good farmer . . ." (p. 58). The dogmatic separation of "fact" and "value" has been most radically developed among twentieth century social scientists. At the end of the eighteenth century, Adam Smith, to the contrary, could still speak of "moral sentiments" as facts of human nature. See E. G. West, Introduction to Adam Smith, *The Theory of Moral Sentiments,* p. xxix.

2

The Emergence of Corporate Constituencies[1]

Less than a decade ago, one American business executive bluntly protested the intrusion of outside groups into managerial affairs:

> They have no right to ask us to do anything. They are not the owners; they do not represent the owners. They are elected by nobody and represent no one but themselves.

Another complained that they are only agitators, irrelevant to a business: "They muddy the waters of decision with foolish arguments and unrealistic objectives." More generally, even if they make a useful contribution, it is at high cost. Business managers can and will, they maintain, act on their own. In a survey of a number of managers, Janger and Berenbeim found that:

> A quarter of the responding executives are convinced that such groups do not even make a constructive contribution. They agreed with the survey statement: "All valid social and economic changes would occur in our company without outsider activity." Almost one-third (31 percent) credited outsider groups with making "valid social and economic changes occur that would not otherwise occur." Almost half (44 percent) rejected both statements, possibly because, as one executive put it: "There are benefits, but they are not worth the costs."[2]

Many business leaders may well find group protests not worth the bother and costs they impose through highly public encounters. The groups challenge managers' judgment and authority in new and unfamiliar ways. In addition, they may be hard to understand; their protests and criticisms may not be credible to hard-pressed, busy, often harried executives who must continually juggle a wide array of invest-

ment, production, and marketing problems within the limits of investment constraints and financial resources.

Furthermore, they may well associate the origins of the groups with
the aftermath of the civil rights movement, the outbreak of student-led
protests against the Vietnam War, the rise of consumerism, and continuing public efforts to preserve the environment and to reduce pollution. Many of the groups involved may appear to managers either as
antibusiness radicals or their successors. Managers have tended to perceive organized groups as hostile to managerial efforts to produce goods
and services effectively and efficiently. Not surprisingly, therefore, many
business leaders have viewed the groups at best as dysfunctional intrud-
ers, and at worst as business enemies. For better or worse, managers
have found many of the groups in the new movement difficult to accept as a part of traditional American business.

That lack of acceptance is based on a misperception of American
history. Below we will first examine the reasons for believing that the
form of these movements, like many other and earlier voluntary efforts
to achieve specific social changes, is as American as apple pie. We will
then consider the major *values* or root social convictions of those who
lead various organizations that make up the movements. We will discover that those who have sought increasing influence on or involvement in business decision making—sometimes known as public interest
groups or as the corporate responsibility movement—are remarkable
for their overall adherence to beliefs that are respected in American
tradition and ensconced in American history. Finally, we shall offer
some opinions about those aspects of the groups that indeed do make
them and their values different from those of earlier times.

THE HISTORICAL CLAIMS OF
VOLUNTARY ASSOCIATIONS IN AMERICA

Change occurs so rapidly and continuously during a lifetime that most
of us have difficulty reckoning with the possibility that all societies, even
those as open to change as ours, are partly products of ancient inheri-
tances. Environmentalists have become justly famous for this perspective. They have called our attention to the truth that many elements of
the human "support system"—air, water, vegetation, and mineral deposits—evolved over millions of years. We are dependent upon this
past even as we go about changing some of it, shaping and using it
according to our sense of values. Building a dam across a river, for
example, takes advantage of all the ancient values of the river, as we
add to the system a few other values. Rivers had their own "value system" long before human beings came along, and woe to those who do
not respect and cooperate with the river's way of running its affairs.

Those who insist upon living on flood plains or in channeling water flow between narrow levees learn through bitter experience the cost and pain of ignoring the river's values.

Social scientists have helped us understand that human societies are a dense mixture of new things—ideas and concepts and goods—piled on top of old things. This image of alluvial layers suggests an important perspective of social change: While the surface of a society may be undergoing various disruptions, some of its traditions, buried deep in its own past, remain undisturbed. In fact, only on the grounds of certain ancient beliefs are some societies able to make some changes in themselves. Observers of modern societies in a state of revolution—contemporary China, for example—testify that the revolutionaries have to appeal to traditions older than the revolution to effect some of their goals.

The United States is not old as societies go. Its founding required the importation of European memories for the making of the dominant American culture. As study of the writings of the Founding Fathers of the Republic shows, they often referred to Greek, Roman, and Hebrew as well as Christian writers, some of whose ideas were a thousand or more years old. To analyze the "values" of any society, therefore, is to be on the lookout for certain roots of change that tap levels of experience and reach strata of tradition below the surface of the present. They may be hard to detect. They are nonetheless there, even in our changeable American society.

There is no more accessible literary door to this truth for most Americans of the 1990s than Alexis de Tocqueville's *Democracy in America,* written more than 150 years ago. Millions of readers of this classic have had the sobering experience of wonderment at how deftly and precisely this French aristocrat, who toured the United States in the 1830s, described social beliefs and habits of Americans that have endured into the twentieth century. The America he described has changed drastically in many ways, but very little in certain respects. One such respect is the remarkable role in our society of a form of organization that Tocqueville called an "association."

With the memory of the French Revolution fresh in his mind, Tocqueville was well aware that the American government was the first modern example of politics organized by the deliberate action of citizens and their leaders. As a perceptive observer of the results of two democratic revolutions, he was profoundly conscious of the *instability* of a society dedicated to "liberty and justice for all." Such a society stands in danger of falling into chaos, as citizens exercise their liberty to pursue personal goals, to define justice according to their own self-interest, and to leave to someone else the definition and pursuit of justice—for all. Such individualism stands in jeopardy of compromising all publicly definable notions of justice, as individuals squabble with their neigh-

bors to obtain personal rights and benefits. The outcome of unre-
strained democratic contending of all with all, as Plato recognized long
ago and as Tocqueville observed in early nineteenth century Europe, is
a return of autocracy, the enforced establishment of order upon a mass
of citizens paralyzed by their individualistic enthusiasm for liberty and
equality.

Americans had developed a remedy for this general danger, he believed. They applied it to a vast profusion of particular problems in
their society, quite apart from direct initiatives by organized government. It is the remedy of the voluntary association, which Tocqueville
described as follows:

> The political associations that exist in the United States are only a single
> feature in the midst of the immense assemblage of associations in that coun-
> try. Americans of all ages, all conditions, and all dispositions constantly form
> associations. They have not only commercial and manufacturing companies,
> in which all take part, but associations of a thousand other kinds, religious,
> moral, serious, futile, general or restricted, enormous or diminutive. The
> Americans make associations to give entertainments, to found seminaries, to
> build inns, to construct churches, to diffuse books, to send missionaries to
> the antipodes; in this manner they found hospitals, prisons, and schools. If
> it is proposed to inculcate some truth or to foster some feeling by the en-
> couragement of a great example, they form a society. . . . As soon as several
> of the inhabitants of the United States have taken up an opinion or a feeling
> which they wish to promote in the world, they look out for mutual assistance;
> and as soon as they have found one another out, they combine. From that
> moment they are no longer isolated men, but a power seen from afar, whose
> actions serve for an example and whose language is listened to.[3]

Tocqueville believed that Americans had learned the art of association from the first European settlements on these shores. From the
seventeenth century on, for example, the descendants of the Puritans
went about organizing towns, churches, and business enterprises with
an intensity hardly matched in any other new society. "I can see the
whole destiny of America contained in the first Puritan who landed on
these shores," not because the Calvinism of the Puritans was to gain
supremacy as the religion of Americans, but because Puritan habits of
systematic organization carried over from their churches into every aspect of the new society. Long before the American Revolution, Tocqueville pointed out, local towns were the scene of people "assembling,
getting excited together, and forming sudden passionate resolves."[4]

The readiness to organize for the pursuit of shared interest affected
the style of politics in the new nation, but it simultaneously affected
economic activity and a broad range of civil (as opposed to political)
purposes. The civil and the political purposes interacted and rein-
forced each other, he observed, with the result that association became

a *social habit* among Americans, a settled routine for effecting social changes. Religious, economic, political, and social purposes were all pursued through various voluntary associations.

The business corporation throughout American history has been a common, though very special example of the voluntary association.[5] No one has more enthusiastically and ably formed associations for the pursuit of this or that social goal than the owners and managers of American business. They have long understood what it means to call the United States a "business society." But it is not merely a business society; the larger truth of the nation is that it is an *associational society*. Tocqueville observed that Americans carried over their associational experience from one sector of the society to other sectors:

> Men chance to have a common interest in a certain matter. It may be a trading enterprise to direct or an industrial undertaking to bring to fruition; those concerned meet and combine; little by little in this way they get used to the idea of association. The more there are of these little business concerns in common, the more do men, without conscious effort, acquire a capacity to pursue great aims in common.

> Thus civil associations pave the way for political ones, but on the other hand, the art of political association singularly develops and improves this technique for civil purposes. . . .

> In this way politics spread a general habit and taste for association. A whole crowd of people who might otherwise have lived on their own are taught both to want to combine and how to do so. . . .

> A political association draws a lot of people at the same time out of their own circle; however much difference in age, intelligence, or wealth may naturally keep them apart, it brings them together and puts them in contact. Once they have met, they always know how to meet again.[6]

The "idea of association" was powerful in the new country because it was pervasive. In his native France, Tocqueville saw no such idea at work on all levels of society. He admired the ubiquity of the idea in America, for he saw it adding practical social meaning and stability to the more general democratic ideas of liberty and equality. Take away freedom to associate, he said, and you take away the chief protection of personal freedom against the powers of government and against freedom's own excess of mere individualism. Take away freedom to associate, and you take away an experience of equality among people who, in many ways, will always see themselves as unequal. Whether on the frontier or in the new commercial cities, the habit of association "draws a lot of people . . . out of their own circle. . . ."

It is no accident, wrote Tocqueville, that the sense of equality is strong among many Americans in spite of many inequalities. "Is there really

some necessary connection between associations and equality?" Yes: "A
people in which individuals had lost the power of carrying through
great enterprises by themselves, without acquiring the faculty of doing
them together, would soon fall back into barbarism."[7] They inevitably
fall prey to aristocracy, too. From time immemorial, aristocrats have
had their own voluntary associations, wherein two or three of them
could control the destiny of many ordinary folk. It is easy for them to
get to know and understand one another and agree on rules of mutual
advantage to themselves.

> But it is not so easy in democratic nations, where, if the association is to have
> any power, the associates must be very numerous.[8]

And, Tocqueville added, as the economy and geographic size of the
United States led to increasing complexity and sense of impotence among
individual citizens, government would tend to override the independence of all sectors of the society. Business and commerce would fall
under the alleged omnicompetence of government if "the idea of association" evaporates from the mainstream of the *total* society.

The freedom of the capitalist to pursue business enterprise is but a
special case of a larger freedom of any and all Americans to associate
for the purpose of pursuing those values that a group of democratic
citizens may wish to pursue. The most influential Americans know this
secret of power and influence, even when they do not use the rhetoric
of political democracy to describe it. "Every rich and powerful citizen,"
wrote Tocqueville, "is in practice the head of a permanent and enforced association composed of all those whom he makes help in the
execution of his designs."[9]

Karl Marx, writing a few years after Tocqueville, identified the important "permanent and enforced association" of a society as social class.
The American of Tocqueville's day made social class a concept difficult
to pin on any group of individuals with certainty over time. The exception, of course, was black slaves. But the fear of many white slave owners that blacks would learn the arts of organized resistance to their
slavery betrayed white Americans' own sense of how changes were most
likely to occur in this society. White slave owners' attempts to cut off
slave rebellion focused always on every sign of organization within the
ranks of slaves.

That fear is ancient in the hearts of totalitarian governments, which
are undisturbed by private personal thought so long as it never seeks
organized social expression. Without the counterpower of some counter
association, organized power runs roughshod over unorganized. It is
in society at large, and not merely in the formal structure of the American federal government, that separation of powers opens the way for

citizens to give concrete meaning to their abstract right to liberty and equality.

To his extensive discussion of the idea of association, Tocqueville appended a perceptive chapter on the role of the press in making associations possible. The organization of a group-interest begins when individuals realize that their private concerns are shared by others in the public. Thus:

> . . . wandering spirits, long seeking each other in the dark, at least meet and unite.
>
> In a democracy an association cannot be powerful unless it is numerous. Those composing it must therefore be spread over a wide area, and each of them is anchored to the place in which he lives by the modesty of his fortune and a crowd of small necessary cares. They need some means of talking every day without seeing one another and of acting together without meeting. So hardly any democratic association can carry on without a newspaper.[10]

If he were writing in the 1990s, Tocqueville would surely add, "Or without a telephone, a fax machine, a television channel, a photocopying machine, or a network of computers."

ASSOCIATION FOR WHAT PURPOSES?

It is clear that Alexis de Tocqueville would chuckle at the suggestion that the corporate responsibility movement began in the America of the 1960s. He recognized that entrepreneurs and businessmen built the business system on foundations of the freedom of individuals to associate themselves for the sake of economic gain. He would have been acutely conscious of the irony involved when business leaders publicly express their suspicions of movements in the society to "call the business corporation to be socially responsible" toward this or that human interest outside its managerially focused operating concerns. From far back in their history, Americans from one sector of society have sought to legitimate their right to call Americans in other, particularly business, sectors to take account of some neglected value, interest, or human purpose. Those engaged in business have not been easily persuaded that nonbusiness groups should enjoy such a right, except for themselves. They have not liked to have people outside business defining public policy toward business, as though such policy should be framed by the public for the sake of the public.

What particular human values, interests, or purposes have these modern expressions of an old democratic impulse organized around during the past twenty years? In one sense this is an uninteresting

question; for, given the pervasiveness of democratic freedom to organize around *any* human interest, it should be no surprise that voluntary
associations, in relation to the affairs of the business corporation, have
sprung up to represent a vast spectrum of such values, interests, and
purposes. The list of such associations in the country runs to hundreds
of thousands. In number, though not in economic resources, they rival
business corporations themselves, which, from a Tocquevillian perspective, constitute only one of many kinds of voluntary associations in the
life of the United States.

He did not foresee the enormous growth of business corporations in
the century following his visit in the first third of the nineteenth century. He would hardly have found surprising, however, the evolution
of managerial power that would render the notion of "stockholder democracy" increasingly difficult to sustain. (It had hardly any reality even
in the early days of business corporations, in any case.) Tocqueville had
a healthy democratic suspicion of the tendency of all leaders to begin
acting like aristocrats. But he would have scorned the notion, promoted
by some corporate critics of the "new" constituencies, that the latter
have no business challenging the wisdom of managers or even of the
interests of stockholders themselves. *The right of any group of people to
challenge any other group on grounds of any value or interest is an unassailable
right in the mainstream American tradition.* It is one of the fundamental
axioms of liberal democracy. Business leaders who think otherwise have
forgotten the cultural, political, and social roots whereby enterprise became "free." Constituencies do not have to await recognition by managers, nor seek first a legal standing in order to assert their existence.
Their essence is a self-awareness among their members that they possess a concern in common and that business policies, decisions, and
activities impinge on it.

To affirm that relationship to business is to identify the first, most
obvious value of the corporate responsibility movement: freedom itself,
the freedom of individuals to organize in pursuit of their perceived
interests in relation to some other organization. The exercise of such
freedom gives far-reaching scope to the "right of assembly" guaranteed
by the Bill of Rights in the Constitution of the United States. In the
early 1970s Philip W. Moore, Executive Director of the Project on Corporate Responsibility, described his understanding of his organization's
purposes *within* existing business structures in terms that fit exactly the
old explanation of Tocqueville:

> . . . the tendency to label the corporate responsibility movement as some-
> thing different from other movements is a fundamental misunderstanding
> of the meaning of corporate responsibility. Corporate responsibility means
> the responsibility that the corporate institution has to all activist concerns,
> and potentially everybody is an activist. . . . And this is true across the entire

spectrum of social action, whether we are talking of war and corporations, the environment and corporations, minorities and corporations, jobs and corporations, consumers and corporations, or people and corporations. Whatever the substantive concern of a person or group, the action of that person or group will inevitably require that the corporation respond to those concerns. And groups like ours are trying to give voice to those concerns—a permanent voice—not because there is anything inherently good about a better board structure or more disclosures, but because those changes in structure and process are a precondition to giving the concerns of citizen activists greater access to corporate power.

Lest you worry about access to corporate power, remember that activism is simply an expression of public concern. The role of activists is to set forth options for society. . . .[11]

"Whatever the substantive concern," the American theory of society gives all its members the right to press that concern upon all the others. In this theory, developed and evolved by judges and justices, no institution is an island; no concern or value is immune to challenge by other concerns and values. The social value of freedom here, then, is its function as a doorway through which a profusion of substantive values can enter the arena of policy debate inside all social institutions. This is formal, procedural freedom that in principle admits anybody to the competition among ideas of what is good and bad for one or another segment of society.

Philip Moore is historically accurate when he characterizes the corporate responsibility movement as hospitable to "the whole spectrum" of human social concerns. The movement has often bewildered the leaders of business corporations because it has seemed to assault them with an array of concerns that are too profuse, contradictory, and noneconomic to be answered with any precision by a mere profit-making institution. But the architects of the movement believed that there were strands of the American tradition that allowed them to claim the right to protect and promote their interests vis à vis any other interest.

THE EARLY MODELS FOR
CORPORATE CONSTITUENCIES

Traditionally corporate managers and their immediate and organizationally related constituencies have dealt with each other through the market. If stockholders did not like managerial policy—such as that on dividend payout—they sold their stock; dissatisfied employees could quit and look for a new job; unhappy suppliers found another buyer; and disappointed consumers switched brands or found a substitute product. Presumably the only bond between managers of corporations and members of constituencies was impersonal and temporary, dependent upon the satisfaction that each party received from the relationship. Its easy breaking assumed readily available alternatives, that is, competitive markets in which the costs of transferring shares, of finding a new job, and of choosing another product were insignificant. As one constituency after another has found the assumptions of competitive markets invalid or the presumption of impersonal, transient bonds unsatisfactory, each has also sought to deal with managers in nonmarket ways. They have sought legal rights, not unlike those long since gained by business corporations through the courts and legislatures, to supplement market power, and they have insisted upon negotiating and bargaining with managers in face-to-face encounters.

Other, more distantly related constituencies—those represented by government agencies, courts, and legislatures or the variety of people, interests, and institutions in communities contiguous to corporate offices and plants—have historically relied less upon the market. Local governments often vie with each other for company investments, offering tax abatements and direct subsidies, but they also impose regulation and fees and charges of various kinds, through nonmarket procedures. Government, at all levels, is more than a constituency, of course, because, being above (in theory, anyway) or often aside from the various parties, it enables each to attempt to define, establish, and enforce rights in law. However corporate managers and constituencies deal with each other, the techniques used and the outcomes reached eventually must be sanctioned by legal right, if the parties are to establish stable and enduring nonmarket relationships. Consequently, the emergence of corporate constituencies has taken place in a political setting of legislative lobbying and court actions, though the immediate drama of demonstration, strikes, marches, and boycotts catches the public eye and lingers longest in popular memory.

The Union Movement

Employees were among the first of the constituencies seeking a nonmarket method of setting their wages, hours, and conditions of work. Early in the nineteenth century they formed unions, seeking to impose terms on their employers that were more acceptable than those offered in the market. They argued in 1806, as they did long afterwards, that all workers possessed natural and inalienable rights, as intended in the Declaration of Independence, to combine for mutual protection. A Philadelphia judge, in the first recorded labor case, ruled that their combination interfered with the natural regulation of supply and demand, thereby injuring the paramount rights of individuals.[12] For a century and a quarter the courts continued to side with employers, most of whom fiercely resisted unions and firmly rejected what gradually became known as collective bargaining. Until the latter half of the century, no union succeeded in sustaining itself more than briefly. The growing economy and fast-paced industrialization after the Civil War encouraged the formation of large companies, and courts found the law flexible enough to legalize the vast combines. They were not combinations of individuals as were unions, but through the magic of legal interpretation they became "persons" and so entitled to full constitutional protection of their property.

Railroads employed just over a million workers by 1900, and more than doubled that number by 1920. Manufacturing firms hired thousands of employees and in many industries impressively concentrated production. By 1901 four rubber companies accounted for all output; in steel, the four largest turned out more than 70 percent of the economy's production, and the top four producers of transportation equipment provided almost 60 percent of the industry total.[13] Unions had begun to establish themselves, but for the most part only among skilled workers, and rarely in the big manufacturing plants. At the turn of the century no more than about one out of twenty had joined; a decade later they had doubled their proportion to one out of ten, which still made them but a small fraction of the labor force.

Managers and courts treated unions as third-party outsiders who endangered corporate property and infringed upon the liberties of individual, nonmember employees. Judges could perceive no legitimate, lawful union function because neither the union nor the employee members possessed any property rights in their jobs. The strikes, picketing, and secondary boycotts thus were at best maliciously destructive of employers' property; at worst they were concerted activities tainted with the dangers of both popularly feared monopoly and violent insurrection.

In fact, violence was an all-too common aspect of employees' struggle to form unions and seek bargaining rights. Employers, seeking state help, often responded with force and intense hostility to any worker attempts to call attention to their condition. Violence was widespread in the great railroad strike of 1877 that began on the Baltimore and Ohio and spread spontaneously throughout the country except for the South and New England. The railroad unions were small and ineffectual; their leaders neither directed nor controlled the outraged, desperate workers. A Philadelphia militia fired upon a crowd of strikers, their wives, and children, killed ten to twenty, and wounded sixty or seventy. The angry crowd then set fire to company property and burned 104 locomotives, 2,552 cars, and the entire depot. In the Louisiana sugar strike of 1887, the peaceful and mostly black laborers met swift and brutal repression. The state militia killed a number of the strikers, and, shortly after, a unofficial force of armed men provoked panic and violence that killed thirty-five unarmed blacks; two of the informal black strike leaders were lynched. At Homestead, Pennsylvania, in 1892, steel workers on the river bank and company Pinkerton guards, ferried up the Monongahela on barges to the struck plant, fought a fierce battle. It ended in defeat for the company forces, but later with 8,000 soldiers of the National Guard to protect the plant, the general manager, Henry Clay Frick, opened the plant and routed the union.

In the Pullman Strike of 1894, which won the support of rail workers across two-thirds of the country, large-scale rioting broke out in Chicago when the rail companies attempted to break the strike. Both federal troops and state militia were called out, and on July 7, four people were killed and twenty wounded in a battle between the mob and the militia. The Ludlow, Colorado, strike of 1913–14 well illustrated the murderous determination of some employers to resist unions and the desperate attempts by workers to sustain them. Though the strikers' demands were moderate—such as enforcement of state safety laws and the right to trade at stores of one's own choosing—the mining company refused to deal with the union in any way. On October 7, 1913, guards attacked a miners' tent colony, killed a miner, and wounded a small boy; several days later they fired into a miners' meeting and killed three men. Miners then retaliated, killing a guard and firing on an armored company locomotive. The next spring the violence began anew. National Guard troops machine-gunned a tent colony; they killed a number of miners, and in burning the tents, killed eleven children and two women. The miners were provoked into a rampaging counterattack. Federal troops finally restored order, but only after seventy-four people had been killed.[14]

Many other examples of violence in labor unions' attempts to organize can be found in our histories. Their numbers diminished after Congress passed the National Labor Relations Act (Wagner Act) in 1935, though the reality of violence lingered on into the 1950s, and for many people the aura of violence around unions continues to color their perceptions of labor and collective bargaining. The Wagner Act required a peaceful selection of unions as bargaining agents, as determined by majority vote of those in a designated unit. Once selected, managers were required to recognize the union and to bargain with it. Essentially the same procedure applies today, and its use by unions and acceptance by managers have contributed to the marked decline of labor violence.

Since the year of the greatest strike losses in American history, 1946, union leaders and corporate managers have learned with increasing success how to resolve their disputes and to enhance cooperation through collective bargaining. Even peaceful (i.e., nonviolent) strikes have declined significantly.[15] Union recruitment of new members requires artful and convincing persuasion, with a vocal, active management usually making a case for no union. Workers choose representatives by secret ballots in elections conducted by the National Labor Relations Board.

Once unions gain employer recognition and the right to represent all employees in a given bargaining unit, they seek, first, a written agreement specifying the pay, hours, and work rules. Second, they almost always establish a multilevel grievance process through which employees can enforce the terms of the agreement and gain a hearing for their complaints and protests. If grievants are not satisfied with lower-level answers, they can usually appeal through the union to a neutral, third-party arbitrator for final adjudication.

Collective bargaining has allowed the unionized part of the corporate employees constituency to enjoy a real measure of industrial democracy at the place of work. It has done so without displacing managers' key function as organizational coordinators or impairing their opportunities for entrepreneurial initiative. It has challenged and checked managers' arbitrary discretion, though. It also continually reminds them that their policies are measured not only by the profit and loss balances of the financial records but also by their effects on human expectations, social values, and institutional loyalties. Each year union leaders and management representatives negotiate tens of thousands of agreements at local plant, company, industry, or national levels. For the most part they perform professionally, carry out their bargaining expertly and efficiently, resolve mutual problems, and exploit common opportunities. For most of them, the labor violence of their great-grandparents' time or even the bitter strikes their grandparents may have known half a century ago, have been lost to memory. They are the stuff of which history is made.

Unfortunately, the precedence of violence in public life has a longer history than people's memories. Angry, threatened, desperate, and fearful groups have, all through American history, used violence to protect "things as they are" or to win changes. It is a tradition to which some members of corporate constituencies and others in recent decades have appealed. Rap Brown, an early civil rights leader, for example, contended that "Violence is necessary and it's as American as apple pie." Business leaders have come to realize that using violence to protect their status is an increasingly poor way to win public approval.

Should a group win legally defined rights, it does not necessarily insure practical enjoyment of them. Immediately after passage of the Wagner Act, for example, many managers and employers doubted its constitutionality and refused to abide by its rules and strictures. They continued to resist unions and to deny good-faith bargaining to their employees. They set up company-dominated unions and were as adamant and dilatory as they could be within—or even beyond—the law. Continued managerial resistance encouraged unionists to innovate, developing new organizing techniques that would both weld the employees into effective fighting groups and also win public notice and support. In so innovating they established precedents to which other constituencies would later appeal.

One innovation was the sit-in strike. The first one that received public attention occurred in an Akron rubber plant in 1935, and it appears to have developed spontaneously over a local shop grievance. Top company managers had changed their inventive pay system after a headquarter's analysis, but had also confused both employees and plant managers about its details. The disgruntled employees complained that they could not figure out how much they were making; local plant managers could not help them, since they did not understand the system either. The workers demanded a return to the old systems, and until their demand was met, they decided to fold their arms and sit by their machines. In a few days, company officials dealt with the grievance straightforwardly, adjusted the methods of wage calculations, and instructed everyone in the complexities of the settlement. The workers returned to work. Nonunion workers in scattered plants of Chrysler, Bendix, Midlands Steel, and Kelsey-Hayes also engaged in brief walkouts and sit-ins. The industrial unrest manifest in these strikes was unfocused and mostly as spontaneous as the strikes of 1877 on the railroads. The strikes usually broke out before union leaders had ever appeared.

At the end of 1935, union leaders made more planned organizing attempts in the vast production plants of General Motors. Faced with management's implacable hostility to unions, the workers borrowed the rubber workers' newly discovered technique of "sitting down." It was so popular that union leaders had to scramble to take control. At the large Flint, Michigan, works, union leaders strongly supported the sit-down strikers when city police attempted to block the passage of food and supplies to those in the plants. Fighting broke out and the event made headlines from coast to coast. The "sit-down" spread across industrial American as the civil rights sit-ins were to spread a quarter-century later. Outraged managers saw basic property rights imperiled; many concerned citizens saw sit-ins as immoral, illegal, and unconstitutional. Some insisted the government should remove the strikers by force. Fearing bloodshed, neither state nor federal officials attempted such removal. Eventually, through negotiations, the sit-ins were relieved and settlements were reached. The dangerous but dramatic sit-in had worked, and most important for the future, it had proved a potent attention-catcher that did not have to involve violence, *if* managers would negotiate and bargain with the protesters.

Union use of sit-in strikes rapidly declined after the mid-1930s, but unionists have continued to rely from time to time upon work stoppages, picketing, and secondary boycotts to exert economic pressure upon employers, when their leaders and managers can not otherwise resolve negotiating disputes. Despite, or because of, the reliance on these techniques, unions have not gone from strength to strength as an employees' corporate constituency. Their membership reached a peak share of the labor force, about 34 percent of nonagricultural employees, in 1954, and has been sliding ever since to 15 to 17 percent in 1989. Even as labor unions peaked in economic power and political influence, other corporate constituencies, long ignored, unorganized, and powerless, began to stir. They had before them the experience of successful union organizing through the 1930s and 1940s, as well as the precedents of various techniques that might be used to dramatize their demands.

The Civil Rights Movement

Among the first to act were African Americans. As citizens they protested segregationist laws that denied their civil rights, and as consumers they sought to overturn those laws that denied their market rights. They sought equality with whites in bus seating, restaurant service, and hotel accommodations as well as equal schooling. They pressed business managers to deal with them and demanded of legislators and courts their proper legal rights.

The most famous action directed at both segregation and business support of it began on December 1, 1955, when the Montgomery, Alabama, police arrested Mrs. Rosa Parks for refusing to sit in the "black" (rear) section of a city bus. She was the fifth person arrested that year for violating segregation rules. Indignation had been rising for some time as a result and suddenly the African American community coalesced; in four days its African American pastors won the support of the community's 50,000 members and began an almost completely effective boycott of all buses that hurt not only the bus company but also downtown businesses. The African Americans, preparing for a long struggle, set up car pools with forty-eight dispatch points and forty-two collection centers. Managers of white-run insurance companies responded by canceling insurance on station wagons used in the pools. The hard-hit downtown merchants remained publicly silent, fearing that any conciliatory approach would produce a white counter-boycott.

The Montgomery boycott continued for over a year, until the Supreme Court invalidated segregated intrastate busing in April, 1956, and specifically ordered, in December, the company to provide unsegregated bus service. In the meantime consumer boycotts spread across the South and white opponents retaliated. Many employers and managers inflicted economic reprisals upon those who supported civil rights. They denied loans, foreclosed mortgages, fired employees, and otherwise harassed activists. It was clear that, despite some important successes, boycotts inflicted losses on those participating as well as upon their targets. In addition, boycotts were difficult to organize and costly to maintain. They required the devotion and active participation of many people to succeed and they took time to inflict their economic costs. Moreover, they did not always pinch the most vocal civil rights opponents. Other, more sharply focused, less costly techniques were needed to back the demands of African Americans for equal treatment by business.

African American activists, particularly in the Congress of Racial Equality (CORE), had long been using sit-ins, throughout the 1943–1959 period, to protest segregated public accommodations in such cities as Chicago, St. Louis, and Baltimore. CORE had received little public attention and only moderate success, but with a change in the Supreme Court's judgment of segregation laws, a shift in national public opinion, and the concentration of African Americans in cities both North and South, after World War II, sit-ins offered increasing advantages. First, they were dramatic and newsworthy in a nation that highly honors property rights as basic to an ordered society. Second, they were efficient, involving only a few demonstrators (compared to a boycott), targeting a specific culprit, and effectively disrupting both production and sales. Third, they could be used against any business that conducted retail operations.

On February 1, 1960, four African American students from the Agricultural and Technical College, Greensboro, North Carolina, entered a Woolworth store and sat at the "white counter" in violation of caste rules. They were refused service. Within eight weeks civil right activists had conducted sit-ins in seventy-eight southern communities, and police had arrested 2,000 of them. In less than two years approximately 70,000 people in both the North and South had actively protested segregated services for African American customers, resulting in changes in at least 110 cities,[16] but mostly on the periphery of the deep South.

MANAGERS' RESPONSE TO MARKET MOBILIZERS

One of the sit-ins that year involved a Maryland restaurant chain and its president, G. Carroll Hooper. He carried the dispute to the U.S. Supreme Court. There testimony and arguments provided insights into what was probably quite common managerial thinking about consumer constituencies at that time. The Court record describes the incident:

> A group of fifteen to twenty Negro students . . . went to Hooper's Restaurant to engage in what their counsel describes as a "sit-in" protest because the restaurant would not serve Negroes. The hostess, on orders of Mr. Hooper, the president of the corporation owning the restaurant, told them, "solely on the basis of their color," that she would not serve them. . . . [The students] refused to leave when requested by the hostess and the manager; instead they went to tables, took seats, and refused to leave, insisting that

they be served. On orders of the owner the police were called, but they advised the manager that a warrant would be necessary before they could arrest [the students.] The managers then went to the police station and swore out the warrants. . . . [The students] had remained in the restaurant in all an hour and a half, testifying at their trial that they had stayed knowing they would be arrested—that being arrested was part of their "technique" in these demonstrations.[17]

The record also provides Mr. Hooper's reasons for the harsh policy he followed:

I set at the table with him and two other people and reasoned and talked to him why my policy was not yet one of integration and told him that I had two hundred employees and half of them were colored. I thought as much of them as I did the white employees. I invited them back to my kitchen if they'd like to go back and talk to them. *I wanted to prove to them it wasn't my policy, my personal prejudice,* we were not, that I had valuable colored employ- ees and I thought just as much of them. I tried to reason with these leaders, told them that *as long as my customers were deciding who they wanted to eat with, I'm at the mercy of my customers.* I'm trying to do what they want. If they fail to come in, these people are not paying my expenses, and my bills. They didn't want to go back and talk to my colored employees because everyone of them are in sympathy with me and that is we're in sympathy with what their objective are, with what they are trying to abolish. . . .[18]

He defended his refusal to serve African American customers with a market argument: As a manager, he was but a passive responder to market forces, providing those services demanded by customers. Even if, as in this case, those demands might deny attractive sales and profit opportunities—serving African Americans—he had to accept the "given" market outcome. Mr. Justice Douglas accepted Hooper's claim that he was not personally prejudiced against serving African Americans. In fact, Douglas noted that "in most of the sit-in cases before us, the re- fusal of service did not reflect 'personal prejudices' but business rea- sons."[19] But if Hooper and other managers were animated by business reasons—to enhance profits and avoid losses—further questions arise.

Throughout the 1950s African Americans were becoming an ever- more attractive market for restaurants and such service businesses; the trend could have been expected to continue, and in fact it did. They were moving into urban areas faster than whites and their income was rising faster as well. (Of course, in absolute magnitude it still lagged considerably.)[20] The business argument is certainly peculiar, for it rests on the notion that passivity is a prominent characteristic of business managers. For market reasons, Mr. Hooper felt compelled to deny his own personal and assertedly unprejudicial racial views, thereby deny- ing his business corporation the possibilities for profits by enlarging his range of customers.

No doubt he feared that opening his restaurants to African Americans would lose him more white customers than the African American customers he might gain. Such a reasonable fear could have been allayed, though, by joining with other profit-seekers in his industry to change community caste rules through legislation outlawing segregation in restaurants; then, all eating establishments would have had to change, and all could have shared in the profit opportunities of serving the previously excluded African Americans. Instead, he fought his case up to the Supreme Court, maintaining that he was protecting his economic interests. Could one conclude that he was using the impersonal, corporate "business argument" disingenuously? Was he actually hiding his own personal predilection for segregation, ironically even at the expense of foregone profits and smaller markets? Constituencies soon learned to take at less than face value managers' ready arguments that because of business concerns they denied the rights sought.

Mr. Justice Douglas, in his dissent, raised another question that bedeviled business managers increasingly as constituencies demanded recognition.

> Who in this situation, is the corporation? Whose racial prejudices are reflected in "its" decision to refuse service to Negroes? The racial prejudices of the manager? Of the stock holders? Of the board of directors?[21]

If business managers were not prejudiced, as they usually maintained, and were sacrificing profits, did they bear responsibility for politically intolerable, socially inhumane, economically wasteful community customs, rules, and traditions?

Do Managers Act by Not Acting?
U.S. Steel in Birmingham

In 1963 Mr. Charles Morgan, a 33-year-old white lawyer in Birmingham, Alabama, raised the question and directed it to the chief officers and board members of United States Steel, then the city's largest employer. Birmingham was a segregationist bastion of the South, and its whites were strongly opposed to change. In one of the worst acts of violence, on September 14, 1963, the bombing of an African American church killed four young girls in their Sunday School rooms. This atrocious act shocked many white leaders in the city and across the nation. Morgan asked U.S. Steel's management, "as the largest employer in the city . . . , the largest owner of real estate, the largest private bank depositor and purchaser and seller of goods," if it did not bear more responsibility for conditions in Birmingham than it had yet shouldered. His answer was "yes." The chairman of the company, Mr. Roger Blough, with his board could "do something about Birmingham." He could call the state governor and the mayor and get action.

Mr. Blough rejected Morgan's answer to his rhetorical question. Speaking in his official position Blough replied that members of management, *as citizens*, should use their influence persuasively to help resolve the problems of communities but

> . . . any attempt by private organizations like U.S. Steel—to impose its views, its beliefs and its will upon the community by resorting to economic compulsion or coercion would be repugnant to our American constitutional concepts. . . . So even if U.S. Steel possessed such economic power—which it certainly does not—I would be unalterably opposed to its use in this fashion.[22]

He did not explain why socially responsible corporate action in the community would be inherently coercive, imposing an unwelcomed view and compelling agreement with its demands. His characterization of any corporate action in such terms appeared needlessly extreme, since he argued that the company did not enjoy such power and influence in any case. He left a small escape door from his condemnation of corporate community action: While he was unalterably opposed to company coercion, imposition, and compulsion, he presented no reason why the large and dominating company might not be able to act, much as managers could act individually, persuasively influencing public decisions and community behavior.

One commentator found Mr. Blough's denial of power less than convincing:

> [His disclaimer is] a new and novel philosophy. What is the history of American business and industry filled with if not rugged chapters of applying every imaginable—or of pressure to establish every possible advantage over the competition, widen profits, corner markets? . . . Under such a doctrine, it would be logical to presume that the next time a U.S. Steel subsidiary was scored by the local civil improvement association for pouring refuse into a river or pollution into the air from smelter smokestacks it would not raise a corporate finger to influence the city fathers in their decision over the alleged nuisance. And of course there would not be an iota of pressure applied at city hall or in the legislature for favorable disposition of a tax case. Maybe a Utopic of selfless corporations is nearer than we thought.[23]

Professor Andrew Hacker, reviewing the controversy some years later, was less stinging in his evaluation of Blough and the managers who borrowed his argument. He admitted that U.S. Steel's willingness to cooperate with the folkways of Birmingham actually served to strengthen those social patterns. The managers were not mere bystanders; by implication they had taken sides on an important and contentious public issue. Corporation leaders can not help but take sides and intervene in community actions. Their looming presence assures them of saying much by being silent, of acting by inaction, and of approving by failure to dissent. The choice of "not acting" is unavailable. He pointed out that if managers went along with a dominant ideology, they did not therefore have to defend it. He strongly insisted that business managers "can and should justify [their] position solely on economic grounds." He warned, though, that community emotions over so contentious an issue as racial segregation might be so strong that public officials and townspeople would ignore economic arguments.[24] On such a matter, managers of large corporations might possess less influence than their size and dominant economic position of their firms would seem to imply.

Most managers probably would agree with Hacker and find comfort in his injunction to define and defend community responsibilities on economic grounds. Such a defense appears businesslike and hard-headed, and it is traditional. Unfortunately, "economic grounds" is neither a self-evident notion nor a clear-cut, substantive standard. In the Hooper case, briefly reviewed above, a manager used "business reasons" (i.e., economic reasons) to argue for segregated restaurants. Since there is good reason to believe that desegregation in 1960 could have opened up wider markets, provided more customers, and enhanced business opportunities, Hooper's "business reasons" were more rationalization and legal rhetoric than a convincing argument.

At the same time that African American consumers were protesting segregation in the late 1950s and early 1960s, a number of African American leaders elsewhere, and with particular success in Philadelphia, were seeking new, additional job opportunities for discriminated-against African Americans. Instead of arguing with business managers over their abstract community responsibility, they chose to apply market (or economic) pressure, through a consumer boycott, to business managers. Like the early civil rights boycotters, they were ready and willing to accept Hacker's and managers' usual position: Business should justify its response to the constituencies and the community on economic grounds. The boycotters would provide the "business reasons" for managerial response by mobilizing individuals in the market. That mobilization enabled the boycotters to wield economic influence through their concerted activities in ways not unlike those of the business corporation itself, a voluntary collective of stockholders acting in concert.

Market Exit to Support Constituency Demands

In early 1960, and continuing beyond 1962, a group of African American pastors organized a Selective Patronage Program (SPP) under the skillful, though hardly public, leadership of the Rev. Leon H. Sullivan. He was pastor of the large Zion Baptist Church in Philadelphia and well aware of the limited job opportunities that confronted many of his parishioners, particularly teenagers and younger members. He was wary of mounting an explicit boycott, for its legality was then still clouded, particularly if used for SPP purposes. He and his fellow pastors stressed the voluntary, but informed, decisions of African American *individual* buyers in the market. They noted that although the city's 534,000 African Americans were nearly 27 percent of the population, they held but 1 percent or less of the "sensitive" jobs in private businesses—professional, technical, and managerial positions, including very junior executive and supervisory posts, clerical and stenographic jobs, and skilled manual as well as driver-sales jobs. They believed that African American consumers ought to know which companies refused to increase their African American hires in these designated positions; the African American supporters, in turn, could then refuse to purchase the companies' products until they were satisfied that the managers had decided to act positively to the SPP demands.

Sullivan said that the pastors were "the only morally organized network of communications" in the African American community. The SPP needed a moral, as well as a mass, base. "To us this is a Biblical—a spiritual—movement. We are saying to our congregation that we cannot in good moral conscience remain quiet while our people patronize companies that discriminate against our people."[25] The pastors had decided that since they held positions of influence in their communities, for them not to speak and work against discrimination was to condone it. The reasoning behind this decision was the opposite of that taken by U.S. Steel's Blough.

The goals the pastors sought were modest but specific. After a determination of the share of "sensitive" positions held by African Americans, they asked managers in formal meetings to recruit and hire African Americans until the percentage figure was significantly raised. They never sought a proportion anywhere close to that of the city's African American population, but insisted upon a good-faith effort, detailing the number or percentage of African Americans in particular jobs and departments. Total numbers did not interest them. If many African Americans were employed in only entry-level jobs and none in higher positions, the company was subject to the SPP. The pastors set deadlines and asked for follow-up reports. If a company did not meet a deadline, then on the following Sunday between 300 and 400 African American pastors reported that failure to their congregations throughout the city. The next day handbills with the same information appeared in beauty shops and barber shops in all Philadelphia's African American communities. As the weeks passed, the number of people involved typically increased as the SPP pastors exerted additional effort to inform more and more African Americans about the situation.

The first company against which the pastors directed their SPP was a baking company; it capitulated in eight weeks. A newspaper settled in seven weeks. A soft drink company came to terms in two weeks, and an oil company in one week. Some companies resisted SPP, complaining that it was both a danger to the free market and very unfair to them. If managers gave in to an SPP, might not another group use the technique to support a blackmail racket—"we won't firebomb your plants if you pay up!"? Others pointed to the dilemmas posed by the SPP demands. The pastors asked an oil company, for example, to hire African American truck drivers during the summer months when regular white drivers were laid-off. The boycott as instituted and continued though the company was legally bound by a collective bargaining agreement to protect the recall rights of the out-of-work white drivers.

SPP continued for several years, directed at about thirty companies. Sullivan claimed that it opened up at least 5,000 jobs to African Americans that otherwise were not then available to them. He made a reputation for himself that later enabled him to play an important role in another constituency movement just beginning to form among stockholders. He and his fellow pastors carefully did their homework, they listened to business managers, and while pushing the managers harder than they cared, and not always agreeing with them, never made outrageous or preposterous demands. They helped white managers in Philadelphia to open up a wholly new, hitherto untapped labor market, that benefitted business indirectly as well as directly. The SPP received little publicity from the regular news media; the parties directly involved proceeded with little drama. Editors and reporters may have believed themselves or their own organizations too vulnerable to encourage SPP even indirectly by covering its progress or reporting its members' activities.

Not all activities in support of opening employment opportunities to African Americans were carried out as quietly as that of the SPP in Philadelphia. Elsewhere opponents often responded publicly and on occasion with legal action. In 1964 the managers of the Bank of America, San Francisco, sought to gain and hold the initiative in dealing with CORE. They had hoped to avoid demonstrations and a boycott by publicly affirming their equal opportunity employment policies and reporting their achievements regularly to the state Fair Employment Practice Commission. Using press conferences and extensive briefings, they secured the publicity they sought, but CORE leaders were still able to discover a dispute over two points: they wanted a specific hiring quota and they insisted that the supporters supply company hiring data to CORE. Because of the dispute, CORE demonstrated at bank branches and picketed them from late May until the end of August. Later, in disputes with other groups, the bank was boycotted and some branches were bombed. Nevertheless the bank managers maintained and argued for a stance far different from that of Blough in Birmingham:

> The corporation, by virtue of its own enlightened self-interest, the conscience of its officers, and expectations of the public has a role to play in solving contemporary ills. Profits are and must continue to be the central concern of any responsible enterprise. . . . But in the long pull, nobody can expect to make profits—or have any meaningful use for profits—if our entire society is racked by tensions.[26]

In Mississippi business managers responded to constituency demands not with initiative in making changes, as did the Bank of America, or even with the passive "neutrality" of U.S. Steel. They resisted the demands, checking them with law suits against the consumer boycotts that accompanied and underlined the demands. One case was carried eventually to the Supreme Court[27] and the resulting decision is a landmark for future constituency actions. We will examine the Court's reasoning and conclusions later.

The case arose out of incidents in Port Gibson, Mississippi, where most of the African American citizens, well over half the population, boycotted white merchants in the spring of 1966 for not supporting African American civil rights and for refusing to hire African Americans in such jobs as clerk and cashier. The boycott continued effectively, and at great loss to local white merchants over seven years, until the Mississippi Supreme Court found it illegal. The court ruled in 1972 that "coercion, intimidation, and threats" were a part of the boycott and that participants had used "threats, social ostracizing, vilification, and traduction" to enforce their demands. It awarded twelve merchants $1,250,599 in settlement against the National Association for the Advancement of Colored People (NAACP), the organization in whose name the boycott operated.

Like union workers before, the NAACP found the boycott a difficult technique to use effectively. First, managers opposed to boycott demands could fight them through the courts, as the Port Gibson merchants had done. The borderline legality of boycotts threatened the leaders and their organizations with punishing penalties. Second, boycotts imposed heavy direct costs on ordinary participants, because they almost always take considerable time and involve inconveniences at best, or second-rate (or even no) alternatives at worst. It was clear to many constituency leaders that they needed to bring to bear upon managers other, more manageable, but still effective pressures and to generate them in ways less costly than lengthy, hard-to-organize and difficult-to-maintain boycotts.

THE SEARCH FOR VOICE
TO SUPPLEMENT MARKET EXIT

In early 1967 an employee and community constituency in Rochester, New York, discovered—or stumbled onto—a way of applying pressure to management, without relying exclusively on a boycott that was difficult to organize and control. It was a technique that a few groups had used from time to time, but whose possibilities few others had explored. Labor unions had never made any effective use of it, though many might have done so. It was attractive because it activated corporate procedures already in place, appealed to American democratic ideals, and allowed many people to show their support of the constituency through quiet voice and vote. Since it had seldom been used, except by contending business managers or dissident corporate investors, its use by broad interest groups generated wide public interest, thereby assuring national publicity. The new way to generate pressure on managers was through voting proxies, introducing resolutions, and speaking or demonstrating at annual meetings.

A constituency solicited stockholders' proxies, and on the basis of their own usually slender holdings protested management policies and/or introduced "broad social" resolutions at corporate annual meetings. Business managers were disconcerted by the new technique; *Baron's,* using the Blough argument of 1963, warned that constituency use of proxies was irresponsible and dangerous.

> Companies best serve their stockholders and communities by sticking to business. . . . For Kodak [the company first involved] and the rest of U.S. industry, it's time to stop turning the other cheek . . . management is the steward of other people's property. It can never afford to forget where its primary obligations lie.[28]

Mr. Saul Alinsky, the consultant and adviser to the African American Rochester group, FIGHT,[29] was as delighted with the tactic as managers were disturbed. He later wrote:

> Corporate executives sought me out. Their anxious questions convinced me that we had the razor to cut through the golden curtain that protected the so-called private sector from facing its public responsibilities. Business publications added their violent [sic] attacks and convinced me further. In all my wars with the establishment I had never seen it so uptight. I knew there was dynamite in the proxy scare.[30]

He exaggerated managerial response. Managers clearly opposed constituency use of the hitherto sedate annual meetings and any rival solicitation of proxies, but their violence was purely verbal. It grew out of an almost aesthetic distaste for uncomfortable, "nonbusiness" issues introduced into a traditional and ossified ritual; it also fed on a fear that they were to be challenged from within their organization by people who did not appreciate the values and attitudes that, they believed, were inherent in the role, position, and status of the stockholder. The Advertising Council and other national business associations had long sought to convince Americans that widespread shareholding was evidence of "peoples' capitalism." However, the appearance at annual meetings of constituencies claiming, not only to be speaking *for* stockholders, but also *to* stockholders, was as alarming to business managers in 1967 as Polish workers' demands for free, open elections were to Communist party managers in the "people's democracies" in 1978. In both cases those in whose name top officials governed sought to realize the substance of a procedurally promised democracy.

Alinsky and the leaders of FIGHT did not originate the use of the annual meeting as a forum or the voting of proxies to influence managers, but they gave it publicity. Other and new corporate constituencies forming across the nation examined closely the possible use of proxies and annual meetings. They discovered that though dissident stockholders had gained some legal rights and had operated for many years as corporate gadflies, they had little to show for their efforts. The Securities and Exchange Commission (SEC), established under a regulatory act in 1934, had long been charged with the duty to compel full disclosure to investors of material facts about securities offered and sold in interstate commerce.

In 1942 it had adopted Rule 14a–8 requiring companies to include shareholder resolutions in their proxy statements *if the resolutions' proposals were proper subjects for stockholder consideration.* A number of professional dissidents, particularly Lewis Gilbert and his brother, John, made use of the SEC rules to seek two goals: First, to expand corporate democracy, by transforming the annual meeting into "a modern extension of the New England town meeting," and second, to increase investors' dividends.[31] Lewis Gilbert was unsparing in his criticism of corporate management:

> It must not be forgotten that we and not management are the employers, but over the years the managers have usurped the rights of ownership through their absolute control of the proxy mechanism. . . . [All the] groups who help make up the segments of business ownership which comprise the American shareholder family are entitled to the opportunity to be fairly represented in the council chamber of owners, just as we have majority and minority viewpoints in the Congress.[32]

The two brothers have regularly framed proposals for inclusion in proxy statements, assured that they would be discussed at annual meetings.

They concentrated upon unquestionably business-relevant topics—managerial compensation and stock options, dividend policies, cumulative voting for board directors (a way of concentrating dissident votes and making them more effective), and location of annual meetings. On a few occasions the Gilberts raised broader subjects; in 1941, for example, they questioned the chairman of Standard Oil of New Jersey about oil sales to the Axis powers, and they protested to the board of Curtis Publishing Company the "scandalous isolationist policy of the editors of *Saturday Evening Post.*" But they never extended their resolutions in the direction that FIGHT's interests lay.

FIGHT and the Eastman Kodak stockholders who sympathized with its goals did not propose any resolutions; they merely withheld their votes at the annual meeting, protesting company policy. The dissidents won most of the publicity, but management took 84 percent of the total vote and all Kodak's officers won easy reelection. Leaders of other interest groups and corporate constituencies decided that more could and should be done with stockholder votes.

During the Vietnam War, a number of shareholders in Dow Chemical Company, a maker of napalm under contract to the Defense Department, had been concerned about the company's contribution to death and destruction in Southeast Asia. Dow's production had become a matter of wide public attention and the company a favorite target for students demonstrating against the war. Between 1966 and 1968, it was the object of 183 major campus demonstrations. In the fall of 1967 one-third of all campus demonstrations protested the presence of Dow's recruiters, and over the academic year 1967–68 alone, its college recruiters were banned by, or had to put up with, demonstrations at 113 institutions.[33] In addition, a number of groups sponsored a boycott of Dow's consumer products, Saran Wrap and Handi-Wrap.

The matter of stockholder rights arose, however, because the Medical Committee for Human Rights had received ten shares of Dow stock as a gift. The chairperson, Dr. Quentin Young, and the committee decided to exercise a stockholder's privilege to propose an amendment to the company charter. The committee's proposal for stockholders' votes was "that napalm shall not be sold to any buyer unless that buyer gives reasonable assurance that the substance will not be used on or against human beings."[34] It was sent to the secretary of Dow Chemical in early March, 1968, but too late for inclusion in that year's annual meeting. Nine months later the committee wrote again and the company replied that after study it had decided not to include it as a stockholder resolution. The reasons given tended to contradict each other. On the one hand, Dow managers argued, the proposal was too specific, dealing with a matter of ordinary business operations and thus within managers' expertise. On the other hand, it was too broad, for it had been submitted for the purpose of promoting general political or social causes.

The committee appealed to the SEC, which upheld the managers' position that the proposal did not need to be sent to all stockholders in the proxy resolutions. It then took its claims further to the U.S. Court of Appeals in the District of Columbia. The majority of the court declared:

> The clear import of the language [in section 14(a) of the Securities Exchange Act], legislative history, and record of administration of section 14(a) is that its overriding purpose is to assure to corporate shareholders the ability to exercise their right—some would say their duty—to control the important decisions which affect them in their capacity as stockholders and owners of the corporation.[35]

The court went on to note that Dow Chemical did not claim to have produced napalm for business reasons but despite them.

> The management of Dow Chemical Company is repeatedly quoted in sources which include the company's own publications as proclaiming that the decision to continue manufacturing napalm was made not *because* of business considerations, but *in spite of* them; that management in essence decided to pursue a course of activity which generated little profit for the shareholders and actively impaired the company's public relations and recruitment activities because management considered this action morally and politically desirable. The proper political and social role of modern corporations is, of course, a matter of philosophical argument extending far beyond the scope of our present concern; the substantive wisdom or propriety of particular corporate political decisions is also completely irrelevant to the resolution of the present controversy. What *is* of immediate concern, however, is the question of whether the corporate proxy rules can be employed as a shield to isolate such managerial decisions from shareholder control. After all, it must be remarked that "[t]he control of great corporations by a very few persons was the abuse at which Congress struck in enacting Section 14(a). We think that there is a clear and compelling distinction between management's legitimate need for freedom to apply its expertise in matters of day-to-day business judgment, and management's patently illegitimate claim of power to treat modern corporations with their vast resources as personal satrapies implementing personal political or moral predilections. It could scarcely be argued that management is more qualified or more entitled to make these kinds of decisions than the shareholders who are the true beneficial owners of the corporation; and it seems equally implausible that an application of the proxy rules which permitted such a result could be harmonized with the philosophy of corporate democracy which Congress embodied in Section 14(a) of the Securities Exchange Act of 1934.[36]

The court remanded the case to the SEC, asking it to rule in a manner consistent with its decision. The commission appealed to the Supreme Court, which dismissed the case as moot. Dow Chemical had already included the committee's resolution in its 1971 proxy statement. In the meantime the proxy resolution had also become moot, because Dow Chemical no longer produced napalm.[37]

The court's decision backed up and provided additional support for an earlier, narrowly approved SEC ruling on stockholder resolutions. In February, 1970, four young lawyers formed the Project for Corporate Responsibility (the Project), promising that through it they would mount Campaign GM, an effort to change the policies and governance of America's prototypical large business corporation. It is notable that three of the four founders had been active in the civil rights movement, one of them having worked with Saul Alinsky and FIGHT in their encounters with Eastman Kodak in Rochester.

A cadre of activist leaders were available. They were professionally well trained and also experienced in the practical organizational work and the managing of publicity necessary for constituency success. They recognized the opportunities that awaited them in directing their efforts toward business firms; they also relished the challenge of dueling with private power after having tested their mettle in contending with public officials over changes in public policy through the 1960s.[38]

The Project owned twelve shares of General Motors stock, and on that basis submitted nine resolutions that its leaders wanted sent to the company's 1.3 million stockholders. They reasoned that "modern notions of the corporation rendered social concerns legitimate areas for corporate activity and for shareholders consideration." They went further to declare, "it was good business for the corporation to weigh social factors in business decisions."[39] General Motors responded with the contradictory arguments noted above that Dow Chemical was to borrow and use later that same year—the proposed resolutions could be omitted because they were at once "ordinary" business matters and also could be ignored because they were "general" public matters. The SEC agreed with General Motors that seven of the proposed resolutions could be excluded but, because of two abstentions, found two-to-one that the remaining two resolutions should be presented to stockholders. One of the resolutions was an amendment to expand the number of directors, and a second was a proposal to create a shareholder committee.

The two 1970 court decisions, the General Motors and the Dow Chemical decision, were important. for they were the first to require corporate managers to include in proxy statements resolutions whose implications were clearly social and not merely financial. Twenty years earlier the SEC had upheld the right of managers of the Greyhound Corporation to refuse to include in the proxy statement a proposal phrased as "A Recommendation that Management Consider the Advisability of Abolishing the Segregated Seating System in the South."[40] The SEC then explicitly changed its rules to exclude matters "submitted by the security holders . . . primarily for the purpose of promoting general economic, political, racial, religious, social or similar causes."[41] Had the group that drafted the Greyhound proposal been more adept— emphasized the economic advantages and business efficiencies to be won by desegregation—perhaps the SEC would have moved more carefully and less sweepingly. As it was, the issue did not again arise until the Campaign GM issue reached the SEC in 1970.

Under the narrow approval given the proxy proposals for GM stockholders, the SEC at last indicated that proposals with clear social purposes did *not* automatically violate the stockholder proposal rules. There was a legitimate area for constituencies to explore through stockholder resolutions, an area within which they could use the massive proxy machinery of business corporations to publicize and pay for their concerns. Through the proxy route they could properly seek business response and legally encourage change.

In 1977 Mary Gardiner Jones, president of the National Consumer League and former member of the Federal Trade Commission, maintained that the bringing of community issues and concerns to managers was valuable for business managers as well as for the public:

> the real problem with too many corporate managements, as I see it, lies not in their lack of concern for public values but rather in their failure to identify the public's expectations and concerns or to recognize or anticipate the public impact of too many of their decisions. For too long corporations have been accustomed to viewing their role as one of producing goods and services desired by the public and to evaluate the public's satisfaction or dissatisfaction with their performance in terms of their profit and loss statements . . . the need exists to strengthen the forces which can influence and control the day to day actions of management so that external long range social and economic consequences of management actions will be as carefully and systematically analyzed during the decision-making processes as are the more traditional, financial and marketing factors which also bear on that decision.[42]

The SEC was not as convinced as Jones of the value of active stockholder participation, but under pressure from the federal court, it had provided at least a small, but significant, new opportunity for constituencies to exploit. They need no longer rely only upon demonstrations, picketing, and boycotts to make themselves heard by business managers. They now enjoyed a legitimacy that entitled them to be heard through regular, well-established and well-financed channels. They could be assured of the rights of due process and recognition *within* the business organization. The opportunity could become, in time, of course, a problem: Through their regular participation in established procedures, constituencies might become an entrenched part of the business system, and feel the need to defend their rights and privileges against those still outside, yet affected by, corporate decisions. This problem was, in the 1970s, far in the future, however.

What Is a Responsible Institutional Voice?

The controversy between FIGHT and Eastman Kodak and the antiwar protests directed at Dow Chemical had alerted stockholders that they were likely to be asked to vote their shares in ways that required a declaration of position on social, political, and moral issues. At the annual meetings of many large business corporations in 1967, critics had protested managerial policies and had begun to solicit proxies, particularly from large institutional investors.

For years institutional investors had followed the "Wall Street Rule," a popular maxim that had been accepted rather than examined. Its advice, "support management or sell the stock," was not required by law or sanctioned by economics;[43] there was no reason to believe that it was an effective method of influencing managers. Stock sold by an investor is usually easily repurchased by other buyers and even large sales, by themselves, are not likely to affect market value significantly. Acceptance of the rule was a denial of owner responsibility for any of the wider effects that corporate policy and activity might have upon various business constituencies. By the late 1960s, however, few institutional stockholders, managers, trustees, and officers in foundations, colleges, universities, pension funds, churches, or union investments, believed they could long continue as passive voters of corporate stock. They had come to realize that the Blough doctrine of "not acting" was in fact "acting."

In voting with managers, under the Wall Street Rule, they were regularly wielding economic power, though passively resigning it to managers. To investigate proposed "social" resolutions on their own merits and occasionally vote against managers when appropriate, would neither increase nor decrease the social and economic power involved; it only assured a more explicit and responsible consideration of that power which was exercised in any case. Consequently, the institutional investors began to explore more carefully and closely the problems and opportunities they would confront as they began to vote, in less passive ways than in the past, the stock for which they were responsible. They moved with great caution, however, and by the end of the 1980s were still primarily passive stock voters.[44]

In the early 1970s many universities issued reports on their responsibilities as ethical investors, and most foundations prepared explanatory guidelines for themselves. In January, 1971, Harvard University published a report of a special committee chaired by Dr. Robert W. Austin, a retired professor from the School of Business Administration. The report examined ways in which universities and corporate enterprise can work together for constructive social purpose; one part discussed the role of the university as stockholder. It easily reached the conclusion that, as an investor, the university "should strive fundamentally for maximum return," though exceptions might be made on the basis of the university's duty to the more or less immediately surrounding community. As an owner of stocks, the university "need not remain passive in the face of evidence that the company is acting in an antisocial way. . . . Certainly the university should vote its stock on occasion in favor of change for symbolic effect of a great university's taking a position on a social problem." It recommended that the Harvard Corporation should collect facts and evaluate information about the effects of its stock, so that it might be voted in an informed way.

The next year, 1972, Yale University published the most detailed examination of the responsibilities of institutional investors yet seen.[45] It offered a Basic Policy:

> The "moral minimum" responsibility of the shareholder to take such action as he can to prevent or correct corporate social injury to the university when it is a corporate shareholder.[46]

Despite many difficulties involved in adhering to such a basic and minimum level of responsibility and in spite of many objections to active institutional involvement in voting shares, the authors of the report strongly urged explicit, open dealing with the issues. They could not accept the long-observed principle

> that the corporate system depends for its health on ignorance and silence rather than on healthy debate—that certain questions ought not to be asked for fear of getting inconvenient answers. While shareholders may prove, in the short run, to be less concerned for the public interest than some advocates of shareholder responsibility hope, in the long run society will benefit from more widespread participation in moral and social issues. Keeping people away from these issues only increases the atrophy of responsibility already pervasive in a highly organized society. The fabric of trust, so essential for a democratic nation, rests on the reciprocal expectation that persons and institutions will take responsibility for the social consequences—intended or unintended—of their acts.[47]

In 1973 the Ford Foundation had reached conclusions about its responsibilities as a large stockholder. First, to vote always with management seemed inappropriate for a public welfare institution with a deep commitment to social issues; second, many different decision-making mechanisms could be developed to take social considerations into account in the investment process, but whichever were chosen required the gathering of essential information. Many other institutional investors had reached similar conclusions. In a 1971 Ford Foundation sur-

vey forty-nine institutions responded that they had initiated reviews of
social aspects of their investment policies. Almost all had abandoned
the Wall Street Rule, and were examining proxy proposals on their
merits.

By 1977 about thirty universities were regularly voting their stock
only after careful consideration. The multibillion dollar College Retirement Equities Fund (CREF) had taken the lead among pension funds
in carefully considering each proxy resolution on its merit before casting its votes; in 1977 it voted against management recommendations
twenty-one times.[48]

If institutional investors were to vote their stock with any competence, they needed far more information than they had hitherto gathered, as most of them recognized. The president of Harvard University
noted in his report of 1972–72 that "Our experience with . . . stockholder issues . . . led us to conclude that our procedures were inadequate to enable us to obtain all the information we needed or to insure
that all points of view within Harvard could be heard." He approached
a number of other universities, and also foundations, with the suggestion that they establish an Investor Responsibility Research Center (IRRC)
to provide timely and impartial analyses concerning corporate social
responsibilities. Enough support was found to fund a modest effort by
the center. Neither it nor the universities it serves have proved themselves to be impressive innovators in using their stockholdings to press
managers for changes in policies.

Some constituency groups were skeptical that IRRC could provide
unbiased information, or if it did, they feared that it would deflect
pressure from institutional officers to consider seriously their social responsibilities. After fifteen years of operations, though, it has proved
itself a modestly useful agency, supplying information and analyses of
many social responsibility issues to some 170 institutional investors. The
IRRC staff and budget have grown to serve roughly equal numbers of
profit and nonprofit institutions. As many banks and insurance companies as universities and foundations regularly use its services.

The large number of church investors has also established an agency
to help secure information about, and to provide analyses of, social
issues raised through proxy proposals. In 1974 several Protestant denominations and Roman Catholic organizations combined their various
informational efforts devoted to examining corporate social responsibilities. The resulting organization, the Interfaith Center On Corporate
Responsibility (ICCR), coordinates the relevant activities of about seventeen Protestant groups and more than 180 Roman Catholic orders
as they exercise their rights over $7 billion worth of investments. ICCR
does not initiate resolutions nor vote stock on its own, leaving those
activities to member organizations, but it has been far more active and
innovative in publicly challenging managers than the IRRC has been.

There are other less prominent groups serving the same functions as the IRRC and the ICCR in providing data relevant to corporate social issues and making analyses of proposed shareholder resolutions for those who need such help; many constituency groups own little or no stock and thus do not need the informational and coordinating services, but nevertheless push their own particular interests for managerial consideration and submit resolutions for inclusion in proxy statements. There are a number of research organizations ready to serve constituency groups and institutional investors by regularly monitoring policy areas and their social, economic, political and technical developments.

Earlier, when constituency groups were just forming, they often had to rely upon "scandals" and search out evidence of bureaucratic business abuses to win notice for their causes. Increasingly, now, they have access to technical information and meet corporate officials on an equal informational basis; sometimes they may even collect better, more timely, and more complete data about operations in the far reaches of the globe than home offices of large corporations.

There is now in place a wide array of permanent support and research agencies, ready to serve all kinds of constituency groups. Some groups that appeared early have died out, and others have been formed. Constituency groups may arise over a particular and dramatic event, such as a plant closing of a large firm that dominates a community or the dangerous pollution of a whole neighborhood or district. Other groups continue for long periods to protest directly the action of a single firm such as Nestlé or to seek remedies for a single problem, such as the siting of nuclear power plants.

The institutional voices examined above were raised in the 1960s and 1970s by stockholding constituency groups whose investments were usually small relative to any company's total shares. They provoked a public debate about the wisdom of "social investing" that continued into the 1980s, but no public consensus developed about the meaning of such investment. A few institutions, such as universities or foundations, owned sizable absolute amounts of stock, but their ownership was an ancillary function of their main activities and interests. Thus the groups were generally considered to be outside traditional corporate boundaries; other large and traditional stockholders feared that they were all too willing to *subsidize* various socially beneficial projects, even if only at the margins. Until both they and the "socially responsible institutions" had enough information to be able to tell what, if any, subsidization was involved they were not apt to receive much support from other stockholders.

By the end of the 1980s, however, new institutional voices joined the debate and shifted attention to a new reality: The emergence of pension funds as the nation's largest source of capital. In 1989 they owned $2.0 trillion in assets, providing retirement security for more than 41 million workers. Having increased by 500 percent in the previous decade, pension funds were expected to double by the turn of the century when they may well hold two-thirds of the equity capital of all U.S. business.[49] So large have pension funds become that the fiduciaries are confronted by a broad array of new needs and constituencies beyond those of retired and current employees. Both the needs and the constituencies have an impact on the welfare of the pension beneficiaries, requiring a broad view of the effects of investing the funds involved.

While no public consensus has yet formed about the role of pension fund managers in exercising their ownership, it appears likely that they will interest themselves in many of the issues that the nontraditional constituencies have raised. No longer can pension fund trustees simply sell their stock, exiting through the market from involvement in a company whose policies and practices they do not like. Increasingly they find that they cannot practically disinvest a company's stock, except gradually over a long period. Because of the enormous size of their assets and long-term liabilities, large pension funds are virtually permanent investors in the market. Certainly as a group they cannot, over the long term, expect to "beat the market" in the same way an individual investor may be able to do.

Harrison J. Goldin, then New York City Comptroller and a cochair of the Council of Institutional Investors (CII),[50] made clear that institutional investors in no way wanted to take part in the day-to-day running of a company. He believed that they would insist, however, upon open and nonadversarial channels of communication between corporate managers and themselves, not as adversaries or distracting transient investors, but as corporate resources and supporters of companies' long-term goals.[51] With their ever enlarging investment in and commitment to corporate businesses, pension fund managers have been urged to consider four new duties. First, they should monitor carefully and fully the performance of the companies in which they invest; second, they should assure a program of reciprocal communications; third, they have an essential obligation to play a constructive part in corporate governance; and fourth, they must be accountable to those whose fund they manage, disclosing how their voice is used and how their duties are fulfilled.[52]

Not all pension fund managers will act alike. Those in charge of public-employee funds will probably be more responsive to current political pressures and interests than will those who oversee private funds. Some may experiment, testing out new areas of investment and push corporate managers to respond more imaginatively than in the past to matters of public concern, such as environmental protection. Others may pursue more cautious, conservative goals, in line with traditional business thinking. The particular message of pension fund managers' voice will be, however, much less important than the use of it.

Constituency Voices and Their Effectiveness

Not all constituency groups espouse what popularly could be labeled as "liberal" causes. For example, Stockholders for World Freedom and Young Americans for Freedom have regularly sponsored resolutions addressed to the stockholders of companies such as Exxon, Dresser Industries, and Gulf, asking for policies that would disallow trade with Communist countries. The issues raised by other groups—resolutions sponsored by the National Wildlife Federation, the American Jewish Congress, or the California Public Employees Retirement System, for example—may not easily be characterized as either liberal or conservative.

Considering the attention and voting of some institutional investors, one may wonder if older conceptions about "liberal" and "conservative" are germane. The administrator for corporate public involvement of the Aetna Life and Casualty Company, Hartford, Connecticut, pointed out that his company in 1982 examined such issues as drug, infant formula, and pesticide exports to Third World countries and nuclear power and arms, and decided to support some of the constituency resolutions. For example, in 1979 and 1980 Aetna voted its holdings in Fluor Mining & Metals, Inc., common stock in support of a church coalition's resolution to halt Fluor's sales in South Africa.[53]

The kinds of issues raised by corporate constituencies and institutional investors change over time. Through the 1970s, and particularly in the early part of the 1980s, equal employment opportunities were a major concern. Many groups proposed resolutions, following on the Watergate revelations in the early 1970s, that dealt with illegal political contributions and questionable payments (bribes) abroad. Bank redlining and the reform of television programming were issues sometimes raised as well.

In the 1980s, resolutions on plant closings, investments in South and Central America and environmental pollution gained prominence. Pension fund investors were concerned with issues of hostile takeovers and "poison pill" defences against corporate raiders. Comparable worth appeared as an issue that women's groups and their supporters raised in the 1980s, and care for employees' children moved to the fore as a lively constituency issue. Throughout the period that constituencies have been using shareholder resolutions, from the earliest days into the 1980s, two issues were prominent. The first was marketing of infant formula in Third World countries, and the second has been American business loans to, and investment in, South Africa. The first appeared to reach final resolution by 1984 when Nestlé managers and constituency boycotters agreed to a settlement, but, the second has continued. That the infant formula boycott continued so long—even flaring up again—and the South African trade issue continues, with no quick resolution, sug-

66

BEYOND SUCCESS

gest that the nature of the problems involved and the solutions required are quite different from those of the other issues that appear to
wax and then wane as managers and constituencies accommodate them.

The raising of issues through proxy resolutions is but one part of a
more complex process of interactions among constituencies, institutional investors, and corporate managers. Often constituency leaders
will approach managers for discussion of particular matters and mutually explore the problems involved, with both parties sometimes modifying their initial positions. If no accommodation can be reached, a
constituency may then propose a resolution, after which more meetings
may be held and discussions conducted. Officers of ICCR report that
church agencies withdraw from one-third to one-fourth of their resolutions because mutually acceptable agreements have been reached before the annual meetings.[54] For example, in 1983 the World Division,
General Board of Global Ministries, United Methodist Church, proposed a resolution on investment in a new Brazilian smelting plant for
the proxy statement of the Aluminum Company of America. Later,
after discussions, the church agency withdrew the resolution, announcing that the company had agreed to issue a requested report concerning the plant; the company also agreed to further meetings with the
agency to discuss data to be used in the report, allowing management
representatives from Brazil to explore the questioned matters with church
officials who had visited the Brazilian project.

Those who seek to change corporation policy by a majority vote of
stockholders have been, and probably will be, often disappointed. The
votes that constituencies win, largely from institutional owners, are a
small proportion of the total. Of course, when they commend management, as church groups did that of Chemical Bank in 1983 on its South
Africa policy, the vote is high. In this case, it was 78.3 percent of the
total. Most other constituency votes seldom are as large as one out of
twenty (5 percent). Some are noticeably higher, though, and managers
no doubt pay more attention. In 1983 a constituency proposal for an
endorsement of the Sullivan principles in South Africa won over 11
percent of the votes at the annual meetings of both the Eaton Corporation and Ingersoll-Rand. Stockholder resolutions on nuclear power secured more than 10 percent of the votes at the annual meetings of
Detroit Edison and Philadelphia Electric. U.S. Steel reported almost a
12 percent vote for stockholder resolutions on plant closings and rein-
vestments in steel. These are all matters to which the particular managements may well have addressed themselves after the votes were taken,
for they indicate a significant, if not yet threatening, stockholder concern with issues that will probably arise again.[55] Mr. Thomas Edwards,
chairperson of Teachers Insurance Annuity Association and College
Retirement Equity Fund, pointed out in a 1982 conference in honor of
IRRC, that institutional investors and the corporate constituencies

can't out-vote management, but even the smallest institutional investors have improved the acoustics for change, whether management admits it or not. . . . In addition the one-on-one exchanges of information generated by investor involvement have revealed at the management level another large reservoir of social concern, or, if not that, at least a wholesome respect for the concern of their shareholders.[56]

Constituencies do not rely merely upon their resolutions and garnered votes to generate and sustain managers' respect, however. Their leaders are well aware of the need to push their concerns by every means available: Political lobbying to change laws and to fund desirable government programs, enforcing and defining legal rights through the courts and regulatory agencies, and developing publicity campaigns for popular support and approval. Managers recognize that the institutionalization of constituency influence affects "voice" rather than directly changing decisions and policies. Proxy resolutions afford managers an additional way of listening. Should they be deaf to the sound of an appealing and widely supported issue, they know that they may well have to deal with it in other forums. Should a constituency group generate enough popular support for a change in company policy, they may be able to mobilize the market, and apply economic penalties— generate market exit to accompany their shareholder voice.

In chapter 9 we will examine a highly publicized example of market mobilization, the long consumer boycott against Nestlé. There have been many other boycotts in recent years, though; they are often recommended, some are actually announced, and a few are implemented, with demonstrations and picketing. One study identified ninety of them, reported in national newspapers, over the period 1970–1980 in various parts of the United States. Although the data are incomplete, they suggest that constituencies may be using boycotts more frequently than in earlier times, but researchers have not yet found ways of evaluating their successes or failures, or even in defining exactly what those terms may mean for the companies involved or the constituencies.[57] It seems likely that few have inflicted very heavy costs upon any of the target companies. Usually, mobilization of the market through boycotts is a dramatic way to emphasize the dissatisfactions that constituencies have already expressed through political and social voice.

The next chapter examines the general goals and purposes of constituencies that may help the reader determine the kinds of standards against which constituency activities, both voice and mobilization of the market, may be judged.

NOTES

1. The term "constituencies" is similar to another term, commonly used in managerial and business literature, "stakeholders." Two scholars who have pro-

vided a theoretical basis for the latter term, define it as "those groups who have a stake in or claim on the firm [and to whom managers bear a fiduciary relationship]. Specifically we include suppliers, customers, employees, stockholders, and local community, as well as management in its role as agent for these groups." See William M. Evan and R. Edward Freeman, "A Stakeholder Theory of the Modern Corporation: Kantian Capitalism," in Tom L. Beauchamp and Norman E. Bowie, eds., *Ethical Theory and Business*, 3rd ed. (Englewood Cliffs, N.J.: Prentice Hall, 1988), p. 97. In this study, "constituencies" refers essentially to the same groups, but implies self-chosen, self-asserting groups, not merely those recognized, chosen, or "managed" by a firm's managers. For further discussion of the difference, see Chapter 3 of this study.

2. The three quotes are from Allen R. Janger and Ronald E. Berenbeim, *External Challenges to Management Decisions: A Growing International Business Problem* (New York: The Conference Board, 1981) p. 7.

3. From *Democracy in America*, vol. II, by Alexis de Tocqueville (New York: Alfred A. Knopf, Vintage Books, 1945), pp. 114, 117–18. [Copyright 1945 and renewed 1973 by Alfred A. Knopf, Inc. Reprinted by permission of the publisher.]

4. Ibid., p. 279.

5. As a voluntary association the business corporation has received special protection. The courts have been exceedingly respectful of it, on the grounds that in reality, it was private property. Few other voluntary associations have won such recognition; not only did business corporations gain constitutional protection of their property rights (Dartmouth case, 1819), they were declared "persons" within the meaning of the Fifth Amendment [*Santa Clara County* v. *Southern Pacific Rail Road*, 118 U.S. 394 (1886)].

6. Tocqueville, *Democracy In America*, pp. 520–21.

7. Ibid., pp. 514–15.

8. Ibid., p. 515.

9. Ibid., p. 514.

10. Ibid., p. 518.

11. Philip W. Moore, "Corporate Social Reform: An Activist's Viewpoint," *Perspectives on Social Involvement*, speech delivered c. 1972 in California according to author, pp. 48–49.

12. Elias Lieberman, "The Conspiracy of the Philadelphia Bootmakers," in *Unions Before the Bar* (New York: Harper and Brothers, 1950), pp. 1–15.

13. *United States Historical Statistics*, Series 197–204, p. 687.

14. For illustrations of violence in American history and fuller details of the examples mentioned here, see Richard Hofstadter and Michael Wallace, *American Violence: A Documentary History* (New York: Vintage Books, 1971).

15. Correcting for the much larger membership and labor force, both the frequency of strikes and members and participation rates in strikes has trended down since the 1930s; in the 1980s they were roughly half the levels common half a century before. Strike losses, as measured by working days idle, did not decline as much because in recent years the average strike has been nearly twice as long as earlier.

16. David Vogel, *Lobbying The Corporation* (New York: Basic Books, 1978), p. 24. Also see Frances Fox Piven and Richard A. Cloward, *Poor People's Movements* (New York: Vintage Books, 1977), pp. 221–22.

17. *Bell* v. *Maryland,* 378 U.S. 227 (1963). In the dissent of Mr. Justice Black.

18. Ibid., pp. 245–46, footnote 2. Emphasis added.

19. Ibid., p. 64.

20. See U.S. Department of Labor, Bureau of Labor Statistics, *The Negroes in the United States,* Bulletin no. 1511 (June 1966) (Washington, D.C.: Government Printing Office). The proportion of Negro urban population in the United States increased from 62 percent of total Negro population, 1950, to 73 percent in 1960. The white urban proportion rose from 48 percent to 58 percent (Table IA-6, p. 67); the share of urban nonwhites with incomes of $10,000 and above was rising rapidly, up from 2.7 percent in 1959 to 8.7 percent in 1964, much faster than among whites (Table IIIA-6, p. 141). Nonwhite wages and salaries, 1956–1963, were also outpacing those of whites (Table IIIA-3, p. 139).

21. *Bell* v. *Maryland* 378 U.S. 277 (1963), Appendix I, Decision of Mr. Justice Douglas.

22. David R. Jones, "U.S. Steel Rejects Birmingham Role," *New York Times,* October 30, 1963.

23. Edward P. Morgan, a liberal radio commentator quoted by *IUE News,* November 14, 1963.

24. Andrew Hacker, "Do Corporations Have a Social Responsibility?" *New York Times Magazine,* November 17, 1969.

25. Quoted by John D. Promfret, "Negroes Building Boycott Network," *New York Times,* November 25, 1962.

26. *Business Week,* May 20, 1972, p. 104.

27. *National Association For the Advancement of Colored People, et al.* v. *Claiborne Hardware Company et al.,* 454 U.S. 1030 (1982).

28. "Who's Out of Focus? A Note on the Harassment of Eastman Kodak," *Barron's,* May 1, 1967, p. 1. [Reprinted with permission.]

29. The acronym stood for Freedom, Integration, God, Honor Today.

30. Saul D. Alinsky, *Rules for Radicals* (New York: Alfred A. Knopf, Vintage Books, 1972), p. 175. [Reprinted with permission.]

31. Lewis D. Gilbert, *Dividends and Democracy* (Larchmont, N.Y.: American Research Council, 1956).

32. Statement on "The Role of the Shareholder in the Corporate World," *Hearings* before the Subcommittee on Citizens and Shareholders Rights and Remedies of the Committee on the Judiciary, U.S. Senate, 95th Congress, 1st Session, Part I, June 27, 28, 1977, pp. 67–68.

33. Vogel, *Lobbying The Corporation,* p. 44.

34. Quoted from the opinion of the U.S. Court of Appeals for the District of Columbia, *Medical Committee for Human Rights* v. *SEC,* July 8, 1970, reprinted in The Extension of Remarks, Senator Lee Metcalf, *Congressional Record,* December 28, 1970, p. E10736.

35. Ibid., p. E10741.

36. Ibid., pp. E10741–E10742. Emphasis in original.

37. Dow Chemical's chairman told the stockholders at the 1969 annual meeting that the company was not going to stop producing napalm. It would bid on government contracts and if successful would continue to produce the substance. Nevertheless, within six months the company had ended its napalm production. Company spokespeople denied that Dow Chemical had deliberately bid high to insure loss of the contract; they did not even announce the

loss for half a year, while campus demonstrations continued. The company maintained that it had handled the napalm contract in normal business ways, unaffected by the protests.

38. Professor James Q. Wilson has pointed out that by the early 1970s the federal government had become an important source of "organizing cadres." Those who had worked with local community action agencies of the war on poverty and the numerous VISTA volunteers were active in forming voluntary associations. The training of the organizing cadres took place more widely than Wilson credits, however. Surely the civil rights movement was as important, if not more important than, the Great Society programs of the federal government. See James Q. Wilson, *Political Organization* (New York: Basic Books, 1973), p. 203.

39. Donald E. Schwartz, "The Public Interest Proxy Contest: Reflections on Campaign GM," *Michigan Law Review* 69 (January 1971):452–453.

40. *Peck* v. *Greyhound Corporation,* 97 F. Supp. 679 (S.D.N.Y. 1951).

41. 17 *Fed. Reg.* 11, 433 (1952).

42. Testimony on "The Role of the Shareholder in the Corporate World," *Hearings* before the Subcommittee on Citizens and Shareholders Rights and Remedies, of the Committee of the Judiciary, U.S. Senate, 95th Congress, 1st session, Part I, June 27, 28, 1977, p. 77.

43. Richard A. Posner, now on the federal bench, argued in his book, *Economic Analysis of Law,* 2nd ed. Boston: Little, Brown, 1977), that it is a legitimate concern that managers not be allowed to substitute personal goals for that of maximizing profits. Presumably he would have raised questions about Dow Chemical's managers who argued that they sought a government contract for the production of napalm for "social" reasons rather than because of the profit to be made. He wanted managers to be held accountable but saw little likelihood that shareholder democracy will ever succeed in enforcing accountability. He emphasized instead the mobility of control, the ability of shareholders to remove managers and the preservation of a competitive market. In Chapter 4 of this study, we will explore this issue in some detail.

44. An acute observer of the financial scene did not find that institutional investors had become very much more active in the 1980s, than in earlier years. Michael C. Jensen, "Eclipse of the Public Corporation," *Harvard Business Review* (September–October 1989):66.

45. John G. Simon, Charles W. Powers, and Jon P. Gunnemann, *The Ethical Investor: Universities and Corporate Responsibilities* (New Haven: Yale University Press), 1972.

46. Ibid., p. 65 [Reprinted from *The Ethical Investor* by Simon *et al.* Permission of Yale University Press.]

47. Ibid., p. 64. [Reprinted from *The Ethical Investor* by Simon *et al.* Permission of Yale University Press.]

48. Ann Crittenden, "Teachers Wield Their Proxies," *New York Times,* March 19, 1978, section 3, pp. 1, 13.

49. The Report of the Governor's Task Force on Pension Fund Investment, *Our Money's Worth,* Ira M. Millstein, Task Force Chair, A Project of the New York State Industrial Cooperation Council, June 1989, pp. 1 and 2.

50. A group of major pension funds formed CII in 1985 to work actively to exercise their rights as shareholders.

51. The Report of the Governor's Task Force on Pension Fund Investment, *Our Money's Worth*, p. 39.

52. Ibid., p. 41.

53. "Religious Group Proxy Power," *New York Times*, April 24, 1982.

54. Patricia Wolf and Timothy H. Smith, "Twelve Years on Corporate Ballot," *ICCR Brief* (April 1983):3A. In early 1990 Lockheed Corporation managers agreed to award three board seats to persons representing pension funds and other major investors after some of the large funds threatened to vote for a dissident, minority stockholder, Harold C. Simmons. See Richard W. Stevenson, "Lockheed's Moves in Proxy Fight," *New York Times*, April 5, 1990.

55. For stockholders to propose a resolution the second year, it must have won 3 percent of the vote the first; the second year it must win support of 6 percent of the shares, and thereafter 10 percent.

56. Lauren Talner, *The Origins of Shareholders Activism* (Washington, D.C.: Investors Responsibility Research Center, Inc., July 1983), p. 48.

57. Monroe Friedman, "Consumer Boycotts in the United States, 1970–1980: Contemporary Events in Historical Perspective," *Journal of Consumer Affairs* 19 (Summer 1985):96–117.

3

What Constituencies Seek:
Their Goals and Purposes

The last chapter, a synoptic history of the development of constituencies, emphasized the thesis of this study that the various groups who might be included as constituencies "are remarkable for their overall adherence to beliefs that are respected in American tradition and ensconced in American history."[1] The history of corporate constituencies reveals groups that have neither pushed revolutionary programs to overthrow capitalism nor promoted radical demands to do away with business corporations, to drastically rework corporate governance, or even to replace managers.

FREEDOM OF ASSOCIATION

The one sure generalization to be made about the values of the various groups that may be counted as corporate constituencies is their dependence on, adherence to, and support of freedom of assembly and speech. They use these freedoms to pursue particular shared citizen interests, to seek access to centers of power, and to exert influence on them. Freedom itself, in short, in an American sense of the word, is a central value of the various constituencies. Their reliance upon freedom to associate and organize—to form communities—has also been a means of protecting their members against freedom's own excess of mere individualism.

Common means to achieve that protection are the use of strikes, picketing and other publicity or demonstration, and boycotts. Many Americans find such means troubling or disconcerting, though they may be seen as ways of mobilizing the market to promote particular interests. Business corporations, simply through their combination of capital and/or number of employees, also mobilize and use market power to enhance their influence in the market.[2] Business use of market power to win terms favorable for a firm is usually considered unremarkable, though constituency use of market power almost always is perceived as invidious.

In part as a consequence of this perception, corporate constituencies have sought other means of exerting influence and more acceptable ways of negotiating with business managers. In contrast to the older unions and civil rights organizations, they have sought a voice and a role as stockholders—as internal corporate participants. In exercising their freedom to associate and seek remedies on their own through the marketplace, they have modeled themselves on those who formed business corporations. Though the courts have long since declared corporations to be, in legal fiction, single persons, within the meaning of the Constitution[3] they are, in fact, associations of individuals, cooperating voluntarily to promote their common self-interest through the market.

"Free" Negotiations

While the members of both business corporations and business constituencies have sought to bolster their claims as recognized legal rights, they display a remarkable distrust of government and a preference for independent, self-reliant activity. Note, for example, the readiness of the National Toxics Campaign, mentioned in Chapter 1, to make agreements, on its own, with managers of various supermarkets and grocery chains. Earlier constituencies, such as unionists, usually joined managers in support of "free" collective bargaining, through which the parties themselves determine the substance of wages, hours, and the terms of employment; both parties have been wary, indeed, of government standards, except as minimums. Those in the civil rights movement lobbied for civil rights and equal employment opportunity laws, but they also insisted upon "free," direct negotiations with employers and business managers. The sit-ins were aimed at opening negotiations, not with government officials but with immediately confronted managers.

The Reverend Leon Sullivan and his fellow pastors in Philadelphia sought to engage corporate managers directly, and used their Selective Patronage Program to secure concessions from them above and beyond whatever legal remedies might offer. Saul Alinsky and the members of FIGHT in Rochester, New York, and various other constituency groups wanted to participate in Eastman Kodak's annual meetings, not to win rulings from government agencies. They wanted the company managers to negotiate terms and provisions with them. The Medical Committee that protested Dow's production and sale of napalm approached and dealt directly with company managers, not legislators; they believed they were free to do so, and expected better results than if they had sought a ruling from a regulatory body.

The willingness of constituencies to act on their own and to use available, nongovernmental, market means suggests that their members' disagreements with business managers are narrowly based. They disagree, not over fundamentals, but over solutions to problems in particular circumstances.[4] Once established, recognized, and regularized, constituencies prefer to eschew even the techniques of mobilizing the market. To contest in the market is by definition competitive, contentious, and costly, as well as difficult, as business managers well know, for they must continually struggle for their sales and earnings in it. On both sides there is considerable reason to seek some kind of mutual accommodation through peaceful, though tough, negotiations and bargaining.

The Power of Metaphors and Labels

The press, certainly managers, and even constituency leaders have portrayed these assertions of constituency rights, and the demonstrations or assemblies to publicize them, as violent, warlike activities. And violence has accompanied some of the demonstrations, as we noted in the review of labor union history; from time to time it flared in the marches, boycotts, and mass meetings of those involved in the civil rights movement. It is worth nothing, however, that *all* parties involved often resorted to violence—business leaders and managers, no less than the government, as well as constituency members. Violence was usually but a passing phase of initial reaction; as soon as managers and government officials accepted and recognized constituencies, providing them with a legitimate place in the social setting and regular role to play in affairs, the violence became rare or disappeared.

The shock of the new and the resulting violence in the early encounters between business managers and constituency members may have misled both parties. The drama of the encounters and the publicity they received, even in the days before television, made the differences on current issues loom larger than the basic agreements to which both subscribed. In the last chapter we called attention to the hyperbole that Saul Alinsky used in describing the response of Eastman Kodak's managers to FIGHT's demands; he described them as "violent attacks" and went on to characterize his disputes with "the establishment" as *wars*. He then described the innovative use of proxy votes as "dynamite," an exaggerated term associated with explosiveness and violence. Business leaders often describe, at least in private, their encounters with constituencies in similar terms—as wars, battles, fights, and struggles. Even an observer as careful as Fred D. Baldwin, a business consultant on management problems, in an otherwise evenhanded examination of business–constituency relationships, described Campaign GM, 1971–72, led by Ralph Nader's group, as "a *battle* against General Motors . . . a *guerilla campaign* deep within what corporate officers had supposed was their own territory."[5]

This casual use of metaphor can hinder understanding and mislead analysis. Of more importance, it may cause the parties to overlook opportunities for joint, cooperative, and collaborative efforts; they may unwittingly trap themselves into confrontational stances and unthinkingly deny to themselves potential resolutions. Consider the progression implied in the relationship of corporate managers and constituencies by the following labels applied to corporate constituencies. They are frequently to be heard in business discourse and read in business journals and news reports: (1) business enemies, (2) antibusiness groups, (3) hostile groups, (4) adversarial groups, (5) business critics, (6) special interests, (7) interest groups, (8) dissident groups, (9) opponents, (10) stakeholders, and (11) constituencies.

The progression imputes to the groups, first, extreme enmity, rancor, and antagonism to business (1–5), then neutrality (6–7). The next two labels (8–9) imply a regularized role, analogous perhaps to contending political parties and factions within them. The last labels (10–11) go further. Stakeholder suggests a position analogous to *share*holder, including those who have a recognized claim on the firm. Constituency, a term we recommend, implies that the groups, because of their own existence, interests, concerns, and activities are—whether recognized by managers or not—an inescapable, necessary part or element of the business corporation.

Constituencies and Stakeholders

Increasingly popular with managers and business scholars, the term "stakeholder" is often used to designate the groups, interests, and organizations who negotiate, confront, and deal with managers. Professors William M. Evan and R. Edward Freeman have proposed a theory of the modern business corporation that makes stakeholders important members of the organization.[6] Among those members they specifically mention are suppliers, customers, employees, stockholders, and the local community, as well as managers in their roles as agents for the others. They argue that the various stakeholder groups possess a general social and philosophical "right not be treated as a means to some end, and therefore must participate in determining the future direction of the firm."[7] A wide definition of stakeholders, they assert, "includes any group or individual who can affect or is affected by the corporation."

The term "constituencies," offered in this study, is similar to and even derived from an earlier business use of the term "stakeholder." We like Evan and Freeman's undergirding of the term with the Kantian injunction that all persons deserve respect. Unfortunately, business managers' long-accustomed use of the term implied no such foundation, but explicitly preserved a special role for themselves—that of trustees balancing the interests of all the other members of the organization.[8] We believe that such a role offers little recognition of the need for managerial accountability instead of a self-defined sense of responsibility. Evan and Freeman's acceptance of that special role for management has made use of the term "stakeholder" widely acceptable to managers, and no doubt has contributed to its growing use in business journals and publications.[9] Managers' easy acceptance suggests to us that they will accept the term, but refuse the substance of it.

Managers are comfortable with stakeholding, for Evan and Freeman offer no change from their own traditional notion of the exalted position of managers:

> Management plays a special role. . . . On the one hand, management's stake is like that of employees, with some kind of explicit or implicit employment contract. But, on the other hand, management has a duty of safeguarding the welfare of the abstract entity that is the corporation, which can override a stake as employee. In short management, especially top management, must look after the health of the corporation, and this involves balancing the multiple claims of conflicting stakeholders.

The authors then make all too clear their agreement with managers' lofty view of their special position. It is certainly undemocratic and can hardly encourage others' participation in organizational policies, decisions, and activities. They use an unfortunate kingly metaphor:

> The task of management in today's corporation is akin to that of King Solomon. The stakeholder theory does not give primacy to one stakeholder group over another, though there will surely be times when one group will benefit at the expense of others. In general, however, management must keep the relationships among stakeholders in balance. When these relationships become unbalanced, the survival of the firm is in jeopardy.[10]

Solomon's harsh, bureaucratic, tyrannical rule provoked much discontent among his subjects, and at his death resulted in a rebellion and the division of his kingdom. Comparison of managers' role to that of Solomon's is hardly felicitous, though it is more accurate than Evan and Freeman may have intended. Institutional investors and shareholders, along with many other constituents, have in recent years made clear their discontent with the autocratic rule of corporate managers.

Although "unbalanced" relationships among constituents may threaten the survival of a firm, we wonder why this is a matter of concern to managers alone and thus a special concern of theirs—many constituents should be, and usually are, concerned as well. The unique managerial role, as we perceive it, is not in providing a particular goal or set of values for the firm, such as assuring survival, but rather in fulfilling a necessary *function:* coordinating all the various demands, negotiations, and settlements so that the outflows of resources match inflows. Managers do not, and should not be expected to, serve as neutral judges or arbitrators among the contending and cooperating constituencies. They are prime actors in the organization, with their own values and interests to pursue. Organizations policy and direction will be determined by the complex of negotiations among all the various groups interested in and affected by the organization's activities. Powerful constituencies able to mobilize market power as investment bankers do or to wield the power of ownership as shareholders may, can reorganize or bankrupt the firm; other constituencies—customers, raw material or parts suppliers, community groups concerned with pollution, education, health, and safety—may exercise their power through political voice, legislative restrictions, and market exit to accomplish the same end. Managers neither deserve, nor can they be trusted with, any special power to override these outcomes, except as it is related to their function of coordination. Whatever the constellation and interplay of constituent forces, managers have to make the best they can of them. They will have to engage in tough bargaining with all constituencies, who are as free to demand and negotiate as the power they can acquire in their defense.

We believe that in today's industrial world, with its highly competitive economy and communications-saturated society, business managers not only must recognize the demands of various groups, but they must also accept their input and contributions, if they are to operate effectively and efficiently. We do not believe that viewing the groups as integral members of the business organization—that is, constituencies—is utopian or romantic. It arises from an appreciation of the difficulties of managing large, complex productive organizations in an intractable, continually changing, and challenging environment.

Asking, or expecting, managers by themselves to anticipate and cope with the multiple social, economic, and political problems arising daily and constantly confronting them, demands too much. They need wider visions, more acute hearing, and more sensitive responses than they are likely to be able to muster without the initiatives of people and groups outside the managerial arena.

Recognizing the various groups as inherent, integral parts of, and contributors to, the future of the firm does not mean that managers should agree to any particular demands or approve all recommended changes. Nor are constituencies likely to develop business managers. The perspectives they encourage, the skills they develop, and the interests they pursue may complement those needed by business managers, but they usually will be too special and tangential to prepare their leaders for the full range of responsibilities that are the lot of corporate managers. The larger truth about professional training, of course, is that leaders in many sectors of American society are likely to develop different, complementary skills. Like Leon Sullivan, pastors of churches often demonstrate managerial skills. They are not strangers to the discipline of budgets whatever may be their unfamiliarity with the discipline of profits.

Since the lives of constituency members are significantly affected by managerial decisions and programs, they therefore have a concern for the future of the business corporations. Their recognition of their inescapable involvement convince them that they have become willy-nilly a part of the corporate community, whatever the formalities of law and convention declare. If and as business managers also recognize constituencies' stakes in the corporations' future, opportunities as well as problems will present themselves. To be sure, conflicts in goals and purposes will still remain, but there will also be possibilities for synergistic approaches to some of them, and opportunities for joint effort in overcoming otherwise intractable obstacles. Beneficial, productive contributions of both managers and their constituencies can be garnered as the parties proceed with their debates, bargaining, and negotiations. A continually changing social scene and economic uncertainties promise that they will always be wrestling with unexpected problems and unanticipated issues.

MARKET GOALS AND SOCIAL PURPOSES

Business managers have long assured themselves and the public that they are the uniquely qualified protectors of corporate business and market values. They have usually disdained help from any other group, usually categorized as outsiders, even when the groups were representatives of their own employees, suppliers, or customers. They have insisted that society's goals and their goals (or those of shareholders) usually coincide. If, however, they come into conflict, managers, in their wisdom, experience, and expertise, acting as trustees for the public, can best balance the interests and offer socially beneficial resolutions. Usually they insist that if production efficiency and economically rational allocation of resources must be subordinated to other, more human values, then government should so decree. Otherwise, they should be allowed to pursue their business—economic values at they define and understand them.

Unfortunately, as we will see in succeeding chapters, managers' understanding of efficiency and rational allocation is often self-serving and myopic; further, they regularly mix other, human values into their notions of efficiency and hardly recognize what they have done. That they might do so is hardly surprising; in large businesses, they are the chosen leaders of vast organizations with many employees. They must win voluntary cooperation by promoting loyalty and devotion to the goals they project. They must promote and maintain a community, which almost surely is held together by more than greed and self-interest, as we noted in the first chapter. In leading the members of their firms they appeal to more than the economic value of efficiency, though it may loom large. They may even sacrifice it to preserve other human values they consider important. Although in general managers possess special skills in such functions as investing, production, and marketing, they usually enjoy no more understanding of social values than their constituencies. At their and society's risk they presume that their special managerial function gives them an exceptional claim to determine values for others.

Economic Efficiency Is Not Always Primary— Even to Business Managers

Consider just two cases already presented in the last chapter. G. Carroll Hooper, the Maryland restaurant owner and manager who had to cope with a sit-in by African Americans wanting to be served. He was so willing to accept what he believed were his current customers' values that he willingly sacrificed the opportunity to increase sales and profits by developing a wider, more inclusive market. He gave precedence to noneconomic values and strongly defended that managerial decision. He appealed his case all the way to the Supreme Court in defense of it. In the case of Dow Chemical and its production of napalm during the Vietnam war, the judge noted that "the management of Dow Chemical Company is repeatedly quoted in sources which include the company's own publications as proclaiming that the decision to continue manufacturing napalm was made not *because* of business considerations, but *in spite of* them."[11] The company managers apparently chose to explain their production decision as one based on patriotic values, not the hardheaded decisions to maximize profits. Few Americans can contemplate the fate of the major manufacturing corporations over the last decade or so in major industries, such as steel, automobiles, rubber, glass, petroleum refining, and paper, and reach the conclusion that their managers were continually and vigilantly seeking economic efficiency and maximum profits. They may suspect that, like Hooper and the Dow managers, they were concerned with a good many other values, including those that various constituencies both approved and sought.

The Relevancy of Social Purposes
to Managerial Efforts

There are good reasons for such concerns. The larger the business corporation and the more intricate its organization and the more complex the community of which the managers are the leaders, the less likely they are to be able to concentrate upon a single economic purpose or business goal. Only an economist could believe that economic constraints press so hard upon managers that they cannot respond to their own humane and human values, which almost surely will be rooted in the same American traditions as those of their various constituencies. Ernest van den Haag, a neoconservative and acutely perceptive social observer, has written:

> Justice is as irrelevant to the functioning of the market, to economic efficiency and to economics, as it is to a computer or to meteorology. But it is not irrelevant to our attitude toward these things. People will tolerate a social or economic system, however efficient, only if they perceive it as just.[12]

While managers generally may not be able to express this notion as succinctly and clearly as the scholar, many, if not most, probably intuit its validity. Managers of almost any large business corporation probably would agree with Irving Kristol's observation that

> . . . the large corporation used to be a single-purpose institution: an economic institution directed toward economic growth. It was very good at this job, would still be good at this job, but its very size and importance have resulted in its job assignment being changed. . . . And so the executives of the large corporation, like the administrators of the large university, have to learn to govern . . . to think "politically"—i.e., institutionally—as well as economically or educationally. That is the price of bigness and power.[13]

Even though business managers may be better trained and experienced to operate efficiently in the marketplace (and more willing and ready to take the risks and suffer the penalties if they do not maintain efficiency), there is no reason to believe that they affirm or promote social values other than those of many of their fellow citizens. We can expect business managers usually to give preference and precedence to the economic values of efficient production and rational allocation of resources; insuring those values is supposed to be a special competence of theirs. When conflict arises between these economic values and other human values, as at times they surely do, both the managers themselves and the society need the balancing effects of public scrutiny and of negotiations between those most directly affected and the managers. That important balancing function should not be—dare not be—left to the managers themselves. No party to any social conflict is likely to

command special knowledge of the just resolution of the conflict, for
no one is a good judge of his own case.

We expect to find differences between the goals and purposes of
business managers and the various constituencies. The disagreements
between them and the constituencies would not be so sharp, nor their
debates so intense, were the differences not substantial. Managers need,
therefore, to comprehend the nature of constituencies' purposes and
goals; they need to understand what they hope to accomplish. With
such comprehension and understanding they should better be able to
formulate *their* strategy—whether to accept constituency demands, bargain for modifications, or reject them completely.

Even a brief consideration of the variety of constituencies will convince the reader that managers' formulation is not apt to be simple.
The very plenitude of constituencies and the variety of their interests
makes the task of determining which deserve serious and careful consideration at once perplexing, arduous, and even tedious. Business demands and market requirements already burden managers and strain
their abilities. However, the nature of their job, as Kristol observes, has
changed in recent decades. They have no choice but to discover the
complexities of corporate constituencies, to come to appreciate them
better, and to *join* them in exploring the new opportunities and problems of changing society.

THE VARIETY OF CONSTITUENCIES
AND THEIR VALUES

A glance at the *Encyclopedia of Associations* for recent years readily turns
up the evidence of the profusion, contradiction, and noneconomic nature of the values promoted by associations that are classifiable as part
of the corporate responsibility movement. The *Encyclopedia* lists some
14,000 associations, a number undoubtedly far below the actual total
for the whole country. Only eight organizations in the 1984 edition are
formally listed under the category of "corporate social responsibility,"
but hundreds of others in fact belong in that category, such as those
listed under "Consumer" and "Environment." Half of American voluntary groups are religious congregations, and the probable total of all
"associations" in the United States is 800,000.

The concerns around which these groups are organized are as various as those Tocqueville found in their early nineteenth century ana-
logues; they range from "serious" to "futile" (a judgment depending
on one's point of view!) and from "very general" to "very limited." Their
membership ranges from "immensely large" to "very limited." In the
late 1970s, Hurst, Texas, was the headquarters of "The Enough is
Enough Club," with a membership of 1,213 fans of national televised

"Monday Night Football" who believed "the program would be more enjoyable without ABC sports broadcaster Howard Cosell." The members may have found other causes to support after Cosell left the ABC position. In Rochester, New York, there sprang up in 1980 an organization, Dignity After Death, whose twenty-one regional groups seek "to boycott and protest the commercialization and marketing of John Lennon after his death."

Quite on the other side of the spectrum from specific to general is the Center for Accountability to the Public (CAP), located in the John Hancock Center, Chicago, Illinois. Its 1981 membership was 100 and its staff numbered two, but its declared constituents and concerns were broad gauged:

> Individuals and organizations interested in making society more responsible and accountable to the public. Examines all aspects of modern society and believes in providing information needed to make basic decisions in order to accomplish accountability. Seeks to provide information which is not publicly known on energy efficiency and air pollution. Publicizes information about solar devices which do not work but are still on the market. . . .

Such a description could be grist for the humor mills of business people who manage by objectives. "Not much focus of objectives here," they might say. But over the past fifteen years, many a business leader has been prematurely amused at the apparent potpourri of objectives proclaimed by such mini-organizations. Plainly CAP has a general value-commitment to an increase in the accountability of all parts of the society to the whole. (That there is such a thing as a whole society is inherent in the meaning of the word *public*.) Plainly, too, the organization's initial specific focus is on a cluster of timely, interrelated concerns: energy efficiency, pollution, and honesty in the marketing of solar devices. Its central strategy is information distribution. And its central strength is organization itself, however weak and unimpressive that strength may look from a cursory reading of an encyclopedia entry.

The combination of initial organizing interest with openness to many future interests is a feature of the corporate responsibility movement that business managers have had to learn to respect, sometimes at the price of much grief. When Ralph Nader, an early hero of the contemporary movement, launched his criticism against a General Motors automobile that he judged "unsafe at any speed," the company managers made the mistake of underestimating the potential power of (1) a single, determined, well-informed individual, (2) backed by at least a small number of organized sympathizers, (3) with access to multiple channels of publicity, (4) focused on at least one value—automobile safety—of relevance to large numbers of consumers in the nation.

One of the organizations spawned by Nader's work, Public Citizen (PC), listed 200,000 supporters. The projects they endorsed through Public Citizen have encompassed targets as diverse as tax reform, health, energy, the courts, and Congress. PC worked to provide effective citizen advocacy on the most pressing problems at the least cost by using the services of volunteers, keeping expenses as low as possible, and hiring dedicated professionals who were willing to work long hours for modest salaries, replicating, in effect the lifestyle of the founder, Nader himself.

Business managers and many of the public probably think of most of the groups listed in the *Encyclopedia* as "liberal." As noted in Chapter 2, however, there are enough groups of all shades of opinion to represent a very wide variety of causes. For example, the Reverend Donald Wildmon of Tupelo, Mississippi, announced the formation of a Coalition for Better Television in 1981. He had earlier founded the National Federation of Decency in 1977, but had discovered many concerned groups that he believed should jointly work together for change in television programs. With a number of other groups cooperating in the coalition, such as the Eagle Forum, the American Life Lobby, the Conference of Women of America, and the Pro Family Forum, the major networks and many corporate advertisers took serious note of their efforts.

The coalition's purpose was to halt the "trend toward increasing amounts of sex, violence and profanity in network prime time programming."[14] In the early 1980s they began a widespread monitoring program with a promise to lead a viewers' boycott of advertisers who sponsored objectionable programs. Some business managers responded quickly to the coalition's demands. Procter & Gamble Company's chairman soon announced that it had withdrawn sponsorship from more than fifty programs, including movies, because they had not met its guidelines on sex, violence, and profanity. However effective the boycott threat, the *Wall Street Journal* announced several months later that "Television Plays Down Sex This Fall, Reacting to a New Public Mood."[15] Later, in 1989, as government regulation lessened and the television networks dropped or cut back their "standards and practices" departments, stations offered advertisers more programs with sexual themes, explicit language, and violence. Still active, Mr. Wildmon, as head of the Christian Leaders for Responsible Television, asserted that he and his group deserved credit for keeping advertisers from sponsoring the new programs. Under threat of boycott from a variety of monitoring organizations, such advertisers as Coca-Cola, McDonald's, Chrysler, General Mills, Campbell Soup, Ralston-Purina, and Sears announced cancellations of commercials in programs that various monitoring groups found offensive.[16]

The Young Americans for Freedom regularly submitted stockholder resolutions forbidding company sales to Communist nations, and other groups with similar goals have, from time to time, instituted consumer boycotts of firms importing goods from Communist countries. The small Stockholders Sovereignty Society, with membership of thirty in 1983, focuses its concern altogether on the internal structures of corporations. It called for "improvement of methods of election of directors, conduct of annual meetings, and stockholder participation in corporate political action committees." It protested all commerce "with Communist bloc countries," membership of Communists on any board of directors, and charitable contributions by corporations "to institutions supporting Communists." These two organizations certainly belong to the right wing of the corporate responsibility movement, but they mean to secure more power for stockholders than they have previously enjoyed in fact. They share this latter purpose with any number of organizations on the left wing; and both agree with the long-time professional corporate critics Wilma Soss and the Gilbert brothers, Lewis and John. All such groups—right, center, and left—are insistent advocates of corporate democracy, even if they limit their social agenda. Their leaders and their supporters make a strong case that corporate owners should enjoy far more voice, influence, and power in determining company policy than they have enjoyed or have yet won.

As the aims of all such groups suggest, whether those of Philip Moore of the Project on Corporate Responsibility or of the Stockholders Sovereignty Society, they share an intent to revise the very structure and workings of business corporations. They want to exercise new power or influence in corporate policy making, and they are ready to prescribe some new distributions of power for the governance of corporations. "What is required," wrote Moore in 1971, "is a shift of accountability from management to the people affected by corporate decisions."

It may have once been accurate for business managers to claim that the people affected by their decisions and corporate behavior were simply the consumers of company products. They exercised economic power over both the corporations and the managers through the market, but only if they acted in concert, or if, through a shift in taste, significant numbers of them chose not to buy the product offered. But the exercise of power by individual consumers is not easily accomplished in the automobile industry, particularly in the years before better quality vehicles from Japan and Korea became available. The industry resisted the notion that it should do anything about urban air pollution, though that problem was well identified and researched by industrial engineers as early as 1951.[17] Only when environmentalists organized in the 1960s did the industry begin to pay attention. Only in response to constituency efforts in support of local, state, and federal regulation, and to direct approaches to the automobile companies did the firms begin to change the design of their products.

The leaders of the various constituency groups are not in agreement about just how the decision-making system in corporations should be revised to take permanent account of a wide spectrum of public and other interests; they all appear to agree, however, that the call for corporate accountability *is* permanent, requiring a search for some long-range internal changes in corporate structures. Such a call and the requirement may sound philosophically dangerous to many business managers. They involve a change in power relations, always a matter of concern to those already exercising power.

The Conference Board concluded, after a wide survey of executives in large business corporations that:

> pragmatically, many . . . have concluded that they have little choice but to develop a broader perspective than in the past. Their attitude is partly defensive. . . . [They] feel that "tall towers attract attention and lightning." Also, the same desire for active participation that has brought outsiders to feel they must insert themselves into the business-decision process has made executives anxious to extend their own reach. "Business has lost too many battles," is the way one executive described the emotions that led to the foundation of the Business Roundtable [a restricted, top corporate managers' public interest group.] [18]

A pragmatic response by managers no doubt saves much ideological turmoil since the pleas for change come from many points on the political compass, as noted already. Not only do liberal groups want dialogue and response from corporate leaders, but so do conservative groups.

All are willing to bring their conflicting political philosophies into debate at annual stockholder meetings. Such a debate may appear at first glance to be unmanageable; in the 1970s it was novel and certainly untraditional in challenging many comfortable definitions of corporate efficiency and profitability. But manageability, precedent, and the sovereignty of profit are, after all, to stockholders and others immediately affected by corporate actions, challengeable values. That is the presupposition of the "democratic way." The right to debate the values that should have priority in policy making is the democratic prelude to policy making. Or so the great majority of these groups claim.

ASSOCIATIONAL FREEDOM TO PURSUE JUSTICE

Once initiated on grounds of associational freedom, what is the debate likely to be *about*? One can answer flippantly, "about anything that comes to the minds of the constituency members." But the answer is not very descriptive of what the new constituents say once they gain access to

boardrooms and managerial suites. Characteristically, they make a case
for some value and related group interest that they believe is threatened, neglected, or misserved by the business managers in pursuit of
their implicit interests or the firm's declared values. When the constituencies assert personal and social interests, seek debate about them—
even deny some of them—and work for a new balance among the goals
and purposes of the corporation, *the overarching social issue at hand is
that of justice.* Any careful survey of the constituencies that might be
included under the title of a corporate responsibility movement will
reveal that, again and again, they are fundamentally concerned with
claims having to do with justice.

Tippers International, founded in 1972 seeks a minor form of justice. Its 30,000 members want to restore an older meaning to service
tips, matching reward with performance. The 10,000 members of the
Environmental Coalition on Nuclear Power, founded in 1970, seek a
wider justice, the implementation of a safe, nonnuclear energy policy
in the United States. Women Against Pornography seeks to change public
opinion about pornography, ending "the degradation, objectivication,
and brutalization of women." The Campaign to Oppose Bank Loans to
South Africa seeks justice abroad, protesting American bank loans to
South Africa, calling attention to the oppression of blacks in that country. Whether the justice sought is small in scope or has an international
range, managers should prepare themselves to ponder the meaning of
this crucial ethical term.

Whose Justice?

In the United States, however, managers most frequently have had to
deal with two kinds of groups whose goals of justice do not always easily fit together, let alone find favor in managerial offices. The first are
those representing minorities and the second, those with environmental concerns.[19] Those representing minorities usually seek as a primary
goal increased opportunity for employment and jobs, with improved
status in the quality of work and the quantity of pay. Environmental
groups usually put the safeguarding of natural resources high on their
goals, even if, and when, it denies opportunities to create jobs and expand work and production. The conflict between these competing groups
and their contradictory goals can pose problems for managers, if they
find themselves seeking to help both at the same time. Of course, there
may be opportunities for them to play one group off against the other,
using the differing approaches to, and definitions of, justice to serve
their own ends.

An example of the conflicting notions of justice developed when the
managers of Consolidated Edison (Con Ed), a power company serving
New York City, proposed in the 1970s to build a pump-storage project

on Storm King Mountain, above the mid-Hudson River, about 80 miles north of the city. Con Ed spokespersons made a convincing case that the project would provide a means of producing electricity considerably cheaper than alternatives, in a region noted for its far-above-national-average power costs. At a time when manufacturing was fleeing the city because of too costly power, the project would have contributed to the saving of thousands of unskilled and semiskilled jobs for a population that desperately needed them.

Conservationists, however, pointed out that the project would also pump brackish water from the Hudson to a reservoir high up on the slopes of Storm King and, in filling and emptying daily, would destroy a large area of natural preserve and drastically alter the ecology of the surrounding area. They argued that the destruction Storm King would suffer outweighed the benefits of more efficiently produced electricity. The conservationists, well organized, vocal, and politically potent, killed Con Ed's proposal; power costs in New York City remained relatively high, manufacturing continued to decline precipitously with a concomitant marked loss of the jobs and skills that it supported.

The conflict among groups over the nature of justice they seek can be sharp indeed, as the example suggests. Bayard Rustin, the veteran civil rights and labor leader, called environmentalists

> . . . self-righteous elitist, neo-Malthusians who call for slow growth or no growth . . . [and who] would condemn the black underclass, the slum, proletariat, and rural blacks, to permanent poverty.[20]

Thomas Sowell, an economist—and an African American—has argued that the environmental regulations of recent years have impeded those who are climbing economically rather than those who are already at the top. He asserts there is a "fundamental conflict between the affluent people, who can afford to engage in environmental struggles and the poor. . . . You don't see many black faces in the Sierra Club."[21] Various surveys tend to support Sowell's observation about the exclusiveness of support for environmental causes. One poll showed that support for such causes is concentrated in the upper middle class.[22]

Two scholars, S. Robert Lichter and Stanley Rothman, surveyed leaders of "public" interest groups and attorneys who practice public interest law in the Washington, D. C., and New York metropolitan areas in 1983. They concluded that those surveyed are an elite—young, well-paid professionals with secular and liberal outlooks and Democratic voting habits—hardly representative of the "public interest" they profess to represent. In particular they concluded:

> If there is one issue that unites this group, it is environmentalism. Protection of the environment provides the *raison d'etre* for many of the groups we

sampled, so it is no surprise that virtually all reject the contention that our environmental problems are not serious. . . . They see American society as dominated by the traditional "power elite" of business, the military, and government, along with the news media.[23]

Other researchers have pointed out that a number of interest groups and constituency organizations do not represent any large number of members. It appears quite obvious that many are not nearly as broad based as their names imply. A study of eighty-three "public" interest organizations in 1977 found that at least one-third of them received more than half of their funds from private foundations. At least 10 percent of them received more than 90 percent of their operating expenses from such sources. Three-quarters of all "public" interest lobbies enrolled fewer than 100,000 members. Fifty-seven percent could boast of not more than 25,000 members, and 40 percent reported no more than 1,000. Thirty percent of the groups studied had no members at all![24] A 1981 study of Washington-based interest groups (and thus a very special sample) confirmed the earlier findings.[25]

It may be worth noting, however, that so famous a group as Ralph Nader's organizations receives wide financial support. In 1988 its total budget was approximately $5 million, of which three-quarters came from individual contributions and sale of publications. The various nonprofit organizations in the group are decentralized, operating in at least 26 states, Ontario, and British Columbia.[26]

With today's means of communication and modern technology, organization, research, and publicity can be secured at low costs; relatively small groups can make themselves heard and gain wide public attention. They can also use that attention, with the legal skills and administrative talent they employ, to help define and claim rights under the law for broad classes of people, including corporate constituencies.

One can view the findings mentioned above in a different perspective, if the emphasis is reversed. About two-thirds of the sample of *Washington-based* groups receive no more than, or less than, half their funds from private foundations (including the Nader organizations); a quarter of them enroll more than 100,000 members—no small number in absolute terms. Viewed this way, there is considerable reason to believe that there are an impressive number of broad-based member-supported constituencies and citizens groups in existence. Corporate managers would be wise not to discount their significance or to denigrate their efforts just because many of them appear to shallowrooted.

As with businesses, many more are formed than succeed in establishing themselves, but there is no end to the numbers that are continually created—more than enough to replenish the number of failures and those that disappear. One should no more expect corporate constituencies to disappear than one should expect business firms to disappear. New ones arise, some flourish, and others die; the same is true of the constituency groups. And some highly organized institutions—churches are the best examples—can be mobilized to support one after another disparate cause when moved to do so by a persuasive small group of people. The Nestlé boycott achieved much of its success by mobilizing churches and labor unions.

The kind of support that the most viable, long-lasting groups attract is basically that of the middle- and upper-income classes. The members of these classes may disagree with some, or even many, business policies, but they accept managers and business corporations as legitimate parts of the economic and political system. In turn, managers will have to learn to accept constituency protests as legitimate, particularly since they will probably be well researched, well reasoned, and institutionalized.

That the leaders of constituency and other interest groups are not thoroughly representative of the American public can hardly be surprising; it does not indicate, of course, that they do not serve their members well. Corporate managers might well consider that in education, income, and age, the constituency leaders with whom they must deal match closely their own characteristics. Both business managers and their constituency counterparts are leaders, both provide services (or goods) to wider publics, and both must prove themselves by securing a continuing flow of income to support their efforts. Constituency leaders may disagree with managers over specific policies to achieve justice, but they share much in common with managers that allows ample room for both compromise and agreement.

PROCEDURAL JUSTICE

"Justice" is old in the lexicon of moral and political philosophy. Debate over its concrete meaning is often the stuff of legislative, judicial, and administrative deliberation in the halls of government. Theologians, philosophers, politicians, jurists, and a host of scholars have done all their part to confuse or illuminate the great debate about justice in the human world, both ancient and modern. The debate has led to uncertain conclusions. Some modern philosophers, such as Robert Nozick, contend that, in the American context, the society has found no way to settle on any official, enforceable definition of justice. Its only official justice is procedural, the freedom of all parties to a controversy to have "their day in court"—or their day in the stockholders' annual meeting. Who can say what version of justice the courts, the legislatures, or the stockholders should vote into policy? Let them enjoy regular and legitimate voice and let them have procedures insuring a proxy resolution and vote. At the same time all can continue their endless debate on "real" (substantive) justice, since no one can authoritatively say what such justice is. Justice is the social search for justice, according to this notion.

However popular this procedural answer may be to the question of justice in America, hardly any citizen can be content with that answer in his or her personal or social life. Concrete self-interest is the beginning of the citizen's stake in the general debate on justice. And, especially in a society like that of America, which depends heavily on procedural safeguards for justice, those procedures are only as valuable as the use that citizens make of them. Such a society naturally becomes litigious, as its citizens discover that to get justice for themselves they have to organize and exert pressure on others to achieve rulings, decisions, or outcomes that provisionally match their standards of justice.

Again, corporate constituencies are characteristically "American" in their search for business responsibility and economic justice: Whether in the interest of one allegedly neglected cause or many, they are usually ready to summarize their disagreements with corporate managers in terms of an appeal to justice. From the Declaration of Independence to the nineteenth-century labor unions to the outcry of environmentalists against toxic waste pollution, individuals, groups, even animals and ecosystems, are alleged to lack something that is "due" from human society.[27]

When leaders of a group speak of "making society more responsible to the public," they are speaking about justice as a social good that all members of a society should enjoy. Some of the movements seek justice along narrow, provincial lines, assuming that it is their right to do so. Who in the country is to say that some group of sports fans had no right to organize for the purpose of getting Howard Cosell off the air? But among constituency groups there is a class of organizations whose leaders would be embarrassed to seem to represent so narrow a band of human interests. Many of these leaders are involved, personally and professionally, in institutions with very broad responsibilities set forth in their own charters.

Two prominent representatives of this segment of corporate constituencies are churches and universities. Business leaders have long exhibited suspicions of these two institutions, though some have made enormous contributions to both. Unlike churches and universities, corporate culture usually calls for relatively narrow human goals: The production of certain commodities or services, the making of profits, the promotion of benefits for some portion of humanity. Business firms that try to do everything or to serve every market seldom succeed. Business leaders have a natural suspicion, therefore, of leaders who pretend to be concerned with everything and everyone.

Awkward and strange as they may seem to the eye of specialists,

churches and universities (as the very name, *university,* implies) have a
self-acknowledged interest in the *comprehensively human.* The religion of
the one and the intellectual tasks of the other incline each to be permeable to "all sorts and conditions" of human need, as a classic prayer
of the churches puts it. It is not surprising that students and faculty of
a university should develop an intense concern for injustice in the relations of the races in South Africa; that nation is a fair focus of the
world-roving intellectual interests of a university. Universities may not
be the first institutions in society to call world attention to social injustice in some part of the world, but there is nothing illogical about their
doing so. Nor is it surprising that the pastors and members of some
Christian churches should organize to show public concern for the welfare of underpaid migrant farmworkers. Their institutional standards—the Bible, for example—long ago gave them reason to be on
the lookout for injustice anywhere in their society, particularly among
the weak and the defenseless, including especially "widows and or-
phans."

One cannot be surprised, then, that churches and universities, when
they have declared themselves as corporate constituencies, appeared to
be unfocused "do-gooders" in the eyes of their corporate-manager critics. For example, a leading church-related organization, the Interfaith
Center On Corporate Responsibility (ICCR), founded in 1974, lists its
concerns as including:

> equal employment opportunity, economic conversion, nuclear weapons pro-
> duction, infant formula abuse, occupational and environmental hazards, agri-
> business, energy, community reinvestment, and nutrition.

Again, only the uninformed critic from business will be inclined to joke
about this wide swath of concerns, for in its first ten years the ICCR
helped to form coalitions of thousands of participants, nationally and
internationally, to bring political, social, and economic pressure upon
business firms, which had to listen, if only because of the numbers and
prestige of the people who represented these concerns of the churches.

The ICCR is one of the most notable church coalitions involving both
Roman Catholic and Protestant members in the United States. It does
not pretend to represent all 130 million people who hold membership
in these churches, but the various churches provide a vast set of possible contacts among some of these millions: Consider church newspapers in light of Tocqueville's discussion of newspapers. Add to them
church telephones, address lists, knowledge of sympathetic supporters
in the national and world settings of organized church life, and one
perceives an impressive array of potential influence and power arising
from these churches. Church members and leaders mobilized themselves as contributors to the success of the civil rights and antiwar

movements of the 1960s. Buoyed by these efforts, the churches presumed in the 1970s to ask many a corporate chief executive officer or president to listen to their critique of corporation policy along a wide front. He (almost always *he*) had to listen. The churches were there in numbers, with expertise and with credible threats of actions—publicity, demonstrations, boycotts, lobbying and stockholder proposals—to make it imprudent for him not to pay attention.

What the manager listened to, in the most general sense, was almost always a claim relating to justice. Somebody was being harmed by corporate policy or action; some human interest was being overlooked; some externally intolerable impact of a corporate program was being internally tolerated. *The corporation is as subject as any other institution of society to the norm of justice in all its social effects and relations.* On this principle, virtually all segments of corporate constituencies have agreed.

WHAT CONSTITUENCY VALUES ARE REALLY NEW?

Was this principle of the applicability of justice new to the business sector of American society? The answer depends on how one reads history, on one's personal philosophy, and on what connections one makes between personal and business philosophy.

The diversity of the movements and the debatable relation of their goals to American history and various philosophical standpoints make difficult any ambitious answer to their question. As students of the social responsibility movement and interest groups, the authors are inclined to believe that in the groups' use of the associational strategy for social change, they are part of a long-standing American tradition. In their appeal to some norm of justice, they are part of the even older Western-Hebrew-Greek-Roman-European tradition, one that stretches back 3,000 years. If the appeal to justice seems new to many business leaders, one reason is that theory, rhetoric, and practice in the free enterprise system have given—can give—little attention to justice; and, as we claim elsewhere in these chapters, justice is as systematically neglected in existing capitalist systems as personal liberties are in existing socialist systems. But neither liberty nor justice as a primary social value is alien to the legal-political institutions or the religio-philosophical culture characteristic of the United States.

Community as a Value in
an Individualistic Society

Closer to alien status in the cluster of reigning social norms in this society is an ideal named "fraternity" by the French revolution and called "community" in American terminology. As Tocqueville observed re-

peatedly, "liberty" and "equality" as American social ideals have a decidedly individualist ring. Freedom, as the right to protest, and equality, as a moral status and as a right to seek social rewards equal to those
of one's neighbors, drive citizens apart as well as together. The associational society has built-in, self-destructive tendency here. Associations,
of which the business corporation is one, come and go, leaving behind
the individuals who composed them. Americans tend to think of society
as just a collection of striving individuals, anyway. Capitalism cultivates
that image, and business practice makes it experiential—as when a 45-
year-old steelworker, who has worked in a mill for 27 years, finds that
the company can no longer employ him, or a 60-year-old executive
faces "early retirement" and a sense of loneliness now that the principal
association of his life is over. Such people, as they become unemployed,
are likely to admit that their work always was for them something much
more than a source of income and individual achievement. It was an
experience of "company" in the human-sense—an experience of community with associates whose support, even friendship, was an indispensable noneconomic value of work.

As our nominee for what is truly new in the galaxy of values among
corporate constituencies and the social responsibility movement, we
suggest *community,* a term whose indistinct meaning for us Americans
is an evidence of its newness. When the leaders of constituencies use
the term "corporate social responsibility," they do indeed mean to call
the managers of the corporate firm to be responsive *to* the interests of
some neglected constituency. But openly or by implication, they seem
also to claim that the corporation, and every other institution of the
society, has a responsibility *for* the interests of an unprecedentedly broad
human neighborhood. A corporation that markets infant formula may
have no direct role in a misuse of its products that leads to the death
of an infant in Africa. But, once discovered, the danger of such a death,
say the advocates in some corporate constituencies, becomes a responsibility of the business firm, which is morally bound to take steps, within
its capability and ambit, that will prevent such death in the future. Are
corporate managers the keepers of their brothers and sisters? Or is
their responsibility limited by some version of the rule, "let the buyer
beware"? Many, perhaps most, corporate constituencies have a clear set
of answers to these two question.

Asking questions and delivering answers to them in a corporate setting may seem to large numbers of Americans something new and
strange. Witness the continuing doubts about the notion of government
welfare payments to the poor and the widespread belief that in this
nation, if one is poor, it is one's own fault. Certainly the sense that
"every person is a keeper of his or her brother or sister" is not as widespread in our society as in some others on the planet. In this respect,
some of the tribal cultures of Africa, for example, may be distinctly

different from the culture of most Americans. But that part of our society called "corporate America" is no place for anyone to preach pure individualism. Corporations, by definition, are social enterprises. In no small way, they manifest *communal* values.

The Corporation as a Voluntary Community

Perhaps the most innovative theme heard among many of the corporate constituencies is the message that business managers must recognize their extensive organizations as communities. Some may find this disturbing to their traditional sense of the firm as simply an economic market entity. Individualism may have worked to build American society so long as the western frontier and abundant resources were open to individuals and small groups for easy development and exploitation. Even during this long period the nation provided an undergirding of locally vital communities that allowed individualism to flourish. Heroic cowboys did their part to "win the West," but federally financed railroads and military troops were probably more decisive.

During all of modern American history a sense of community and cooperation was so common, and so taken for granted, that Americans could easily fail to recognize its omnipresence. An ancient rabbinical question is, "who discovered water?" The answer: "We do not know, but it was not the fish!" For the same reason, Americans find it difficult to discover cooperation, for the reality of their daily lives is that they "swim" in it, surrounded in every direction, enveloping almost all that they do. Even the westward expansion and the world-market expansion of the economy have been more a collective social achievement than the symbol of the bold, individual entrepreneur implies. Individualism, indeed, had a basis in the reality of American history. Its presence and effects are undeniable. But, as these first three chapters imply, changing social and economic conditions are changing our recognition of, and the composition of, society itself. Free enterprise and entrepreneurs flourish, but individualism in our modern, industrial, corporate, urban nation is imposing narrower limits, tighter constraints, and more checks than ever known in the past 300 years.

The gradual realization of these unexpected, untraditional curbs upon individualism, often expressed forcefully, dramatically, and publicly through constituencies, enter the experience not only of business managers but also of Americans generally. Many find it a disturbing experience. In a complex and increasingly interdependent world, a decision to abandon sheer individualism and to shoulder responsibility for one's neighbors can be costly. As a nation we live in a mixed economy, where responsibility for the needs of the society as a whole is spread among a great variety of systems and institutions. This diversification of responsibility expresses the American preference for both freedom and justice. We are wary of loading all responsibility for all our needs on any one institution, such as government or the business corporation or religious bodies. Americans have not searched vigorously for systems of collaboration that will serve the interests of literally everyone in the society. We are a people who expect each person to assume responsibility for him- or herself. It may be that we have now pushed this individualistic notion of responsibility inhumanly far. We may be at the point in our history when it would be wise to discipline our historic individualism with wisdom like that of Lewis Thomas when he wrote

> The urge to form partnerships, to link up in collaborative arrangements, is perhaps the oldest, strongest, and most fundamental force in nature. There are no solitary, free-living creatures: every form of life is dependent on other forms. The great successes in evolution . . . have [been] so by fitting in with, and sustaining, the rest of life.[28]

NOTES

1. See Chapter 2.
2. If a large retailer like Sears or Montgomery Ward cannot reach a satisfactory agreement with a furniture supplier, it may simply refuse to conclude a deal, taking its business elsewhere. Such an act has much the same effect as the strike of a group of employees or as a consumer boycott; in each case a party is using market pressure in support of more favorable terms.
3. See *Santa Clara* v. *Southern Pacific*, 118 U.S. 394 (1886).
4. Leonard Silk and David Vogel make the same point:

> Yet the very values that keep critics of American business from formulating an ideology that explicitly challenges capitalism—a pragmatic philosophy that respects what "works," an upward-striving ambition for limited goals (especially a higher standard of living), a privatism that regards politics as a game or a racket, or in any case a matter of limited importance, and a resentment toward government control—are shared by the business leadership itself. Neither business nor its critics are able to stand sufficiently apart from the American orthodoxy to appreciate the relative narrow parameters of their disagreement.

See Leonard Silk and David Vogel, *Ethics and Profit: The Crisis of Confidence in American Business* (New York: Simon & Schuster, 1976), p. 28.
5. Fred D. Baldwin, *Conflicting Interests: Corporate-Governance Controversies* (Lexington, Mass.:Lexington Books, 1984), p. 2. Emphasis added.
6. William M. Evan and R. Edward Freeman, "A Stakeholder Theory of the Modern Corporation:Kantian Capitalism," in Tom L. Beauchamp and Norman E. Bowie, eds., *Ethical Theory and Business*, 3rd ed., (Englewood Cliffs N.J.: Prentice Hall, 1988), pp. 97–106.
7. Ibid, p. 97.
8. Committee for Economic Development, *Social Responsibilities of Business Corporations*, June 1971, p. 22. Several years later, The Business Roundtable endorsed the same notion of business responsibility, insisting that managers

have a special, neutral, balancing role to play. See "Statement on Corporate Responsibility," in Thomas G. Marx, ed., *Business and Society: Economic, Moral and Political Foundations* (Englewood Cliffs, N.J: Prentice-Hall 1985), p. 155. See also Chapter 5 of this volume.

9. See, for example, Charles E. Exley, Jr., "The Stakeholder Concept Creates Value for All," *Financier* 13 (February 1989):20–25; Stephen R. Covey, "Universal Mission Statement," *Executive Excellence* 6 (March 1989):7–9; and William H. Gruber, "Planning for the Next Market Break," *Institutional Investor* 22 (December 1988): 14–15.

10. Evan and Freeman, *Ethical Theory and Business,* p. 103.

11. See Chapter 2, from the opinion of the U.S. Court of Appeals for the District of Columbia, *Medical Committee for Human Rights* v. *SEC,* July 8, 1970, reprinted in The Extension of Remarks, Senator Lee Metcalf, *Congressional Record,* December 28, 1970, p. E10741. Emphasis in original.

12. Ernest van den Haag, "Confusion, Envy, Fear and Longing," in *Capitalism: Sources of Hostility* (New Rochelle, N.Y.: Epoch Books, 1979), p. 19. [Reprinted with permission.]

13. Irving Kristol, "The Corporation and the Dinosaur," in *Two Cheers For Capitalism,* New York: Basic Books, 1978, p. 75.

14. Mel Friedman, "Will TV Networks Yield to New Pressure Groups?" *Television/Radio Age,* May 4, 1981, p. 38. For a full history of the use by various groups of consumer boycotts to influence television programming see Kathryn C. Montgomery, *Target: Prime Time; Advocacy Groups and the Struggle Over Entertainment Television* New York: Oxford University Press, 1989.

15. "Television Plays Down Sex This Fall, Reacting To A New Public Mood," *Wall Street Journal,* November 5, 1981.

16. Bill Carter, "TV Sponsors Heed Viewers Who Find Shows Too Racy," *New York Times,* April 23, 1989.

17. Cf. A. J. Haagen-Smit, "Chemistry and Physiology of Los Angeles Smog," *Industrial and Engineering Chemistry* 44 (June 1952):1342–46.

18. Allen R. Janger and Ronald E. Berenbeim, *External Challenges to Management Decisions: A Growing International Business Problem* (New York: The Conference Board, 1981), p. 11. Note that the authors, no doubt reflecting managers' own perception of the challenge, perceive it as a problem, not an opportunity. While constituency challenges to managers certainly have, do, and will create problems, neither the reader nor managers should assume that there have not been, are not, and will not be new possibilities opened up.

19. See Janger and Berenbeim, *External Challenges to Management Decisions,* p. 5.

20. William Tucker, *Progress and Privilege* (Garden City, N.Y: Anchor Press, Doubleday, 1982), p. 37.

21. Ibid.

22. Robert Cameron Mitchell and J. Clarence Davies III, "United States Environmental Movement and Its Political Context: An Overview," Discussion Paper D-32 (Washington, D.C.: Resources for the Future, 1978). Also, Mitchell, unpublished data.

23. "What Interests the Public and What Interests the Public Interests," *Public Opinion* (April/May 1983):47.

24. Jeffrey N. Barry, *Lobbying For The People* (Princeton, N.J.: Princeton University Press, 1977), p. 72.

25. Jack L. Walker, "Origins and Maintenance of Interest Groups in America," Paper delivered at the annual meeting of the American Political Science Association, New York, September 2–6, 1981.

26. The Nader organizations include Public Interest Research Group, Center for Study of Responsive Law, Corporate Accountability Research Group, Citizens Utility Board, Buyer's Market, and public interest research groups operating in twenty-six states, Ontario, and British Columbia, as well as seventeen "Public Citizen" groups. See Douglas H. Harbrecht and Ronald Grover, "The Second Coming of Ralph Nader," *Business Week* (March 6, 1989):28.

27. See Christopher Stone, *Where The Law Ends* (New York: Harper Colophon Books, 1976). As a legal scholar, he argues that the claims of justice may be made for many nonhuman creatures, and even for trees, stones, valleys, and mountains. All have a legitimate claim as part of an environment whose completeness and wholeness need consideration and, often, preservation.

28. "On The Uncertainty of Science," *Harvard Magazine* (September–October, 1980):21.

4

Business as a Source
of Social Discontent

Business managers have not found congenial the notion that the organizations that they head are—or should be—collaborative partnerships. Commonly, those who directed large corporations have pictured themselves as field marshals and generals whose orders were to be obeyed. To be sure they expected cooperation, but on *their* terms, for *their* designated goals and purposes. Were employees, contributing suppliers, or small retailers to dispute or protest company policies or programs, they were condemned for being disruptive, destructive, undesirable, and antisocial. The pyramidal, hierarchical organizational chart, widely used to picture the corporation's internal structure, was more than merely a sketchy outline of authority and reporting relationships; it was also a symbol of the power, status, and special position that managers claimed for themselves.

That they would make special claims for themselves is not surprising when one considers the privileges under the law they have enjoyed from the earliest days of the Republic, and the wide public acclaim they and their firms received, particularly during periods of economic prosperity and rapid economic growth. More than half a century ago Calvin Coolidge summed up their pretensions when he declared that "the business of America is business."[1]

The Great Depression seriously eroded managerial claims, but by the end of World War II and in the quarter-century afterward, the amazing productive achievements of American business raised managerial prestige to new levels. Their position became the model for, and their claims the envy of, business around the world. Most business managers, like most Americans, probably believed that the unprecedented production record was an obvious justification for managerial command; the rapid economic growth and the rising standard of living were reasons for the honors they gave themselves. The economic success over

which managers presided was so impressive that they convinced themselves it could continue only if they were left free, without either the
obstruction or hindrance of distracting constituencies, including stockholders.

Not all Americans were convinced that business managers alone were
responsible for the economic accomplishments or that the rapid economic growth was an adequate measure of success. Most managers,
however, tended to dismiss such doubts and qualifications as to be expected from unappreciative, anti-American dissidents; they often characterized those who held such opinions as irrational opponents of a
productive economy and a beneficent society whose flood of goods and
services could not be ignored. A large portion of the public appeared
to agree with managers, ready even to accept profits they won as indicators of public good rather than of private greed. As late as 1965
more than two-thirds of those polled agreed with the statement that
"the profits of large companies make things better for everyone."[2]

In justifying corporate business power and their own presumed prerogatives on the basis of a quarter-century's "performance," managers
were leaning on a weak reed, however. First, the performance could
have—and as we shall see, did have—unintended, paradoxical, adverse
effects, and second, the uncertainties and instabilities of a capitalist system provide no assurance that a given level of performance can be
maintained for very long. Irving Kristol warned managers that:

> Life has its ups and downs; so do history and economics; and men who can
> only claim legitimacy *via* performance are going to have to spend an awful
> lot of time and energy explaining why things are not going as well as they
> ought to. Such repeated, defensive apologias, in the end, will be hollow and
> unconvincing. Indeed, the very concept of "legitimacy," in its historical usages,
> is supposed to take account of and make allowances for all those rough pas-
> sages a society will have to navigate.[3]

The very success of large business corporations and their managers in
the decades after World War II created problems whose size, complexity, and intractability, few Americans, and even fewer managers, fore-
saw at the time.

THE CHANGING ENVIRONMENT FOR
CORPORATE MANAGERS AND CONSTITUENCIES

The daily routines of organizational life, the deep involvement in the
progress and perturbations of professional careers, and the diversions
of dramatic events and highly publicized developments distract managers and the public from gradual transformations of the economy and

fundamental trends in the social setting. Year-to-year social changes usually appear almost glacial, even when over a life time they produce radical transformations. An acute awareness of change requires both time and effort; most busy people, including business managers, seldom have the luxury of examining events carefully and knowledgeably enough to become aware of the deeper currents that change their condition.[4]

It is not surprising, therefore, that managers misunderstood or overlooked the social transformation that their very success involved. As a consequence, most of them mistook the rise and organization of their constituencies as an attack upon themselves, their businesses, and the economic system. They did not easily accept the new constituencies nor did it cross the minds of many that the constituencies might be organizations modeled after their own corporations, entrepreneurially directed toward creating *and serving* new markets.

Business's Delivery of Prosperity

In the quarter-century from 1947 to 1972, Americans were well aware of their prosperity and the economic rewards that were accruing at a rapid rate. Gross National Product (GNP) increased at an average rate of more than 3.2 percent a year, raising the total from $1,4408 billion to 3,443 billion (in 1989 dollars). Personal consumption expenditures—the income people spent on durables, nondurables, and services—increased at an even faster rate, 3.6 percent, allowing them to double within twenty years!

Population growth was also unprecedentedly high, 1.5 percent annually, providing a net addition of nearly 66 million people. Despite this increase, economic growth was fast enough to raise per capita disposable income from $6,363 to $11,302 (in 1989 dollars). Other obvious indicators of economic success were the threefold increase in industrial production and the net creation of more than twenty-five million jobs for a rapidly expanding workforce. That number, alone, was one-half the total workforce at the beginning of World War II; it amounted to a million net new jobs each year!

Business leaders trumpeted their accomplishments and congratulated their success. A typical statement was that of the Committee for Economic Development (CED) in 1971. After surveying the business scene, the manager members concluded that the corporate business system had treated and was treating Americans very well:

> . . . business on the whole has done its job remarkably well. Since 1890, the total real national product has risen at an average of more than three per cent a year compounded, almost doubling every 20 years. . . . Real disposable income per person has more than tripled and work time has declined

by a third over the past 80 years. . . . All other major institutions of society, including government, have been sustained in substantial measure by the wealth produced by a business system which provides a strong economic foundation for the entire society.[5]

The Loss of Public Confidence in Business and Its Managers

Despite the attention managers drew to their accomplishments and although the economy had performed quite remarkably in the two and one-half decades after World War II, the public was expressing more and more doubts about the contributions of business. Seymour Martin Lipset and William Schneider, two experts in public opinion surveys, concluded that "the period from 1965 to 1975 . . . was one of enormous growth in anti-business feeling." (See Table 4.1.)

The public loss of confidence in business was shown in a variety of other surveys and public opinion polls. The average favorability toward both companies and industries declined markedly from 1965 to 1975; the former dropped to 73.5 percent to 48 percent, and the latter from 68 percent to 40 percent. Some surveys reported even steeper declines in probusiness sentiment, from around 70 percent to a low of 15 percent.[6]

Remarkably the period of declining public confidence in business was also the climax of the rapid growth and business success. Business leaders were shocked and dismayed at their loss of standing. Given the economic success to which they had contributed greatly, how could the public lose confidence in them and their firms? Why would people so well provided fail to appreciate their efforts?

Since the loss of confidence was accompanied by the appearance of newly organized and vocal constituencies, perceived as critics and anti-business organizations, managers at first found in them the cause and source of the public's loss of confidence. Certainly there was a surface plausibility in their perceptions. If the business environment was relatively stable and if economic progress offered unquestionable benefits, then it followed that some groups, for their own hostile reasons, were out to destroy business.

HOSTILE CONSTITUENCIES AS CREATIONS OF THE PRESS AND INTELLECTUALS

By the time the Committee for Economic Development issued its congratulatory report, a portion of which was quoted above, corporate managers were already familiar with hostile criticism from many different groups. Some had to contend with public demonstrations outside

Table 4.1. Attitudes toward Business, 1965–1975

Statement	Percentage Agreeing	
	1965	1975
"As they grow bigger, companies usually get cold and impersonal in their relations with people."	58	79
"There's too much power concentrated in the hands of a few large companies for the good of the nation."	52	78
"In many of our largest industries, one or two companies have too much control of the industry."	58	82
"The profits of large companies help make things better for everyone."	67	41
"For the good of the country, many of our largest companies ought to be broken up into smaller companies."	37	57

Source: Opinion Research Corporation as reported by Lipset and Schneider (see endnote 2).

their offices and inside stockholders' meetings; a number had been threatened by nationwide consumer boycotts and a few had actually endured them; others had felt constrained to meet and even negotiate with local groups of citizens aroused about various business actions such as plant closures, investments in South Africa, pollution of air or water, or alleged discriminatory hiring and promotion. Such unprecedented criticism appeared exceedingly unfair to many business managers. They looked for causes outside the corporation; almost all managers were convinced that neither they nor their organizations could be the cause. After all, were they not still producing as efficiently and effectively as ever, carrying on the great tradition so clearly established during World War II? Business performance, as perceived by managers, had been too impressive and consistent to be the source of the rising hostility. Therefore, the cause had to lie outside business.

The CED pointed the finger of blame at intellectuals, for unrealistically raising the expectations of the younger generation:

> . . . a highly idealistic and restless generation of American youth; a cultural leadership class of writers, filmmakers, artists, and intellectuals which is exerting considerable influence through communication's media, literature, theaters, and universities; and numerous citizens' groups which are crusading for . . . [various] objectives.[7]

Business managers blamed college teachers not only for inculcating unrealistic expectations among their students but also for projecting an unflattering, damaging image of business, creating it out of their general antibusiness bias. Typical of the managerial comments were these:

There is no comparison between free enterprise and other systems if they understand the facts. But few young people know what free enterprise means. That's why they don't like it.

The young know more about the Cuban and Chinese systems than they do about ours. Students are being deprived of a freedom of choice. They don't know how our system works, so how can they choose it?[8]

Some business leaders accused the causal culprits more sharply. The then-chairman of Citicorp asked in 1974, "Why is the future, in our eyes, not what it should be?" and answered, "One reason is the omnipresence of the media . . . the profession as a whole lacks a sense of history which is essential to balance, to perspective and to optimism."[9] After sitting in on a series of business conferences and conducting many interviews with business leaders, Leonard Silk and David Vogel found this perception to be one held generally:

The business community feels extremely hostile these days toward the press and the electronic media, which it blames for the low public esteem of business. No one theme was so consistently mentioned . . . and few themes enjoyed such unanimous support. Executives are firmly convinced that "a majority of citizens would support the free enterprise system if they understood it," and that the press denies the public that knowledge.[10]

Even at the end of the decade business leaders were repeating these conclusions, reassuring themselves that the public would appreciate them and their efforts, if only intellectuals and a critical press—and "liberal," unsympathetic television—would not undermine them. Herbert Schmertz, vice president for public affairs at Mobil, repeated the charge in a speech before the American Iron & Steel Institute in September, 1978.

Through the 1960s and right to this day, a strong undercurrent of anti-business sentiment has made it difficult for corporations to get a fair hearing in the forum of national debate. The academic community has trained nearly a generation of young Americans in a catechism which holds that "big is bad." . . . Self-anointed defenders of the public interest—whatever they may be—have gained status as quasi-official watchdogs and they have the ear of the government. . . . Sad to say, the free American press—print and electronic media alike—has been caught up in this anti-business maelstrom. Usually, the media will more readily believe those who oppose business than those who support it.

While Schmertz took aim at the rising numbers of corporate constituencies as well as academics and the media, Donald U. Honicky, director of university relations for the American Telephone & Telegraph

Company, in late 1978 limited his criticisms to the latter, and in a spirit
of generosity assumed, for business, part of the blame.

> . . . it's no wonder that so many college students have such strong anti-
> business views. Business has apparently been doing a poor job of communi-
> cating with the educational community, and that clearly includes faculty as
> well as students, because it's from faculty, presumably, that students are being
> misinformed.[11]

Walter H. Annenberg, a leading publisher and close friend of incoming President Ronald Reagan, nearly two years later in full-page advertisements in newspapers across the country expressed his agreement
with Schmertz and earlier business spokespersons: "Provoked by chronic
faultfinders, distrust of publicly owned companies permeates not only
Government but the media and the public."[12] He went on to argue that
the widespread mistrust was hampering business and making it ever
more difficult for domestic producers to compete in world markets.

Does the Public Know Only What
Professors Teach and Media Report?

In blaming intellectuals, college teachers, and news reporters for the
criticisms directed toward them, business managers imagined an influence for their putative detractors that was undoubtedly inflated. After
all, the public did not have to rely upon teachers or the press (or television) for their knowledge about how business and their markets operated. The public learns about, and knows business from, its own intimate experience; it does not have to read news stories or study textbooks
to discover how managers deal with employees, investments and plant
closures, disposal of wastes, quality of products, and many of the other
activities of industrial organizations. First, a sizable portion of the labor
force works for large corporations. Perhaps a quarter of it, more than
twenty million, hold jobs in the 1,000 largest firms and three-quarters
of the labor force work in the private, for-profit sector; they and their
families are quite familiar with the daily routine of business, at least at
the place of work. As insiders, they observe and learn much about how
business is conducted; they see managers firsthand or at least feel the
immediate consequences of their decisions and directives. If workers
do not know much, or if their information is wrong, the fault might
seem to lie more sensibly with business managers than with outsiders.

Second, almost the whole American population, as consumers, has
firsthand experience with business. Consumers soon discover the quality of the goods they purchase, and they understand very well the kind
of service they receive through the market. Those who bank in large
cities, for example, know well the efficiency of long lines at noontime

bank counters and the indignity of being told that their balances are so small they must deal with machines rather than tellers or that they cannot draw upon checks that they have just deposited.

Customers know how the market operates, for they are involved with it daily and weekly as they shop; they continually test the claims and advertisements of sellers against the use of the goods they buy, and they know directly the degree of satisfaction they receive in the services they regularly purchase. Many have initially believed the promise of quality in American-made automobiles, for example, only to be disillusioned by defective motor parts that may be replaced after legal action; not a few car owners have discovered the exasperating, small-print exceptions to the warranties of a major auto producer. And what consumer has not bought a frozen food dish, attracted by the carton's colorful picture, to discover within a dull-looking, salty, hardly edible concoction? Can there be much doubt that when Americans give a low rating to business performance in specific areas, such as those in Table 4.2, that they are drawing upon firsthand, direct experience, not what they heard professors say or read what news columnists wrote?

For business managers to blame those who bring them the news that their public standing has fallen sharply may be to confuse the messenger with the message. As in the case of the Persian king who beheaded the heralds who brought him evil tidings, business managers do not change events by denouncing the news, and should they succeed in silencing those who bring the news, they may lose a most valuable source of information. Business managers who accuse the public, consumers, intellectuals, and reporters of misunderstanding them and their system might remember Bertolt Brecht's mocking of the Communist rulers during the 1953 workers' uprising in East Berlin: "The government has lost the confidence of the people. So it is necessary to elect a new people." [13] Business managers are not going to get a new people, and the public's loss of confidence in business may be more solidly based than business managers realize; it may reflect more difficult problems than they have generally cared to contemplate.

We believe that there were sound reasons for the loss of confidence during the late 1960s and through the 1970s, without attributing the cause of the loss to "outside" scapegoats and hostile, ignorant critics. A more substantial cause, ironically, is to be found in the success with which the business system had fulfilled its promise of providing economic growth, and so enriching Americans and others around the world. Few managers imagined that in furthering economic progress they were social revolutionaries, creating new social forces and undermining the older economic order of the post–World War II period. The productive results they achieved were so remarkable that the unintended consequences and unforeseen outcomes escaped their attention. They were too busy with their regular, routine work, too distracted by immediate

Table 4.2. American's Attitudes about Business

Statement about Business	Percentage of Polled Americans Agreeing
Business is	
Producing good quality products	64
Paying fair share of taxes	42
Controlling pollution	38
Advertising honestly	37
Setting reasonable prices	36

Source: Adapted from Joseph Nolan, "Business Beware: Early Warning Signs in Eighties," *Public Opinion,* April/May 1981, p. 206a. (Survey by the Roper Organization, 6-13-80). [Reprinted with permission of the American Enterprise Institute for Policy Research, Washington, D.C.]

problems either to comprehend the opportunities opening for them, or to understand the nature of the problems confronting them.

THE FAILURES OF SUCCESS

One of the unforeseen outcomes of economic growth that gradually came to wide public attention in the late 1950s and early 1960s was the accumulation of externalities—the costs of production or economic change not reflected in the prices of particular goods or services but nevertheless paid, as costs, by the people whose lives were adversely affected. The degradation of air and water quality by pollution is an easily recognized example; and road or beach congestion, when too many people attempt to use them at the same time, is another. Galbraith described the paradoxical situation when the beneficiaries of affluence find themselves surrounded, at the same time, with disagreeable externalities.

The family which takes its mauve and cerise, airconditioned, powerbraked car out for a tour passes through cities that are badly paved, made hideous by litter, blighted buildings, bill boards, and posts for wires that should long since have been put underground. They pass on into a countryside that has been rendered largely invisible by commercial art. . . . They picnic on exquisitely packaged food from a portable icebox by a polluted stream and go on to spend the night at a park which is a menace to public health and morals. Just before dozing off on an air-mattress, beneath a nylon tent, amid the stench of decaying refuse, they may reflect vaguely on the curious unevenness of their blessings.[14]

Many environmental externalities can be eliminated or greatly lessened through appropriate pollution charges, taxes, or perhaps by government regulations. Most economists believe the latter are far less effective or efficient than the former if consumers can be charged the full social costs of their purchases. In recent years Americans have made measurable advances in cleaning up their environment,[15] though much remains to be done, since continued economic growth increases old kinds of pollution and creates new ones.

There are other kinds of externalities arising out of affluence that are not as amenable to solution; in fact some of them may have no solution at all. Some of them have contributed to the public's loss of confidence in business and have encouraged members of the public to join constituency organizations to press for changes in the way business managers conduct affairs. In the following sections we examine four important externalities that have confounded managerial expectations that economic progress would increasingly satisfy and please the public. The appearance of these externalities contributed to the public's loss of confidence in business noted from 1965 on and the rise of corporate constituencies more than any hostility of intellectuals and the media. Such hostility may exist, but to assume it is a major source of difficulty for business over simplifies a complex set of causes and ignores the radical social effects that business itself has caused.

Happiness Can Be Added, But Not Summed

As incomes rise people probably expect increasingly to emulate the lifestyles and pleasures of the rich. They buy more goods and services for private use—automobiles, color televisions, video home recorders, and vacations at "in" spots—and thus pattern their consumption to match those with high incomes. If millions of consumers make such purchases, they cannot gain the full enjoyment known by the wealthy, however, when they alone could afford them. The wealthy consumed the goods in the privacy, seclusion, and spaciousness that even upper middle class families are not apt to enjoy. The wealthy command the personal services of butlers, valets, and maids, which the ordinary consumer seldom knows. The attempt by persons of average income to improve their relative status by buying goods once reserved for the rich almost surely will produce disappointment as long as such purchases are based, not on their intrinsic satisfactions, but on social comparison. If the mass of people advance themselves, raising the average, their relative position does not change. It is analogous to the situation of one who stands in the stadium to improve one's view of the players on the field; if all stand, however, no one can see any better.

The continuing scarcity of relatively high positions, even in an economy of growth, tends to frustrate economic expectations. It belies what may be widely perceived as an implicit promise of economic growth: getting ahead of everyone else. Certainly business feeds the notion that every man can live like a king—or at least a Robert Redford—and every woman like a queen—or a Dolly Parton.[16] The resulting frustration may be a major cause of the happiness paradox Professor Richard Easterlin identified some years ago.[17] He found that opinion surveys in various countries regularly show a strong, consistent, positive association between income and happiness. At any particular time many more of the rich report they are "very happy" than do the poor, as might be expected. It accords with common experience that higher income and wealth help in achieving happiness. The association is more apparent than real over an extended period, however. Surveys taken roughly at the same time in countries with widely different income levels reveal but slight differences in the degree of happiness reported, much smaller than the differences of incomes among the countries. Furthermore, in any given nation over a period of years when incomes have increased markedly, there is no associated rise in reported happiness. Total happiness does not seem to increase with gains in absolute income over time.[18]

Abraham H. Maslow long ago suggested that human wants form a hierarchy. As a lower order want is satisfied, it becomes less important, with the next highest order want becoming more and more important. He perceived that wants are not absolute, but relative.[19] Want satisfaction is more complex than Maslow's hierarchy, however. Wants may indeed follow some hierarchical order at any given time, but they and their order may change as satisfaction is gained. Increasing economic rewards may become less and less satisfying, but not less important. In fact, as the ability of economic reward to provide a positive incentive diminishes, its capacity to create dissatisfaction, increases rapidly.[20]

If, in fact, rising income does not always and directly increase the sum of happiness in society, the relative contribution to human welfare that business can make through economic growth diminishes significantly. People may enjoy the more comfortable life of an advanced economy when they compare it with the scarcities of a former time, but they may also feel trapped in a "rat race." They work hard, buy much, and yet find themselves *relatively* where they have always been. People whose incomes rise may even feel worse—more harried and less satisfied—because the demands of owning more goods and choosing among more services effectively reduces the time available to enjoy the benefits of a richer lifestyle.[21]

Relative deprivation, the invidious comparison of one's own position, consumption, or acquisitions, may be a more widespread cause of discontent with business in an advanced economy than absolute deprivation.[22] A social critic once asked, "What is poverty?" and answered, "If all but you can buy a bottle of beer, you feel poor." Insofar as business managers believe they have satisfactorily discharged their social obligations by providing more "beer" and making it available for anyone and everyone, they may misperceive the nature of the faith the public has in them. Individuals may want to rise relatively, whereas business aims to provide "only" the means for all to rise absolutely.

The Inequality of More Equality

Not only does economic growth problematically affect the general well-being, as suggested by the Easterlin paradox, but it may also paradoxically sharpen the issue of inequality, to the discomfiture of defenders of the business system. The sharpening has occurred even though, by many measures, the economy has moved the nation well along the path toward equality. While the gross distribution of income has moved over the last half-century toward equality, it varies from period to period.[23] If one considers only a variety of basic goods and services—shelter, clothing, household appliances, and private transportation—historic differentials have shrunk surely and noticeably.[24]

In 1942, for example, just as wartime rationing began, the highest third of urban households consumed more than twice as much meat per person as the lowest third; a quarter-century later the highest third consumed only 16 percent more. Consumption of sugar and sweets had reached equality as early as 1948. The distribution of owner-occupied housing has also widened markedly over the last forty years. In 1940 it accounted for 44 percent of the total, but by 1980 had reached 66 percent. In the half-century prior to 1940, the owner-occupied share changed very little. Housing units with central heating increased from 50 percent in 1950 to 83 percent in 1980. In a marked increase over the 1940s, today 97 percent of all dwelling units enjoy piped water and inside plumbing. Practically all households now own a telephone, a vacuum cleaner, a toaster, a refrigerator, at least one television set, several radios, a steam or spray iron, a coffeemaker, and a food mixer. In 1960 one-quarter to one-half the households did not possess those useful appliances. Refrigerators and television sets had already become popular possessions by 1960, with 98 percent of households having the refrigerators and 89 percent television sets. In 1948, 54 percent of all families owned an automobile, and more than 80 percent had one by 1970.[25]

Certainly the lower one-third of families today can afford less expensive cuts of meat than the top one-third; the houses they can buy are smaller and probably not as well made or as fully furnished; their appliances may be more basic models and their cars are more likely to be secondhand than those of the rich. Yet the mass production and wide distribution of such goods, basic to modern existence, are more than a casual achievement. It is indeed impressive to those old enough to have been reared on farms where outdoor toilets, water pails, stove-heated flatirons, hand-powered washing machines, and once-a-week meat for Sunday dinners were standard parts of daily life for whole communities. Even more dramatic have been the changes in dress, with the poor as well as the rich wearing "designer" jeans, well-made boots or sneakers, and easy-to-care-for fashionable shirts, blouses, and jackets. There is also the startling change in entertainment, having moved from locally produced recitations or musicals at the Grange hall and church to the bright, colorful professional offerings on radio and television.

Professor Paul Samuelson has suggested that an observer ought to look beyond merely the official data on inequality of incomes, which had shown considerable constancy over the forty years from 1945 to 1985. He declared,

> I believe that true inequality, measured over people's entire lifetimes, has significantly decreased in the mixed economy. More people went to bed hungry, more children had rickets because their parents were unlucky or feckless, when I began my economic studies than is the case today in North America and Europe.[26]

We agree with Samuelson—the gap between rich and poor has unquestionably narrowed over the past generation, yet few readers will be as impressed by its narrowing as they will be conscious of the gap itself and the remaining distance yet to go.

A variety of interest groups remind the public of and express uneasiness about the inequality that continues to distinguish rich and poor throughout the economy. Although inequality of possessions has significantly declined, many members of the public are concerned because the distance to go toward equality seems so small compared with the distance already covered.[27] They readily believe that in an affluent society, the existence of any poverty at all is a moral scandal. But others besides members of interest groups, who are often perceived to be socially liberal, are also concerned about poverty, homelessness, and illiteracy. Conservatives, too, are offended by the deficiencies of a nation as rich as the United States.[28] Adding to the problem is the shift toward a widening of the distance between rich and poor that began in the early 1970s, reversing a trend of more than thirty years.[29]

Professor David M. Potter suggested that American success in reducing many of the outward forms of class differentials through economic abundance may have stimulated new discontents.

> In diminishing the physical differentials, the social diversity, and the real economic disparities that once separated classes, it has made any class distinction or class stratification seem doubly unfair and discriminatory.[30]

If he is right, business managers' contribution to long-term, widely diffused economic growth may be forgotten as the public focuses upon concerns over remaining inequalities. Economic growth, thus, may have

to move at a fast pace, indeed, to outrun the grievances and resent-
ments that arise from the effects of the growth itself, We may conclude
that a rising standard of living, based on continually improving business performance, may neither offer increased, overall happiness for a
society nor adequately satisfy the egalitarian thrust of a democratic
people. The result may be increased discontent even as standards of
living rise.

The Rising Dependency of
the Increasingly Independent

There are other dissatisfactions that may also arise from the successes
of a business system. Consider the ambiguous independence successes
offer. As incomes rise, employees win more leisure in which to pursue
individual interests. In addition, the increased income makes available
for their own and their family's enjoyment an even wider range of goods
and services. Thus, economic growth has freed most Americans from
the tight constraints of dire necessity and has given them more and
more scope for independent choice of lifestyle by seeming to offer wider
options in their personal lives. Out of such variety, Americans may pursue their ideal of social individualism.

Unfortunately, the independence and individualism that is realized
has been accompanied by an increased dependency in other ways. Most
Americans earn their livelihood only as employees. More than 90 percent of the labor force do not work for themselves, but are dependent
upon others for their wage or salary. Almost all in the population of
an industrial society depend upon others—many of them usually unknown, faceless, distant producers—for all that they consume. Few consumers are able any longer to make for themselves the necessities of
life. The dependency upon others for the basics of life—job, income,
food, shelter, and clothing—and even for less essential comforts creates
a vulnerability that can infect the public with a general unease and
insecurity not easily remedied.

Consumer dependency. The scope of consumer dependency has increased
generally with the decline of the American farm population and the
growth of urban centers. Compare the situation of the population a
generation or two ago with that of 1970. In 1920 just under half of the
population lived on farms in rural areas; in 1940, 46 percent still lived
in rural areas or towns under 5,000 population.[31] By 1970 the farm
population had fallen to less than 5 percent.[32] Farmers, rural folk, and
small town people did not enjoy easy access to the well-stocked markets
of the city and had to rely extensively upon local merchants or more
importantly, their own production. Railroads, then paved roads and
automobiles, and finally the postal service (mail-order catalogues) pro-

gressively made available wider markets and a greater variety of goods and services. Gardening, home canning of vegetables, preserving meats, and raising cows, pigs, and chickens continued as major contributors to the larders of millions of families. Housewives sewed many of their own and their children's clothes, quilted bed coverings, and fashioned kitchen towels, linens, and even dresses out of flour and sugar sacks. Farmers continued to build their own houses, barns, and outbuildings. Church members and community groups furnished most of the entertainment available for special occasions. Only slowly were they replaced by store-bought goods and commercial services.

As the farm population dropped sharply from about 25 percent in 1940 to 15 percent in 1950 and then to less than 5 percent over the next three decades, the whole nation came to rely upon distant producers and manufacturers for both the basics and the trivia of daily life. By 1967 surveys showed that American consumers were beginning to feel their remoteness from those upon whom them depended.[33] They reported a widening gap between themselves and the businesses from whom they bought. In four surveys made by Cambridge Reports, Inc., between 1975 and 1977, 69–79 percent agreed with the statement, "Big business doesn't care whether I live or die, only that somebody buys what they [sic] have to sell." Fewer than 20 percent disagreed. The Harris poll in December, 1976, found that 71 percent agreed that "businessmen will do nothing much to help the consumer if that reduces profits—unless it is forced to do so." Only 11 percent disagreed. In surveys taken in 1975, 1979, and 1981, the Roper Organization reported that 64–65 percent of those polled approved the statement that "American business and industry has lost sight of human values in the interests of profits." The proportion disagreeing was 22–23 percent. In four Harris polls from 1966 to 1978 a majority of respondents gave negative marks to business on "better," (as opposed to good) quality products. There appeared to be little relationship to political inclinations; liberals, moderates, and conservatives responded almost identically, though more of the poor than the wealthy reported that business firms were increasingly distant and unresponsive.[34]

Technological change has probably contributed to a sense of consumer dependency. As services as well as products proliferate and become increasingly complex, consumers find comparisons difficult among close substitutes and different forms or models.[35] Furthermore, owners of sophisticated products must depend on trained service technicians for repair even if the malfunction potential is lower than with less sophisticated products. The spread of self-service among retailing establishments and the use of computer-generated communications with customers removes personal encounters in consumer transactions. The widespread and immediate coverage of news emphasizes consumer dependency as well. In recent years news alerts have warned the public of the dangers of poisons in food—botulism in canned soup, pesticides on fruits, lead in canned fish, and PCBs in fresh fish. The public frequently reads stories about dangerous products—new model cars with unprotected rear-end gas tanks, faulty steering mechanisms, or slip-prone gearshifts—that cause injury and death to the unsuspecting or careless user. Congressional studies and reports by interest groups identify companies resisting governmental attempts to safeguard the public and remove unsafe products from the market.

In 1982, a startling incident starkly revealed to the public its vulnerability in the modern market. An unknown, presumably mad person inserted cyanide into Tylenol capsules before they were sold, killing seven people in the Chicago area. The poisoning highlighted the trust a public must put in those who produce for it. Consumers must rely upon the manufacturers to maintain quality and safeguard health, for they literally put their lives in the hands of those upon whom they depend for the goods they purchase. The necessary reliance of consumers upon the good faith and goodwill of others—distant, unknown, and impersonal—may well contribute to their feeling of remoteness from business firms and their managers. It may also contradict and undermine their sense of individual independence by injecting a sense of unease about the business system and creating an underlying discontent with its benefits.

Employee dependency. There are indications that many of the 90 percent of employed Americans find a dependent working relationship less than satisfactory. The Harris polls have regularly found that a majority of those interviewed in recent years agreed that business did not allow "people to use their full creative abilities."[36] A survey conducted by the Public Agenda Foundation of New York in 1983 showed that many workers distinguished between *agreeable* jobs and jobs that *motivate* them. They would work harder, they said, if there were potential for advancement, a chance to develop abilities, and a challenging job. Most workers reported, however, that employers had little knowledge of how to motivate workers and less than a quarter of those surveyed said they were performing on the job to full capacity. Sixty-one percent identified "pay tied to performance" as a feature they wanted most in their work:[37] So consistently do surveys report such findings that some industrial relations scholars wonder why managers do not act upon the information. More than half of workers surveyed regularly indicate an interest not only in work that allows them to expand and fully use their skills, but also in a greater say in job-related decisions.[38] That employees might want more say at the place of work is hardly surprising, since the conditions under which many work contradict fundamentals of the American mythos. Managers typically assign workers to groups rather than allowing independent work. They are required to merge their efforts with those of others in joint activities. Few readers will find strange that most workers hanker after forms of remuneration and restructured jobs that might offer at least the illusion, and perhaps some of the substance, of traditional American self-employment, independence, and self-reliant individualism.

As long as managers of large corporate bureaucracies do not, or cannot, fulfill such hope for their employees, they can expect widespread worker discontent. The traditions are kept alive in workers by the frontier myths with which Americans grow up. The discontent may well find generalized public expression, blending with consumer dissatisfactions and other disappointments that will bring about unexpected and unintended consequences. A part of the problem of lagging productivity in the United States may be found in the poor motivation managers provide for their employees.[39]

If employees do not find satisfaction at work, there are few alternative venues in which to find it. Although forty years ago one out of every five persons in the labor force was self-employed, since the early 1970s fewer ethan one out of ten has been able to claim that independence.[40] Those who find self-employment may be exemplars of the nation's mythic rugged individualism, but they pay an increasingly high price for it. The shares of national income accruing to them appear to have declined markedly over the past forty years. Certainly proprietors have lost economically. Their shares of national income, *both nonagricultural and agricultural,* have dropped even faster than their representation in the work force. Whereas in 1948 they received nearly one-fifth of national income, in 1989 their share was just a bit more than one-twelfth. If self-employment or setting up a small business are still options for a few Americans, they are pinched and relatively unrewarding opportunities. Probably they are chosen more out of desperation than in pursuit of independence or as a declaration of self-reliance.

BUSINESS SUCCESS AND
THE ENABLING OF DISCONTENTS

Managerial rationalization of business in pursuit of efficiency has contributed to continuing high levels of economic growth, but it has not increased general happiness as much as managers have probably expected. Nor has it appreciably lessened public concern over disparities in the distribution of income. Furthermore, managerial rationalization has produced a widespread dependency among consumers and has shrunk the opportunities for self-employment and proprietorship. The development of a dependent citizenry, however improved their economic well-being, denies Americans' traditional image of themselves. Improving economic well-being paradoxically calls attention to the contradictory effects of economic growth. Business success thus has created for many Americans double-edged, paradoxical outcomes.

Not only has the success fostered discontents and unease among many people who have enjoyed the material benefits of success, but is has also enabled those same people to express their discontents coherently and increasingly effectively. It has provided a burgeoning new technology that makes the organization of constituencies both easy and inexpensive; and in encouraging the bringing forth of the Immense Generation, the post–World War II baby boom, it provided the talent and leadership for them.

The Contribution of New Technology

The business success after the end of World War II brought forth a communications revolution, introducing new means of travel and remarkable electronic technologies. Corporate constituencies were quick to exploit them in their organization of the discontented. Business firms, of course, also exploited the new technologies to offer a wide array of ever less expensive, more accessible services. In doing so they have helped constituencies greatly. The recent decades' great technological achievements—from facile air travel to inexpensive facsimile transmission— have increased by orders of magnitude the ease, rapidity, and economy of travel and communication. Groups whose members were once too scattered to be effectively united can easily stay in touch across the continent; associations with few resources of their own have discovered how inexpensively they may get their messages on television news programs, presented at legislative hearings, or covered by the press. Legal suits provide a lively forum, well reported in the press and on television, for publicity and recognition of their causes. Leaders of interest groups must still skillfully craft their appeals to members, and they must carefully match their efforts to the concerns of members. If they do so, however, they purposefully reach a wide audience, and effectively and efficiently mobilize their adherents.

The Contribution of Demography
and Education

Demography and education have joined together to supply a cadre of constituency leaders for the discontented and dissatisfied. One of the responses to the revivified economy after World War II and the unexpected prosperity it brought to many young families was the "baby boom." Between 1945 and 1960 the birth rate was over one-third greater than it had been during the decade of the Great Depression. It produced the Immense Generation, seventy million persons born between 1946 and 1964, large enough to create and sustain a "youth culture," a new social phenomenon. That generation questioned and disturbed the major institutions through which it flowed—suburban family, church or temple, elementary and high schools, and then the universities, culminating in the student riots and rebellions that swept across the country in 1968.

The sheer number of youths was larger than the nation had ever beheld. As they moved through their stages of life, they crowded every entry port from hospital maternity wards to positions as bank tellers and jobs on auto assembly lines. The members of the Immense Generation had to compete fiercely among themselves for preference. Graduating from colleges and professional schools, seeking always to gain ascendancy over their competitors and enhance their chance of success, they have honed their instincts for opportunity. Competition has encouraged those with ability and imagination to try new and innovative services. Some were able to discover "needs" that the public had not previously recognized; they proceeded entrepreneurially, creating new markets as earlier business leaders had done.

The successful and innovative among the Immense Generation were better equipped to lead than any of their predecessors. It was the most schooled generation in American history. In 1960 fewer than 4 million students had enrolled in colleges and universities; by 1970 they had more than doubled to 8.6 million and increased to 11.2 million by 1975, and the numbers continued to rise. The educational level of the total population rose markedly, with an overwhelming proportion of youth (85 percent by 1980) completing high school. Almost half of them entered college, and a quarter graduated.

So rapid an expansion of schools, colleges, and universities may have lowered the quality of instruction and perhaps the quality of students, but, nevertheless, the educational effort revealed new and untraditional notions about the world to millions of Americans. They were generally inculcated with a more rational and skeptical view of the nation's institutions and a more critical approach to values than were people with little or no schooling. They probably understand better how government operates and are less mystified by business operations than their parents were. Lipset and Schneider found that, indeed, the level of schooling adversely affects people's *confidence* in institutions. They conclude that;

> the less well educated and the less sophisticated are more likely to express a "naive faith" that the people running our major institutions know what they are doing. Those at higher educational levels are less likely to say they have "a great deal of confidence" in the people running things. This relationship did not show up at first because the better educated also show higher personal satisfaction and greater trust in others. These factors increase confidence in institutions and therefore counteract the negative effect of educa-

tion. . . . when all three variables are taken into consideration simultaneously, each has a distinct impact: education (negative), and personal satisfaction and trust (both positive).[41]

Better educated Americans may still use the rhetoric of flag, patriotism, glory, myth, or ideology to describe their nation and economy, but they are not apt to mistake soothing assurances for remedies to their perceived injuries and expressed discontents. They know how to raise their voices in protest; when they appeal through broadcasts and broadsheets for public support, they can sometimes mobilize millions. From among the ranks of the well-schooled Immense Generation there are sufficient numbers of young professionals of all kinds to provide an ample supply of leaders for all the various corporate constituencies. There has been no shortage of those who can furnish leadership for the discontented.

LEADERSHIP CHALLENGES

There should be little wonder that the number and variety of constituencies have impressed themselves upon business managers. Without seeking causes in business shortcomings or faults of greed, one can find in business success and managerial triumphs ample reasons for public discontents of the last generation and their organization into constituency form by able young professionals. The appearance of modern constituencies marks a further development of the business system, a consequence of its success in enriching the society and producing as plentifully as it has. Business leaders have responded to their constituencies as if they were necessarily hostile groups, organized to hamper, restrict, and damage business. It is our judgment that the new corporate constituencies were not so much hostile to business as they were—and remain—*hostile to the notion that they should be excluded from the world of business.* They view themselves as integral parts of a capitalist business system, useful contributors to a more effective, and even a more productive, economy than the nation has yet known. In short, they consider themselves to be constituents of the business system and its firms.

Valéry Giscard d'Estaing, president of France from 1974 to 1981, reflecting on his experience as well as on modern social trends, called upon political leaders to respond differently from the ways they had in the past. His call may also be addressed to business managers who increasingly must respond to their new constituencies:

[The] new dimension of the citizen (better informed, more educated and with greater aspirations to the full recognition of his individual rights and individual expression) have to be taken into account by the political [and business] leadership in its three main functions, which are for me the three functions of modern statesmen: to listen, to decide and to communicate.[42]

He warned that listening is not easy, and that the amount of knowledge received by managers is vast. Sorting the fundamental message from the transitory and contradictory requires careful discrimination. Nevertheless, both must be attempted, for the intervention of the media, the wide use of polls, and constant monitoring by interest groups promote a kind of direct, continuous referendum on every type of subject. Furthermore, the referenda are largely outside regular institutional forms and channels.

Giscard also noted that government leaders have to deal with the public and interest groups *in real time*. The same is true of business managers dealing with their constituencies. Decisions or activities, the reports about them, and the reactions to them, with comment by outside commentators, are simultaneously transmitted to the public. Since the public may be continually and immediately assessing almost any business decision or action, managers must pay close attention to the mutuality of communications. They must do it ably and well. Mr. Giscard points out, however, that;

> This function is in itself complex, diversified, frustrating and evolutionary. It will dominate the procedures of selection of political [and business] leaders, who will become increasingly judged according to a dual standard, with different criteria: The ability to conduct society [or business] through permanent changes and also the ability to communicate to the citizen the goals to be reached and the path to be followed to reach those goals. Communication . . . is not only a question of words to be transmitted but also of acts and decisions which are perceived as symbols of the direction in which our societies have to engage themselves.[43]

NOTES

1. The American passion for business was hardly a post–World War II phenomenon. Francis J. Grund, a nineteenth-century Austrian immigrant noted that "There is probably no people on earth with whom business constitutes pleasure, and industry amusement, in an equal degree with the inhabitants of the United States of America. . . . Business is the very soul of an American." Quoted by George E. Probst, ed., *The Happy Republic* (New York: Harper & Row, 1962), p. 7.

2. Reprinted with permission of The Free Press, a Division of Macmillan, Inc. from *The Confidence Gap: Business, Labor, and Government in the Public Mind* by Seymour Martin Lipset and William Schneider, reporting the polls of the

Opinion Research Corporation, p. 30. [Copyright © 1983 by Columbia University in the City of New York.]

3. Irving Kristol, " 'When Virtue Loses All Her Loveliness'—Some Reflections on Capitalism and 'The Free Society,' " in *On The Democratic Idea In America* (New York: Harper Torchbooks, 1972), p. 100. The chapter from which the quotation is taken appeared first as an article in *The Public Interest* (Fall 1970). [Reprinted with permission.]

4. Two recent scholars have noted the same difficulty in perceiving social and economic trends that usually appear to move slowly even though most people understand that they move with almost unstoppable force. See Alan M. Kantrow, *The Constraints of Corporate Tradition: Doing the Correct Thing, Not Just What the Past Dictates* (New York: Harper & Row, 1984), p. 2; and James R. Beniger, *The Control Revolution: Technological and Economic Origins of the Information Society* (Cambridge: Harvard University Press, 1986), p. 2.

5. *Social Responsibilities of Business Corporations,* A Statement on National Policy by the Research and Policy Committee of the Committee for Economic Development, June, 1971, pp. 11–12. [Reprinted with permission.]

6. Lipset and Schneider, *The Confidence Gap,* pp. 36–39. See p. 183 for results of a survey by Yankelovich, Skelly & White, which suggests a greater decline in response to the statement, "Business tries to strike a fair balance between profits and the interests of the public."

7. *Social Responsibilities of Business Corporations,* p. 13–14.

8. Leonard Silk and David Vogel, *Ethics and Profits: The Crisis of Confidence in American Business* (New York: Simon & Schuster, 1976), p. 118, quoted excerpts of managers' comments. [Reprinted with permission.]

9. Walter B. Wriston, "Even The Future Is Not What It Used To Be," remarks before The Commercial Club of Boston, November 12, 1974, p. 2.

10. Silk and Vogel, *Ethics and Profits,* pp. 108–9.

11. Donald U. Honicky, "It's Time We Got Together," *AGB Reports,* January/February 1979, p. 43. The article was first presented as a speech, October 30, 1978. [Reprinted with permission.]

12. "What's in it for us?" *Wall Street Journal,* July 14, 1980, p. 9.

13. Quoted by Flora Lewis, "Hannah's Distant Cousins," *New York Times,* January 30, 1987.

14. John Kenneth Galbraith, *The Affluent Society* (London: Hamish Hamilton, 1958), pp. 196–97.

15. These advances may have little to do with the introduction of new government regulations, though such public attention indicates an increased public awareness of pollution. Studies find that the new government regulations of the 1960s and 1970s made no significant change in the generally upward, improving trend in quality of product and externality effects of American industries. See Paul W. MacAvoy, *The Regulated Industries and the Economy* (New York: W. W. Norton & Co.), 1979.

16. Alexis de Tocqueville in his study of *Democracy In America* [Vol. II. New York: Alfred A. Knopf, Vintage Books (1945), p. 146] noted the paradox that increasing equality brought to Americans. "The same equality that allows every citizen to conceive these lofty hopes [of an unbounded career and no common destiny] renders all the citizens less able to realize them; it circumscribes their

powers on every side, while it gives freer scope to their desires. Not only are they themselves powerless, but they are met at every step by immense obstacles, which they did not at first perceive. They have swept away the privileges of some of their fellow creatures which stood in their way, but they have opened the door to universal competition; the barrier has changed its shape rather than its position. When men are nearly alike and all follow the same track, it is very difficult for any one individual to walk quickly and cleave a way through the dense throng that surrounded and presses on him." [Copyright 1945 and renewed 1973 by Alfred A Knopf, Inc. Reprinted by permission of the publisher.]

17. Richard A. Easterlin, "Does Economic Growth Improve the Human Lot?" in P. A. David and Melvin W. Reder, eds., *Nations and Households In Economic Growth* (New York: Academic Press, 1974), pp. 89–125.

18. For another discussion of the paradox, see Moses Abramovitz, "Economic Growth and Its Discontents," in Michael J. Boskin, ed., *Economic and Human Welfare,* (New York: Academic Press, 1979), pp. 8–16.

19. See Abraham H. Maslow, *Motivation and Personality* (New York: Harper & Row, 1954).

20. Peter F. Drucker, *Management: Tasks, Responsibilities, Practices* (New York: Harper & Row, 1974). Also see Frederick Herzberg, *The Motivation to Work* (New York: John Wiley & Sons, 1959), and *Work and the Nature of Man* (Cleveland: World Publishing Company, 1966).

21. Staffen B. Linder, *The Harried Leisure Class* (New York: Columbia University Press, 1970).

22. See Albert O. Hirschman, "The changing tolerance for income inequality in the course of economic development," in *Essays in Trespassing: Economics to Politics and Beyond* (Cambridge: Cambridge University Press, 1981), pp. 39–58. The chapter was first published as an article in 87 *The Quarterly Journal of Economics,* (November 1972):544–65.

23. Isabel V. Sawhill reported that "poverty has increased over the past two decades; that is, a higher fraction of the population had incomes below one-half the median income in the mid-1980s than in the 1960s. Measured in absolute terms, poverty is almost as high now as it was two decades ago." "Poverty and the Underclass," in Isabel V. Sawhill, ed., *Challenge to Leadership* (Washington, D.C.: The Urban Institute Press, 1988), p. 217.

24. There may be an exception to the narrowing in the case of the unemployed, urban "underclass." The breakdown of the family, particularly the high rates among black families pose a new, difficult challenge to historic income trends. See William Julius Wilson and Kathryn M. Neckerman, "Poverty and Family Structure: The Widening Gap Between Evidence and Public Policy Issues," in Sheldon H. Danziger and Daniel H. Weinberg, eds., *Fighting Poverty: What Works and What Doesn't* (Cambridge: Harvard University Press, 1986), pp. 232–59, and Isabel V. Sawhill, "Poverty and the Underclass."

25. Data taken from the following sources: Food, housing, and automobiles—Bureau of Census, *Historical Statistics of the United States* (Washington: Government Printing Office, 1975), Tables G866–G880, p. 329; N238–N245, p. 646; and Q175–Q186, p. 717. Heating systems and appliances—*Statistical Abstract of the United States, 1983* (Washington: Government Printing Office, 1984),

Table 1358, p. 758; Table 1359, p. 759; Table 1349, p. 751. Water and plumbing—*Social Indicators III, 1980* (Washington: Government Printing Office, 1981) Table 3/21, p. 155.

26. Paul Samuelson, "Has Economic Science Improved the System?" in Winthrop Knowlton and Richard Zeckhauser, eds., *American Society: Public and Private Responsibilities* (Cambridge: Ballinger Publishing Company, 1986), p. 311.

27. Changes in the distribution of income appear to have moved toward a bit more equality since 1929, but they show a cyclical variation. See *Survey of Current Business* (October 1974):27, for data from 1929–1971, and various *Current Population Reports* for later data. A comparison with other advanced, industrialized countries is given by Malcolm Sawyer, *Income Distribution in OECD Countries*, Occasional Studies, OECD Economic Outlook (July 1976):14, Table 3.

28. Karen Pennar, "The Free Market Has Triumphed, But What About the Losers?" *Business Week* (September 25, 1989):178, quotes Dr. Herbert Stein to this effect. He was chair of the Council of Economic Advisers under President Richard Nixon and an articulate conservative.

29. Joseph J. Minarik concluded in 1988 that "incomes have been distributed less equally—both at the individual level, as earnings from labor, and as total family incomes. These unfavorable shifts have followed a period of unprecedented favorable movement toward both faster growth and greater equality." Minarik, "Family Incomes," in Sawhill, ed., *Challenge To Leadership*. Also, Stephen J. Rose, *The American Profile Poster: Who Owns What, Who Makes How Much, Who Works Where, & Who Lives with Whom* (New York: Pantheon Books, 1986).

30. David M. Potter, *People of Plenty* (Chicago: University of Chicago Press, 1954), p. 103. Michael Kammen and others have also noted that a spreading of equalities appears to sharpen the concern over the remaining inequalities. "[E]conomic abundance has reduced older norms of inequality but created new forms as well. By eliminating the traditional system of class distinctions, abundance has created a situation in which almost any social differentiation seems invidious." Michael Kammen, *People of Paradox* (New York: Vintage Books, 1973), p. 276. See also Sidney Lens, *Poverty: America's Enduring Paradox* (New York: Crowell, 1969).

31. *Historical Statistics of the United States, I,* (Washington, D.C.: Government Printing Office, 1975) A52–72, p. 11.

32. See *Statistical Abstract of the United States,* 1987 Edition (Washington D.C.: Government Printing Office, 1986) Table No. 1094, p. 620. The rural population did not fall nearly as fast or as far. In 1980 more than 26 percent of the population still lived in such areas.

33. Lipset and Schneider, *The Confidence Gap,* pp. 36–37.

34. Ibid., pp. 170–71.

35. See Kenneth Arrow, "Social Responsibility and Economic Efficiency," *Public Policy,* 21 (Summer 1973):303–19, for a discussion of the problems arising when there is a disproportion in the knowledge between seller and buyer; also see Richard Nelson and Michael Krashinsky, "Public Control and Economic Organization of Day Care for Young Children," *Public Policy* 22 (Winter 1974):53–76, for an analysis of situations where consumers can be assumed to be expert

in the service purchased and those in which quality is difficult to define, let alone measure.

36. Lipset and Schneider, *The Confidence Gap*, p. 167. The negative opinions were consistent over four surveys, 1966–1978.

37. William Serrin, "Study Says Work Ethic is Alive But Neglected," *New York Times*, September 5, 1983, p. 8.

38. See Thomas A Kochan, "Adaptability of the U.S. Industrial Relations System," *Science* (April 15, 1988):290. For data on the surveys, see M. L. Weitzman, *The Share Economy* (Cambridge: Harvard University Press, 1984).

39. Alan S. Blinder, "Want To Boost Productivity? Try Giving Workers A Say," *Business Week*, April 17, 1989, p. 10.

40. In 1962 over 15 percent of all the employed still worked for themselves, one-third of them in agriculture. By 1985 less than 10 percent were self-employed, and those in agriculture accounted for fewer than one out of six.

41. Seymour Martin Lipset and William Schneider, *The Confidence Gap*, p. 120.

42. Valéry Giscard d'Estaing, "The Opportunities and New Challenges," *Foreign Affairs* (Fall 1983):194. [Reprinted with permission.]

43. Ibid., p. 195.

5

The Socially Responsible, Autonomous Corporation

A rapidly changing industrial technology, an ever more schooled (possibly more educated) workforce, and the paradoxical discontents of relative prosperity—the failures of business success—have combined, as we saw in the last chapter, to create and to enable new corporate constituencies. Increasingly, these constituencies are asking business managers to provide more than simply defined and measured output. Business managers are well aware of constituency expectations. In a 1981 statement the Business Roundtable, whose members include the CEOs of the largest corporations declared;

> . . . it is clear that a large percentage of the public now measures corporations by a yardstick beyond strictly economic objectives. People are concerned about how the actions of corporations and managers affect them not only as employees and customers but also as members of the society in which corporations operate. . . . More than ever, managers of corporations are expected to serve the public interest as well as private profit.[1]

We believe that the various constituencies are seeking a regular, legitimate role in the business system through which they seek their own goals and purposes. If managers do not voluntarily recognize the role sought by constituencies, the lessons of history and current practice suggest that the constituencies will learn how to create it for themselves. In Chapter 2 we noted that constituencies have been experimenting with ways of using the levers of economic and corporate power that managers have long manipulated themselves. In this chapter we will examine how constituencies, old and new, are asserting their claims against corporate managers more successfully than at any time in the past.

Since the rise of large business corporations nearly 120 years ago,

123

managers have been exceedingly reluctant to recognize either internal or external constituencies, though they realize, as noted above, that the constituencies reflect a new and changing public attitude toward business. Stockholders, as legal owners, possess the best claim to recognition, but until the 1980s professional managers offered them only perfunctory de facto attention, while in form and rhetoric they usually insisted that stockholders were their main concern. Managers were long adamantly opposed to unions, an early constituency; for several generations they brushed off consumers with the ancient common law injunction, "let the buyer beware"; and environmentalists and conservationists were dismissed as impractical visionaries. Certainly in the decades after World War II, managers in large corporations believed themselves too secure in their position of command and authority to perceive any reason for responding positively to constituents. They assumed a condescending manner generally—paternalistic at best and hostile at worst—toward any group that sought to protect its interests.

As the decade of the 1980s opened, some business managers were ready to accept the advice which Giscard d'Estaing offered to his fellow political leaders: *to listen,* to decide, and to communicate through acts and decisions as well as words. Yet few among them were willing to recognize constituencies formally and to grant them a formal place and position in corporate governance. Most of them had not yet realized how rapidly and massively their environment had changed since business's heydays of the 1950s. As we have seen, leading spokespersons such as Herbert Schmertz of Mobil and Walter Annenberg, the publisher, even in the late 1970s continued to argue a devil theory to explain business problems: The media and intellectuals maliciously disparaged the great accomplishments of managers.

BUSINESS MANAGERS IN
THE NEW INDUSTRIAL STATE

The accomplishments of the past—the production miracles of World War II and the amazing rise of per capita income through the 1950s and 1960s—were hardly matched by those of the later 1970s and the 1980s, however. The shock to the economy and the public of the steep rise in oil prices, a declining rate of productivity, high inflation, decreasing real earnings for millions of employees, and rising unemployment, all made managers' claims of success unconvincing and their denunciations of critics all too self-serving. Business managers were nostalgically looking back to a time when they occupied a high position and enjoyed a style of operations, at least among the upper ranks of business corporations, well described by John Kenneth Galbraith in a popular, widely read book, *The New Industrial State.*[2] Prototypical of the climactic industrial organization, the culmination of forces and tendencies in a capitalistic economy, were the large American manufacturing firms.[3] The managers of such firms as DuPont and General Motors were answerable to almost no one and directed their company's activities through a decentralized system of command and control, but relied upon central service staffs, which Galbraith dubbed "the technostructure," to oversee personnel practices, define budgets, and set overall policy.[4] At the time, such firms and their management style appeared to be manifestations of a mature, complete industrial economy; in fact, they turned out to be a transient form of capitalism, able to flourish only under the very special circumstances of the time.

The Public Characteristics of the Galbraithian Corporation

The mature business corporation of the 1950s and its managers could be easily identified by four widely discussed and obvious characteristics: (1) the pursuit of economic growth, with stability; (2) an embracing sense of social responsibility; (3) a bureaucratic, hierarchical organizational structure; and (4) an authoritarian relationship with employees, involving both an adversarial stance toward unions and often a paternalistic attitude toward employees.

Stable economic growth. Few managers had any doubt that a primary purpose of their firms was the pursuit of economic growth, though it was to be carried out in a regular, controlled, and stable way. While managers sought constantly to enlarge production, expand sales and increase revenues, they also realized the benefits of stabilizing the market environment within which the firm had to operate. They needed to extend their control of all developments and activities that surrounded and affected production and sales.

Galbraith noted the imperatives of the massive technologically driven demand for capital and the long lead times required to transform big capital investment into productive operations like a new steel mill or a modern automobile assembly plant. Uncertainties in costs and fickleness in demand endangered the stability of the firm and the position of managers. Costs and demand had to be controlled somehow, to allow detailed, long-run planning to be carried out. The control and planning function that managers assumed was daunting and audacious, and eventually, it became presumptuous. Social, political, and economic circumstances are unpredictably changeable so that even very powerful organizations cannot be reliable instruments of control, of either their internal or external activities.[5]

For half a century, 1920–1970, however, many of the large industrial corporations occupied special, protected positions that permitted a good deal of control. In the auto industry the few large domestic firms constituted an oligopoly; in steel the huge United States Steel company was of such a dominating size, with more than one-half of the industry capacity, that the smaller companies operated in its protective shadow. Firms in communications, transportation, and financial industries benefited from a government-regulated, monopolistic status, like that of the American Telephone & Telegraph Company. Given the protected status they possessed, their managers were able, in remarkable degree, to plan and control the environment of their firms. Unsurprisingly, managers whose whole professional careers were played out in such organizations came to accept economic growth—the production of more and more through larger and larger firms—as an unquestioned, unquestionable organizational purpose. If, as they came to assume, growth could be assured only through the stability brought by the continued, even permanent existence of their organizations'[6] overall planning, centralized control, and managed demand, then the necessities of the situation defined their rights and legitimated the ways they had developed to exercise their power.

Social responsibility. In an attempt to win public acceptance of the power they necessarily wielded, managers developed a second corporate characteristic, a credo of social responsibility. The leaders among managers of the large business corporations recognized that the control they exercised and the planning they deemed essential had to encompass more than the immediate interests of the firm. Their organizations were so large, with "millions of employees, stockholders, customers, and community neighbors in all sections of the country and in all classes of society—that they actually constitute[d] a microcosm of the entire society," as one group of prominent business managers put it.[7] In 1978 the chief executive officer of Exxon, concisely expressed a widely held view of the corporation's social responsibility:

> Corporations are part of the life of the communities in which they do business. Their operations affect not only the economic and physical well-being of such communities, but the social and cultural environment as well. Consequently, corporations carry responsibility well beyond conducting their primary business.[8]

The managers of General Electric introduced, in the early post–World War II period, the concept of corporate *stakeholders:* all those affected by the policies and programs of the firm. Managers were responsible to all of them, and thus had to assume a wide social responsibility, far beyond that owed to stockholders, the legal owners of the firm. Implicit in the notion was the possibility that managers might properly subordinate stockholder claims to those of other stakeholders. By the 1980s many managers, threatened by hostile takeovers and corporate raids, explicitly declared the desirability of that subordination. The members of the Business Roundtable wrote in 1981;

> Balancing the shareholder's expectations of maximum return against other priorities is one of the fundamental problems confronting corporate management. The shareholder must receive a good return but the legitimate concerns of other constituencies also must have the appropriate attention.[9]

When taken to task by Paul MacAvoy, a prominent economist, for such a stand,[10] Andrew C. Sigler, chairman of Champion International Corporation and president of the Business Roundtable, responded by arguing that too few shareholders are now long-term, personally involved individual investors. Large numbers of them are institutions, "unidentified short-term buyers most interested in maximum near-term gain. Such interests must be balanced with a long-term perspective. The simple theory that management can get along by considering only the shareholders has been left in old economic dissertations."[11] In 1987, Hicks B. Waldron, chairman of Avon Products, Inc., boldly proclaimed his belief that managers not only *might*, but have a positive duty to subordinate the maximizing value for stockholder.

> We have 40,000 employees and 1.3 million representatives around the world. We have a number of suppliers, institutions, customers, communities. None of them have the democratic freedom as shareholders do to buy or sell their shares. They have much deeper and much more important stakes in our company than our shareholders.[12]

Sigler agreed with Waldron. He put the matter as a rhetorical question, with his own answer:

> What right does someone who owns the stock for an hour have to decide a company's fate? That's the law, and it's wrong.[13]

Managers insisted that *they* and they alone, were capable of choosing among the various competing demands put forth by the many stakeholders. The managers of General Electric had long made clear *their* claim of autonomy in the company motto repeated in advertisements and company literature all through the post–World War II decades: "Do Right Voluntarily in the Balanced Best Interest of All." In 1971 the Committee for Economic Development described the full meaning of social responsibility:

> The modern professional manager also regards himself, not as an owner disposing of personal property as he sees fit, but as a *trustee balancing the interests of many diverse participants and constituents in the enterprise,* whose inter-

ests sometimes conflict with those of others. The chief executive of a large corporation has the problem of reconciling the demands of employees for more wages and improved benefits plans, customers for lower prices and greater values, vendors for higher prices, government for more taxes, stockholders for higher dividends and great capital appreciation—all within a framework that will be constructive and acceptable to society. This interest-balancing involves much the same kind of political leadership and skill as is required in top government posts. The chief executive of a major corporation must exercise statesmanship in developing with the rest of the management group the objectives, strategies, and policies of the corporate enterprise. In implementing these, he must also obtain the "consent of the governed" or at least enough cooperation to make the policies work. And in the long run the principal constituencies will pass judgment on the quality of leadership he is providing to the corporate enterprise.[14]

This remarkable description of social responsibility placed corporate managers in a very special position, particularly in a nation that values democratic procedures. And a decade later the members of the Business Roundtable described their role very much as the CED had earlier:

Carefully weighing the impacts of decisions and balancing different constituent interests—in the context of both near-term and long-term effects—must be an integral part of the corporation's decision-making and management process. Resolving the differences involves compromises and trade-offs. It is important that all sides be heard but impossible to assure that all will be satisfied because competing claims may be mutually exclusive. . . . Balancing the shareholder's expectations of maximum return against other priorities is one of the fundamental problems confronting corporate management . . . some leading managers have come to believe . . . that by giving enlightened consideration to balancing the legitimate claims of all its constituents, a corporation will best serve the interest of its shareholders.[15]

In both the earlier and more recent statements, the managers made clear they were not to be *accountable* to any of their various constituents, not even stockholders. On their own, in their wisdom, and from their positions of high status, they would set social priorities, choosing which interests to favor and which to slight. They would decide for themselves, in light of the general welfare, what, if any, resources might be devoted to community activities, social needs, or constituency demands.[16]

The statements of the CED and Business Roundtable were typical descriptions of corporate social responsibility as it had come to be accepted by the managers of most large business corporations since the late 1950s. By 1958, three-quarters of 700 companies queried had formulated statements of managerial responsibilities.[17] Business managers may have felt that they had to advertise publicly their responsibility because, in fact, they were *not* responsible. More accurately they were not easily accountable to anyone or any constituency for many of their decisions and actions, except through the market. They could close a factory, mine, or mill and move an operation to Tennessee, Texas, or Taiwan without securing permission from any group but their own colleagues and officers.[18] They might tear out the economic roots of a well-established community and build a new city in the open countryside, without having to account to anyone directly or immediately. Within the market's often wide limits, they were responsible only as they chose to define responsibility.[19]

Even those who otherwise endorsed managerial direction and the business system had doubts about such a sense of responsibility. Philip Sporn of American Electric Power and Reed O. Hunt of Crown Zellerbach dissented from the CED statement. It was entirely too ambitious, they argued; managers' responsibilities should be limited to discharging their business responsibilities as business managers.[20] Milton Friedman had denounced such formulation some years earlier. He found so broad a definition of corporate social responsibilities to be a "fundamentally subversive doctrine." He asked how business managers could determine, on their own, the best balance among the contending interests. He wanted to know how they would justify the balance selected. In a democratic society, he noted, elected officials and their civil servants set the taxes and allocate public revenues for the general welfare, but they are accountable to the electorate. Should self-selected managers of private firms act in a similar way, pricing their products or services to raise revenue for distributing among the interests they rate highest, they arrogate a public function for themselves, and "sooner or later, [they will be] chosen by the public techniques of election and appointment."[21]

Paul MacAvoy argued that the Business Roundtable statement "implies that the large corporation is a political entity subject to the votes of interest groups, rather than an economic organization subject to the market test for efficient use of resources."[22] Later Peter Drucker warned that the social responsibility enunciated by managers in the last few decades was incompatible with the realities of American society. He compared the business managers who espoused the doctrine to the Enlightened Despots of eighteenth century Europe. They gloried in their "good intentions" as they removed the last obstacle to an "enlightened" rule—a truly independent board of directors.[23]

The managers' definition of their and their firms' social responsibility, of course, allowed managers great leeway to do as they wished, and offered its own justification. Some business critics did not accept the notion that managers should be able to define their responsibilities for themselves.[24] Yet, from the perspectives of the late 1980s, the social responsibility of the large business firms over the preceding quarter-century had been quite remarkable. In the automobile industry, for example, the major producers raised their workers' wages steadily up through the 1970s. They assured a steady, and profitable market for thousands of parts suppliers, particularly in the MidWest, who in turn were able to provide steady jobs and good pay to thousands of blue-collar industrial workers. As a result, many small communities, as well as larger manufacturing towns throughout the country, flourished and supported what most Americans took to be the good, middle-class life. Plant closings were rare, there was no "Rust Belt," and the stable prosperity of middle America commended itself to most people; in retrospect it was a time when corporate social responsibility appeared to be working reasonably well.

As might be expected, however well-off the various constituencies were, they were seldom satisfied with their benefits. Each complained that managers should have been far more generous and wiser in their enlighten choices. Economists viewing the scene from their special and limited perspective sourly described managers' sharing of earnings among constituencies as merely shared gains of quasi rents or oligopolistic gains won in protected markets. Indeed, with the coming of more and widely felt competition, managers found they could no longer protect their various constituencies as they had over the decades after the end of World War II.

Bureaucratic hierarchies. To handle the burgeoning work of planning and the ever more complex problems of control, managers of the large industrial corporations created arrays of staff offices to offer specialized advice, propose policy, collect data, and monitor performance in all divisions and parts of the firm. They followed the hierarchical structure that early corporate managers in the mid- and late-nineteenth century corporations had first used, probably borrowing from the command-organization of the armies in the Civil War.[25]

The size and extent of business firms' bureaucratic hierarchies were remarkable. By 1975 one-fifth of all industrial employees in the United States worked in firms with at least *six levels* of managers.[26] As late as 1987 General Motors, one of the largest business corporations, still maintained as many as fourteen management levels. Its organizational form contrasted strikingly with that of a chief world competitor, Toyota Motor Company, which managed with only five levels of managers.[27] Firms built up their professional and technical staffs at a particularly rapid rate through the 1950s and 1960s. The various levels of their hierarchies swelled enormously. The manufacturing nonproduction labor force increased far more rapidly than did employment in general. In the years between 1950 and 1970 firms in the durable goods industry, where most of the large corporate firms were found, hired managerial, professional, and technical employees in such numbers that their share of the industry labor force increased from about 17 percent to more than 28 percent. Though employment in the industry accounted for less than 4 percent of total private employment, the hiring in these occupations was so large that it accounted for 13 percent of the net employment gains for the *entire economy* over the period![28]

The ever longer span between top and bottom managers produced large-scale bureaucracies, far bigger than those in comparable European and Japanese firms, and rivaled those of the Department of Defense and the U.S. Post Office in the federal government. The bureaucracies of the latter two organizations had long been infamous for their sluggishness and inefficiencies, but the same adjectives were seldom applied to corporate bureaucracies at the time. Of course, such large bureaucracies, growing as fast as they did, were evidence of slack and even spawned new structures within which more slack might grow. They also encouraged another invidious corporate development, the pushing of responsibility for performance downward and of rewards upward; it encouraged a dangerous split between performance and recognition of achievement and thus a pathological "game playing" among managers to the detriment of the firm.[29] For example, General Motors' managers around 1960 established an index for tracking the quality of autos made in their various plants. The standard was 100, with a point taken off for each defect found in the cars chosen at random for inspection, and a "passing grade" was fixed at sixty. Too few plants were able to receive a "pass," however, so an adjustment was made. The standard was raised to 145 points, so that the measured cars with more than forty-five defects easily won "passing" scores.[30]

Authoritarian management. Few American employers or corporate managers have believed that there either was or should be any balance in bargaining power between them and their employees. Managers generally have assumed the proper relationship of employer and employee was the one defined in the old common law—that of masters and the servants. It is a belief that appears to sustain the continuing, virulent antiunion stance of many American managers. Even in the late 1980s most managers had not perceived any reason or even questioned why they should not command a right to uncontested, unilateral authority within their organizations. Managerial attitudes toward employees have changed in their style of expression over the years, but their substance is remarkably similar for many managers. Consider the bluntly worded declaration made in 1851 by the editor of the *New York Journal of Commerce:*

Who but the miserable, craven-hearted man. would permit himself to be subjected to such [union] rules, extending even to the number of apprentices he may employ and the manner in which they shall be bound to him, to the

kind of work which shall be performed in his own office at particular hours
of the day, and to the sex of the persons employed, however, separated into
different apartments of buildings? For ourselves, we never employed a fe-
male as a compositor, and have no great opinion of apprentices, but sooner
than be restricted on these points, or any other, by a self-constituted tribunal
outside of the office, we would go back to the employment of our boyhood,
and dig potatoes, pull flax, and do everything else that a plain, honest farmer
may properly do on his own territory. It is marvelous to us how any employer, having the soul of a man within him, can submit to such degradation.[31]

Although few managers would use his words, the typical manager of
the last eighty years was probably sympathetic to the paternalism expressed by George F. Baer, president of the Reading Company, during
the 1902 anthracite strike:

. . . the rights and interests of the laboring man will be protected and cared
for . . . by the Christian men to whom God in his infinite wisdom has given
the control of the property interests of the country.[32]

Managers of J. P. Stevens, who fought unions for twenty years from
the 1950s into the 1970s, lawfully if they could but unlawfully as well,
argued in the same vein, as did the antiunion managers of Texas Instruments and Litton Industries. By 1920, the managers of most large
corporations had learned to express themselves more moderately than
Baer and to act less openly hostile to unions than Texas Instruments
and Litton. Whatever their words or actions, though, most were still
firmly opposed to unions, unless they could choose the workers' representatives. Royal Meeker, commissioner of the U.S. Bureau of Labor
Statistics, reported on the President's First Industrial Congress in 1920:

The employer group in the conference must be taken as representing the
majority of employers the country over. The speeches made by these representative employers were often difficult to understand, but their attitude of
mind was never for a moment in doubt. They had been driven by hard
experience [during World War I] to abandon individual bargaining, but they
vigorously maintained their right to *dictate* the terms of the collective bargaining.[33]

Though less than one-fifth of the labor force works under the terms
and provisions of collectively bargain agreements today (down from
nearly one-third in the mid-1950s), most American managers find even
that much unionism a standing threat to managerial authority. In the
late 1970s the National Association of Manufacturers created the Council
on Union-Free Environment, with 450 member companies joining together to discourage unionization and collective bargaining. The announced purpose of the council was to benefit themselves and their

employees.[34] Corporate antiunion programs are widespread and used by the largest business firms. The AFL-CIO reported in 1983 that 401 consulting firms had conducted antiunion programs in the preceding three years. An additional 126 firms had conducted seminars on union avoidance.[35]

Not all the managers of large business corporations are antiunion, but the perception has been widespread among them that unions are a necessary evil rather than possibly useful contributors to the firm or a complement to management. They are seen as a rival authority at the place of work, diluting the power managers must necessarily exercise to control and direct the work force, as well as to maintain their superior position in the firm. Few managers ever question such a view of their power. They assume, with William Graham Sumner, that "Industry may be republican; it can never be democratic."[36] Some managers recognize the nature of the power they defend. Robert E. Wood, when president of Sears, Roebuck, noted that,

> We stress the advantages of the free enterprise system, we complain about the totalitarian state, but in our individual organizations we have created more or less a totalitarian system in industry, particularly in large industry.[37]

A factory manager at Unilever writing for the company magazine in 1974 described what he had observed to be a prevailing managerial approach to employees:

> It is my submission that the worker in the large industrial enterprise is accorded the status of a child. . . . When he walks into the factory he is given a number; "punches the clock"; is closely supervised; and is assumed to be capable of accepting no more than a bare minimum of responsibility. If he then reacts to this in either an apathetic or aggressive manner, management reacts by tarting up the physical environment or putting another quid in the wage packet, and is comfortably confirmed in its stereotype of the average worker as solely motivated by irrational and childish impulses. If, in desperation, the worker then latches on to the "militant" who appears to offer hope of escape from this morass, he is branded a sheep.[38]

The indictments made by Wood and the Unilever manager are probably too sweeping; most Americans do not work on assembly lines and probably only a minority still punch time clocks. In many firms managerial styles are more sophisticated and benign. They go by many names: some scholars have identified autocratic, democratic, and laissez-faire management,[39] while others have specified authoritarian and participatory styles.[40] Nevertheless, the substance of power beneath the cover of style is often authoritarian, widely exercised in the hierarchical business corporation.[41] By 1975, one of every five industrial workers in the United States and Europe was employed by a large, and therefore hierarchically structured, company. A far larger proportion are employed in lesser but still multilevel hierarchies.

Few scholars have explored the reasons that managers, operating in a democratic, individualistic society, adopted an authoritarian employment policy.[42] Competitive market theory does not predict either hierarchical or authoritarian management, nor do practical market operations require those kinds of management. Why then are they so common among large American business corporations? Alternative ways of coordinating production and distribution were and are possible. Small companies, with little or no hierarchical structure and with managerial authority exercised in diverse ways, could have coordinated their activities through trade associations, interest groups, loose combines, or even cartels, negotiating among themselves the details of their membership and relationships. In fact, business managers experimented with such arrangements in many industries through the nineteenth century. Popular dislike of cartels, trusts, and monopolies in the United States, however, made these arrangements politically suspect and ultimately illegal.

Alternatively, managers might have coordinated their major activities through the competitive market of small units, guiding their production and marketing decisions according to price information supplied through market exchange. In such a case, most firms probably would not have grown to the size of the large corporations today, hierarchies might have been avoided or greatly flattened, and managerial power could have been more easily defined and exercised in any number of ways.[43]

On the basis of their prevalence and little or no other evidence, economists conclude that the hierarchical, authoritarian corporate form must have been a more efficient way to coordinate production and distribution than either federations or market coordination. Assuming that business managers are always profit maximizers, the record implies that corporations were particularly efficient in capital- and energy-intensive, high-volume, standardized production typical of mass-marketing industries. But managers' choice of the large corporation is not necessarily proof that it was made primarily to secure efficiency. The record may indicate only that the hierarchical, authoritarian corporation is *efficient enough to sustain itself*. After all, hierarchy and the bureaucracy it creates pose barriers to both rational decisions and efficient operations. As we will see in the next chapter, business managers in such industries as automobile manufacturing, steel production, and air transportation have not always acted to gain and maintain the most efficient production. They are often quite willing to settle for as efficient an operation as will produce satisfactory profits. Other values than those of efficiency influence managers' perceptions, and other goals than those of maximum profits motivate their actions.

Furthermore, if hierarchical, authoritarian corporate businesses were efficient enough to have maintained themselves in the past, that is hardly proof that those characteristics will serve them well in the future. Corporate managers employ workers whose skills, schooling, and specialties are very different from the workers of even a generation ago. Employees in the 1980s were, and those in the 1990s will be, knowledgeable to a degree unimaginable to managers of earlier years. They are and will be better able to insist upon the right to participate in decisions that affect their lives. In addition, employees' skills, training, and education have prepared them increasingly to work on their own, thereby reducing the need for supervision at the levels past workers were thought to require.

The Ideological Source of Authoritarianism

That American managers have long continued to consider themselves an elite in their large organizations, deserving of high status and unilateral authority, may be as much a matter of ideological preference as of economic necessity. An exploration of the various ideologies that they have used to bolster their claim as employers will help the reader understand as well their hostility to constituencies other than employees. Those ideological roots feed the same managerial sense of superiority and autonomy.

Over the century or more that the large business corporation has grown to its present predominance in many industries, managers have advanced one defense after another for their special claim to elite status. None of them has rested upon market performance, but rather upon various social claims and assumptions about human nature. As each has been undermined or proved to be untenable, they have sought another. They have seldom carefully or objectively examined the costs of maintaining their claims, as profit maximizers might be expected to do. Only the exceptional manager has contemplated other methods of exercising authority or experimented with nonhierarchical ways of organizing those who make up the corporation.

In the generation before the turn of the century, successful American business leaders celebrated their careers by emphasizing the virtues of character and the religious mission of their work. They generally regarded their success and wealth not only as proof of capitalist progress but also as evidence of their own fitness in the struggle for survival. Their gains unfortunately, but inevitably, were accompanied by the terrible poverty of those who failed. Popular journalists and scholars alike developed the notion of "social Darwinism" to explain and to honor the titans of industrialism. The sociologist C. R. Henderson explained:

> . . . the "captain of the industry" . . . has risen from the ranks largely because he was a better fighter than most of us. Competitive commercial life is

not a flowery bed of ease, but a battlefield where the "struggle for existence"
is defining the industrially "fittest to survive." . . . The successful man is
praised and honored for his success. The social rewards of business prosperity, in power, in praise, and luxury, are so great as to entice men of the
greatest intellectual faculties. Men of splendid abilities find in the career of
a manufacturer or merchant an opportunity for the most intense energy.
The very perils of the situation have a fascination for adventurous and inventive spirits. In this fierce, though voiceless, contest, a peculiar type of
manhood is developed, characterized by vitality, energy, concentration, skill
in combining numerous forces for an end, and great foresight into the consequences of social events.[44]

Social Darwinism. The achievements of business managers were founded,
according to business apologists, in the laws of evolution, and thus confirmed by the new science of the age. The business managers' success
justified the authority that they wielded in factory, mill, and mine (and
also confirmed their basic virtue). They were entitled to command by
the workings of Nature herself. Workers, by the mere fact of their lack
of success, had proved themselves unfit. Experience showed they lacked
the wit, imagination, and direction possessed by managers. Without the
creative contribution of the naturally superior managers, they were
doomed to live in worse squalor than they did. A "how-to" book of the
times pointed out that not everyone could expect to win in the fierce
struggle for riches and wealth:

> Many a man is entirely incapable of assuming responsibility. He is a success
> as the led, but not as the leader. He lacks the courage or willingness to as-
> sume responsibility and the ability of handling others. He was born for a
> salaried man, and a salaried man he had better remain. If he goes into busi-
> ness for himself, the chances are that he will fail, or live close to impending
> disaster.[45]

If industrial success was to be won in a fierce struggle, and if one
proved fitness by mobilizing one's "vitality, energy, concentration, and
skill," then workers could enter the struggle, and in their own ways,
with their own means, attempt to prove they were the equal of their
managers. Social Darwinism was an attractive doctrine to the down-
trodden; they could improve their lot and raise their station through
their own efforts. They began to organize unions in large numbers,
membership rising fivefold in the seven years after 1897, for example.

Managers were not consistent in applying social Darwinism. However
uplifting, productive, and virtuous their own struggle had been, the
struggle of workingmen was dangerous, a savage conflict that could not
be tolerated. The president of the National Association of Manufacturers in 1902 warned against it:

Organized labor knows but one law and that is the law of physical force—the law of the Huns and Vandals, the law of the savage. All its purposes are accomplished either by actual force or by the threat of force. . . . It extends its tactics of coercion and intimidation over all classes, dictating to the press and to the politicians and strangling independence of thought and American manhood.[46]

Increasingly, managers came to understand that social Darwinism as a defense of their own privileges served them badly; its glorification of struggle, conflict, and competition gave too much aid and comfort to those they were determined to keep in subordinate roles. They needed another and more amenable argument to bolster and justify their authority over workers. The National Metal Trade Association and the National Association of Manufacturers shortly after the turn of the century supported an "Open Shop Campaign" to emphasize the absolute authority of employers and to resist unions at all costs. Though there were managers such as those represented in The National Civic Federation, organized in 1900, who advocated a conciliatory approach to unions, most managers strongly supported the declaration that they would "not admit of any interference with the management of our business."[47]

Scientific management. Fortunately, Frederick W. Taylor proposed a new rationale that could be used, suitably modified to bolster the managerial case for their authoritarian, elitist image of themselves. Taylor offered Scientific Management as a way of increasing production, securing greater productivity, and ending the conflicts between capital and labor. Best of all, it proposed a method of ascertaining scientifically "the one best way" of performing any job. With careful analysis and measurement, Taylor could determine how employees should work, with no personal prejudice or arbitrary judgments involved. He distrusted managers, however, believing that only engineers could determine "the one best way." Managers as well as workers need to learn new methods and rethink what they were about.[48]

Many managers ignored his distrust of them and interpreted Scientific Management as evidence that only managers could and should direct employees at work. In increasingly complex industrial organizations, managers could appeal to science to support their contention that their absolute authority was both good and required. In their scientific wisdom they could place the right worker in the right job and specify the correct methods of carrying it out.

Not all managers were entranced with Scientific Management. It raised presumptive questions about their natural good judgment and superior abilities. Special study and new methods of analyses were required to discover the best way to manage. Managers did not become superior, skilled, or all-knowing merely because of their position. Thus, Scientific Management was a doctrine subtly subversive of the prerogatives that many managers claimed as theirs by right of possession, promotion, and position. Such a claim does not fit well in a democratic society where individuals define their market relationships through voluntary contracts. Prerogative has a medieval connotation quite out of keeping with a free-market system, conjuring up arrogating presumptions of kings or the special privileges accorded senators in the Roman Empire.

A new scientific management—The consent of the governed. In the 1950s Harold Smiddy, an executive in General Electric and an ardent advocate of Scientific Management, attempted to reconcile it and free market contracts.[49] He was a strong believer in the absolute authority of managers over employees. Cooperation at the place of work was necessary, but it was possible, he maintained, only on managers' terms; workers' values, wants, and needs were to be considered only insofar as they might be expressed in explicit employee contracts.

> [Managers] need ability to motivate so they [will] make the organizational goal their own and will voluntarily want to accomplish the task and timetables for which they are personally responsible, and in the context of the jobs and relationships of the others on the team. . . . Anyone who does not care to discipline himself to adopt the organization's objectives, in addition to and in synchronism with his personal, professional, or other work aims— and thus to live harmoniously so that normal organizational restraints are mutually effective and least chafing—can, and in due course he normally should, switch jobs to be able to operate in the context he most prefers.[50]

When employees joined a company, they agreed to managers' terms for their work in return for their job; they thereby voluntarily accepted a modification of their natural liberties. Smiddy insisted: "It is the modification of the liberties that justifies joint teamwork." The team cooperated in attaining managers' goals, of course, not those of the workers.

His reconciliation of managers' absolute authority at the place of work with contractual liberty did not receive wide acclaim in business corporations. The defenders of management prerogatives increasingly were specialists in industrial relations and personnel and human resources concerns. Smiddy's reworking of Taylor's ideas to fit more accommodatingly the constraints of market contracts left open the possibility that employees might someday choose to demand new terms, restricting managerial authority. Managerial power was all too vulnerable if subject to such possible restraint. It was too dependent on and thus too vulnerable to the contingencies of bargaining power.

As professionals who dealt with employment problems of complex, technologically oriented organizations, managers preferred farther reaching and more basic underpinnings for their pretensions. They had come to realize that the scientific solutions offered by engineers were far too simple to help them deal with tantalizingly difficult employment problems like motivation, loyalty and performance evaluation in the absence of tangible output. The insights and wisdom that psychology and sociology provided were examined in the hope that they would help managers improve and sustain their control of their work force.

Human relations—Managers as social governors. The research findings of experiments conducted by Elton Mayo, Professor of Industrial Research at the School of Business Administration, Harvard University, provided a welcome base for managerial ideology. They were published through the 1930s and 1940s. He and his associates reported in the famous Hawthorne studies[51] that workers spontaneously formed groups and enjoyed a social life at the place of work. They were social creatures who, in his interpretation of their activities, achieved complete freedom only in submerging themselves in the group. Taylor was quite wrong in treating each worker as a separate, isolated person, responding only to individual incentives. Workers naturally rely upon social routine rather than logical thinking and systematic analysis to deal with the daily problems on the job. If they were unproductive and uncooperative, they probably suffered from low morale. Managers controlled the industrial environment, so that if workers were bewildered, lost, isolated, and discontented, managers were not managing properly. They had in all likelihood stripped away the reassuring customs and traditions that made work meaningful for employees. It was management's responsibility to make the workplace both human and humane.

Mayo suggested that managers were more than mere employers. They were neither managing workers nor work; they were administering a social system. Their goal should not be a bit more efficiency or a bit less waste, but social stability; production was not as important as harmonious relations in the factory and office. Managers, as rational directors of the industrial civilization, carried the burden of interpreting the emotional conditions of workers, molding their feelings, and mastering the facts of human social life at work.[52] Managers had to learn the art of "human relations" to eliminate workers' misunderstandings and to provide useful facts to those in the shop, all in the promotion of teamwork necessary to the achievement of managerial goals.

Mayo's contribution to managerial ideology lay in his emphasis upon managers as an elite—trained, rational, and possessed of a kind of industrial noblesse oblige. He saw no need for unions, government regulation, or outside pressures to guide managers in the fulfillment of their social responsibilities. He assumed that managerial goals were and ought to be the basis for social and industrial cooperation. The effective working of the social order thus depends upon the full, unquestioned authority of managers. Since the 1950s, American managers had not developed their managerial ideology much beyond that based on Mayo's human relations. Most have been unwilling to contemplate sharing authority with employees or any other constituency. Cooperation has almost always been defined as employees joining managers in pursuit of managerial goals, never a joint exploration of joint goals. By definition managers are assumed to be superiors; only managers are the source of ideas, suggestions, or production know-how. Business leaders still defend managerial prerogatives without embarrassment.[53] Managers have been intent upon maintaining their "right to manage," as if the basis of that right was obvious and necessary. Those who negotiate with unions have almost universally demanded and secured a "management rights" provision in the labor agreement. It has become a symbol of managers' reach for authority, but at the same time an indicator of their fears.

Managerial rights extend just as far and no further than the cooperation, respect, and willingness of workers who choose to recognize them. In a free society no person has special claim over another, even though one in a superior economic position can sometimes dictate choice to one in an inferior economic position. Managers' claims to hierarchical, authoritarian autonomy are strange, indeed, in a democratic, contractual market society. Of course, autonomy ultimately rests upon the power of economic coercion. It is a power that fits American traditions and ethos, however, no better than does prerogative. Economic coercion has always generated resentment and fear, hardly a good base on which to build loyalty and cooperation at the place of work.[54]

IS MANAGERIAL AUTONOMY OBSOLETE?

In the quarter-century after World War II, managers of large corporations were able to conduct business affairs without having to account to many of their constituencies, except for employees who had gained certified union representation. They had won legal right in 1935 to organize and bargain collectively. With this one exception, managers had neither to listen to nor to communicate with their various constituencies, most of whom were still unorganized. Out of a self-defined sense of social responsibility, however, they often did listen carefully, seeking through opinion surveys to discover the desires and preferences of those who could be presumed to be members of informal constituencies. Many of the large firms communicated regularly and at length to the public at large through newspaper, television, and radio advertisements, company brochures, magazines, and pamphlets. Much of the listening and communicating, though, were more a public relations exercise than a truly responsive encounter with constituencies. Many Americans were persuaded that the means used were all too self-serving, and "PR," for public relations, came to be an invidious term, not one of serious responsibility.

While corporate managers increasingly demonstrated their awareness of the social impact their firms had upon community life and the effects they produced for different interests, they continued to insist that they alone could make operative decisions. Few considered the possibility that they might formally recognize and use constituency groups as sources of information or as aides in dealing with the complex problems confronting them. They presumed that it was their right, even their duty, to fashion *their* firms to serve community needs as they defined them and constituency preference, as they chose to interpret them; they saw nothing odd about such autonomy in a democratic society.

The technology of the period and the sheltered markets for which they produced coincidentally worked together for more than two decades, enabling managers to insulate themselves from the demands of shareholders, the grievances of unorganized employees, and the complaints of consumers, suppliers, and other constituencies whose number and variety increased rapidly after the mid-1960s. The technology and competition of the 1950s and 1960s gave way to new forms, however. New technological discoveries appeared, changing industrial advantage, and markets continued a slow but massive shift. Technology and competition created an environment increasingly inhospitable to both existing managerial style and continued corporate success. As the 1980s approached, business managers were discovering that the assumptions of and the expectations generated by the Galbraithian corporation were not sufficient to support their claims to either social autonomy or organizational authority.

Economic growth slowed significantly; economic instability, with its rising unemployment and plant closings as well as the experience of recessions *with* inflation, provoked public disillusionment with business accomplishments. Managers could maintain their policies of self-defined social responsibility and authoritarian direction of the nation's productive effort only if public acceptance was continued. As Irving Kristol had warned earlier, however, that acceptance implicitly rested upon managers' ability to produce the flood of goods and services that American consumers had come to believe was their due. Only ever increasing rates of production and a rising standard of living had allowed managers to respond as they saw fit to the demands of both old and new constituencies.

As industrial productivity declined and many industrial workers realized their standard of living had stabilized or fallen, managers found themselves confronting more challenges than at any time since the Great

Depression. By the 1980s, it was apparent to all that the economy and the society had changed markedly, though gradually, in the decades since World War II. As stockholders increased their demand for a larger say in business decisions and the distribution of corporate earnings, managers began to recognize that the autonomy to which they had long made claim was isolating them from potential allies and supporters. A few managers began to understand that their interests and those of many, if not most, of the constituencies might be complementary rather than conflicting. Both managers and some of the constituencies needed to unite against the insistent demands of shareholders. How and why shareholders became as influential as they did is the story of the next chapter.

NOTES

1. The Business Roundtable, "Statement on Corporate Responsibility," in Thomas G. Marx, ed., *Business and Society: Economic, Moral, and Political Foundations* (Englewood Cliffs, N.J.: Prentice-Hall, 1985), pp. 152–53. [Reprinted with permission.]

2. John Kenneth Galbraith, *The New Industrial State* (Boston: Houghton Mifflin Company, 1967). This volume had been preceded by two other influential studies of the modern corporation and its considerable, but not necessarily benign contributions to society. Galbraith's first was *American Capitalism: The Concept of Countervailing Power* (Boston: Houghton Mifflin Company, 1952) and the second was *The Affluent Society* (London: Hamish Hamilton, 1958).

3. Adolf Berle and Gardiner Means, *The Modern Corporation and Private Property,* (rev. ed.) (New York: Harcourt, Brace and World, 1968), first examined the new industrial corporation that had emerged after the turn of the century. They emphasized the separation of ownership and control, with managers effectively exercising power.

4. Peter Drucker, "The Coming of the New Organization, *Harvard Business Review,* (January–February 1988):53, argued that this form of organization is increasingly obsolete.

5. S. Frederick Starr, "Soviet Union: A Civil Society," *Foreign Policy* 70 (Spring 1988):26–27, has noted that even so centralized, powerful, and all-controlling an organization as that of the Soviet government has not been able to control its external and internal environment. It is ironic that the large American business corporation should suffer some of the same troubles that the centralized state planning system of the Soviets has encountered. As the reader will see in the analysis of this chapter, the source of those troubles are remarkably similar; both capitalism and socialism are subject to them. Or more accurately, one can characterize the large, Galbraithian business corporation as the closest Americans have come to a socialistic system.

6. See the description of the modern corporation given by the Committee for Economic Development (CED): "One of the most important change is that the corporation is regarded and operated as a *permanent institution* in soci-

ety. . . . [the] aim is to further the continuous institutional development of the corporation in a very long time frame. . . . As a permanent institution, the large corporation is developing long-term goals such as survival, growth, and increasing respect and acceptance by the public." *Social Responsibilities of Business Corporations,* A Statement on National Policy by the Research and Policy Committee of the Committee for Economic Development, June 1971, pp. 21–22. [Reprinted with permission.]

7. *Social Responsibilities of Business Corporation,* p. 20.

8. Clifton Gavin, "Exxon and the Arts," *The Lamp,* 1978, p. 10. Mr. Gavin's declaration of social responsibility in Exxon's corporate publication sounds hollow to Americans in 1989, who have watched on their television screens the *Exxon Valdez* polluting Prince William Sound in Alaska by a massive oil spill. Exxon's managers, more than a decade after Gavin's assuring declaration, hardly gave serious consideration to the well-being of the community within which they carried out their operations.

9. Business Roundtable, "Statement on Corporate Responsibility," in Thomas G. Marx, ed., *Business and Society: Economic, Moral and Political Foundations* (Englewood Cliffs, N.J.: Prentice-Hall, 1985), p. 155.

10. Paul W. MacAvoy, "The Business Lobby's Wrong Business," *New York Times,* December 20, 1981.

11. Andrew C. Sigler, "Roundtable Reply," *New York Times,* December 27, 1981.

12. Bruce Nussbaum and Judith H. Dobrzynski, "The Battle For Corporate Control," *Business Week,* May 18, 1987, p. 103. [Reprinted with permission.]

13. Ibid.

14. Ibid., p. 22.

15. Marx, *Business and Society,* p. 155.

16. The reader should note that business corporations have hardly been generous even in their charitable giving. In absolute amounts, their gifts appear large, compared to family incomes. In 1979 dollars, they provided between $1.0 billion and $2.3 billion in the years from 1955 to 1980; however, as a share of their own pretax net income, they have contributed from a low of 0.86 percent in 1955 and 1956, to a high of 1.26 percent in 1969. "Lend A Helping Hand," *Public Opinion,* (February/March 1982):24. In 1988 business firms, including their foundations, gave $4.7 billion to charitable causes. It was a large sum absolutely, but only 4.5 percent of total American giving; for every $1 corporations gave, all individual contributions equalled $20. See Katherine Teltsch, "Americans Donate $104 Billion in 1988," *New York Times,* June 7, 1989.

17. See Earl F. Cheit, "The New Place of Business: Why Managers Cultivate Social Responsibility," in Earl F. Cheit, ed., *The Business Establishment* (New York: John Wiley & Sons, 1964), p. 157–58.

18. The Business Roundtable statement, Marx, *Business and Society,* however, mentioned plant closing as a socially difficult problem, requiring a careful balancing of various constituency interests. It did not call for the participation of constituencies in arriving at a decision, however; managers both could and should make the decision by themselves, in the interests of the various constituencies involved.

19. The limitedness of such responsibility may provide the economy with a flexibility that is far more productive than those adversely affected realize. Compare the ease with which a business firm closes an out-of-date plant with the difficulties encountered by the Department of Defense when it attempts to close unneeded military bases, obsolete supply depots, and inefficient facilities.

20. *Social Responsibilities of Business Corporations,* pp. 63–64.

21. Milton Friedman, *Capitalism and Freedom* (Chicago: The University of Chicago Press, 1962), pp. 133–34.

22. Paul W. MacAvoy, "The Business Lobby's Wrong Business," *New York Times,* December 20, 1981.

23. Peter Drucker, "Corporate takeovers—what is to be done?" *The Public Interest* 12 (Winter 1986): 20. In later writings, Drucker explores the wider meaning of social responsibility in a pluralistic society, for constituents as well as business managers. See Peter Drucker, *The New Realities: In Government and Politics/ In Economics and Business/ In Society and World View* (New York: Harper & Row, 1989) pp. 86–105.

24. Paul Weaver scoffs at the notion that the corporate social responsibility discussed in this chapter is an expression of any altruism. It is, he asserts, "actually a reflection of its aggressiveness. The idea that business has social responsibilities has been a weapon wielded by the corporation in its war to wrest advantage from customers and the political system." "After Social Responsibility," in John R. Meyer and James M. Gustafson, eds., *The U.S. Business Corporation: An Institution in Transition* (Cambridge: Ballinger Publishing Company, 1988), p. 133.

25. Peter Drucker, "The Coming of the New Organization," *Harvard Business Review* (January–February 1988):45. Professor Richard A. Gabriel notes that the military, during World War II, borrowed the same hierarchical, bureaucratic model back from the large corporations. It had become the recognized standard of what "a large organization should be," if one wanted to gain both efficiency and control. See Richard A. Gabriel, "What the Army Learned From Business," *New York Times,* April 15, 1979.

26. Alfred D. Chandler, Jr., and Herman Daem, *Managerial Hierarchies* (Cambridge: Harvard University Press, 1980), p. 1.

27. William J. Hampton and James R. Norman, "General Motors: What Went Wrong?" *Business Week,* March 16, 1987.

28. See *Employment and Training Report of the President, 1982* (Washington, D.C.: Government Printing Office, 1982), Table C-2, p. 241. "Administrative and managerial personnel" made up over 10 percent of all nonagricultural employment in the United States in 1980, compared to 4.4 percent in Japan, 3 percent in West Germany and 2.4 percent in Sweden. See Mark J. Green and John F. Berry, "Taming the Corpocracy: The Forces Behind White-Collar Layoffs," *New York Times,* October 13, 1985, p. F3.

29. Robert Jackall, *Moral Mazes: The World of Corporate Managers* (New York: Oxford University Press, 1988) and Michael Maccoby, *The Gamesman: The New Corporate Leaders* (New York: Simon & Schuster, 1976).

30. Reported by Dean Foust, "A Tough Look At General Motors," *Business Week,* September 18, 1989, p. 12, in a review of Maryann Keller, *Rude Awakening: The Rise, Fall, and Struggle For Recovery of General Motors* (New York: William Morrow, 1989).

31. From George A. Stevens, *New York Typographical Union No. 6,* New York State Department of Labor, Annual Report of the Bureau of Labor Statistics, 1911, part I, pp. 239–41.

32. Quoted in Lewis Corey, *The House of Morgan* (New York: G. Howard Watt, 1930), p. 213.

33. Royal Meeker, "Employees' Representation in Management of Industry," *Monthly Labor Review* 10 (1920):7. Italics added.

34. "Taking Aim at 'Union Busters,'" *Business Week,* November 12, 1979, p. 98.

35. Carey W. English, "Business Is Booming for 'Union Busters,'" *U.S. News & World Report* 94 (May 16, 1983):61.

36. Quoted by Earl Latham, "The Body Politic of the Corporation," in Edward S. Mason, ed., *The Corporation In Modern Society* (Cambridge: Harvard University Press, 1959), p. 223.

37. David W. Ewing, *Freedom Inside the Organization: Bringing Civil Liberties to the Workplace* (New York: E. P. Dutton, 1977), p. 21. [Reprinted with permission.]

38. Ian Cameron, "In Defense of Conflict?" *Unilever Magazine* (September–October 1974):30.

39. Harry Levinson and Stuart Rosenthal, CEO: Corporate Leadership in Action (New York: Basic Books, 1984), p. 4.

40. Allen Weiss, *The Organization Guerrilla: Playing the Game to Win* (New York: Atheneum, 1975), Part III, pp. 91–124.

41. Hierarchies are so integral a part of modern business enterprises that Alfred D. Chandler, Jr., defines business firms as economic institutions, owning and operating multiunit systems, dependent on multilevel managerial hierarchies for administration.

42. Though few studies have explored the matter, we suggest that many managers were greatly influenced by their war experiences, first in the Civil War, then in the Spanish–American War, and most recently (and perhaps most strongly) in World War I and World War II. Several generations of managers learned as young men that the "proper" way to command large organizations and direct many subordinates was through a rigid, hierarchical bureaucracy. Having been exposed to the military life ourselves while young, we can testify to its powerful, though subtle power to legitimize such organizational structure.

43. An economy of small business units would require a great many more market exchanges than at present. Such exchanges or transactions are costly; about one-fifth of the labor force is now employed in wholesale and retail trade. Many of the workers store, move, and package the goods with which they deal, but a sizable portion of them merely administer, carry out the exchanges, and record them.

44. C. R. Henderson, "Business Men and Social Theorists," *American Journal of Sociology* I (1896):385–86. [Reprinted with permission.]

45. N. C. Fowler, *The Boy, How to Help Him Succeed* (New York: Moffat, Yard and Company, 1902), pp. 56–57.

46. *Proceedings of the N.A.M.,* 1903, p. 17.

47. From the principles of the National Metal Trade Association, frequently reprinted in its journal, *The Review.*

48. Taylor's suspicions of managers were clearly evident. See his comments

in F. B. Copley, *Frederick W. Taylor* (New York: Harper and Brothers, 1923), vol. I, p. 417.

49. Harold Smiddy, "The Search for a 'Science of Managing,' " in Melvin Zimet and Ronald G. Greenwood, eds., *The Evolving Science of Management,* (New York: American Management Association, 1979).

50. Ibid., pp. 33–34. [Reprinted with permission.]

51. See Fritz Roethlisberger, *Management and Morale* (Cambridge: Harvard University Press, 1943).

52. See Elton Mayo, *The Social Problems of an Industrial Civilization* (Boston: Graduate School of Business Administration, Harvard University, 1945).

53. Peter Drucker roundly condemns managers' invocation of prerogatives. "This is a singularly unfortunate phrase. A prerogative is a privilege of rank. Management has no claim to any such privilege. It exists to discharge a function. Its job is to make productive the resources in its trust. A prerogative is never based on responsibility or contribution. . . . Management has authority only as long as it performs. To invoke management prerogatives undermines managerial authority." Drucker, *Management: Tasks, Responsibilities, Practices* (New York: Harper & Row, 1973), p. 301. [Reprinted with permission.]

54. The legacy of adversarial-union, paternalistic-employee relations has greatly impaired American managers' ability to respond to the challenges of their rapidly changing environment. A number of firms have tried new, more participative programs, but they are still few and experimental. Changes will have to be much broader, and more basic than have yet appeared, if employee relations are to be transformed. See Harvey Brooks and Michael Maccoby, "Corporations and the Work Force," in John R. Meyer and James M. Gustafson, eds., *The U.S. Business Corporation: An Institution in Transition,* (Cambridge: Ballinger Publishing Company, 1988), pp. 113–32.

6

The Competitive Corporation
and Its Constituencies

The prestigious and influential Committee for Economic Development (CED) asserted in 1971 that the corporate business system had treated and was treating Americans well.

> . . . business on the whole has done its job remarkably well. Since 1890, the total real national product has risen at an average of more than three per cent a year compounded, almost doubling every 20 years. . . . Real disposable income per person has more than tripled and work time has declined by a third over the past 80 years. . . . All other major institutions of society, including government, have been sustained in substantial measure by the wealth produced by a business system which provides a strong economic foundation for the entire society.[1]

The assertion certainly had a factual base, but after the early 1970s, the American economy slowed and business managers' boasts of ever-increasing accomplishments were belied. Although managers could point to absolute gains, the rate of increase in real GNP declined markedly and the change in disposable income (income after taxes) fell even more. On a per capita basis the change in basic disposable income dropped sharply and hourly earnings actually declined! The data suggest that the expectations of a lot of ordinary Americans were being disappointed. (See Table 6.1 and 6.2.)

Public approval of business began its sharp decline before the economic slowdown of the 1970s, probably as a consequence of the failures of success, as noted in Chapter 4, but the sluggishness of the economy after the oil shocks of the 1970s neither inspired nor renewed confidence.

147

Table 6.1. Selected Indicators of Economic Progress in the U.S.

Indicators	1947	1960	1973	1988
Gross national product[a]	$1,066.7	$1,665.3	$2,744.1	$3,996.1
Disposable income[a]	694.8	$1,091.1	$1,916.3	$2,788.3
Per capita disposable income[b]	$4,820.0	$6,036.0	$9,042.0	$11,337.0
Adjusted hourly earnings index[c]	58.5	81.4	101.1	94.6
Productivity index[d]	51.4	71.0	96.4	111.2

[a]Billions of dollars (1982 constant).

[b]Thousands of dollars (1982 constant).

[c]Private nonagricultural production workers; 1977 = 100.

[d]Output/hour of all persons; 1977 = 100.

Source: Economic Report of the President, 1989.

THE RISE OF EFFECTIVELY COMPETITIVE MARKETS

The very foundations of the economy were shifting, transforming the environment within which managers functioned. As never before they had to cope with an unprecedented degree of competition, not only from abroad but also at home. Effective competition began to spread first in the construction and transportation industries after World War II; then through the 1960s, it began to expand in almost all the other industrial sectors, especially trade, finance, and the services. Even in manufacturing, the home of the largest, most prominent industrial corporations, competition increased at an accelerating pace. Over the period from 1939 to 1958 it rose by less than 10 percent, then by nearly 25 percent from 1958 to 1980, and through the 1980s it has certainly increased even faster.[2]

As the European and Japanese economies began to recover from World War II, they proved to be formidable competitors, producing goods of higher quality and lower price than American manufacturers were able to deliver. In addition, a number of other developing economies, particularly those on the Pacific rim—South Korea, Taiwan, Singapore, and Malaysia—began to serve the American market.

As foreign competitors became a force, there appeared a new and widely supported movement to deregulate industries that had long enjoyed government protection against rampant competition. Under both Republican and Democratic administrations, Congress weakened or eliminated the long-established forms of industry-by-industry regulation, particularly those in transportation, communications, and finance.

The enormous success of the large business corporation in the quarter-century after World War II convinced many Americans, and certainly many members of constituencies, that these firms were and would continue to be ample treasuries, able to finance almost any and all de-

Table 6.2. Average Annual Percentage Changes in
Economic Progress in the U.S.

Indicators	1947–1960	1960–1973	1973–1988
Gross national product[a]	3.49	3.92	2.54
Disposable income[a]	3.53	4.43	2.53
Per capita disposable income[b]	1.75	3.16	1.52
Adjusted hourly earnings index[c]	2.57	1.68	−0.44
Productivity index[d]	2.52	2.38	0.96

[a] Billions of dollars.

[b] Thousands of dollars.

[c] Private nonagricultural production workers, 1977 = 100.

[d] Output/hour of all persons, 1977 = 100.

Source: Economic Report of the President, 1989.

mands made upon them. The absolute size of corporate profits and the evident influence of corporate managers in politics and on the economy convinced many people that corporate managers should share their profits and use their influence "to accomplish good deeds." It is not difficult to find corporate constituencies and their leaders, even today, who still share this conviction.

The rising tide of competition that engulfed the economy after the early 1970s weakened managers' defense of their autonomy and accustomed status. The swift decline of manufacturing employment especially contributed to a disturbing erosion of the vast, stable, blue-collar middle class. Consumers discovered new measures of quality in European and Japanese goods that found American products wanting. The increased competition also impaired the ability of managers to respond easily to their critics simply by giving money to "good"—or importuned—causes. Through most of the 1980s, profits were often too limited and prices too squeezed for managers to grant unions' requests for higher wages, for example. More and more managers insisted that they could not even maintain the existing, generous work rules or sustain the high hourly wage rates agreed to in more robust years; they demanded concessions. Other constituencies also discovered that they, as well as managers, had to give careful consideration to the business values of efficiency and profitability if the firms were to continue their operations. For both managers and their constituencies, the competitive changes in the economy offer opportunities and problems. Intense competition disallowed some constituency demands, but it also made business managers more vulnerable to other kinds of demands, if constituencies could mobilize their members' economic power.

Both parties discovered that their usual views of each other were partial and incomplete. Their conventional dealings with each other hardly sufficed to treat either the problems or the opportunities as they arose. On the one hand, constituencies discovered that in the new competitive economy profit making was exceedingly chancy. They began to appreciate the high order of skill required for efficient production as well as the contingent and problematic nature of markets. Business managers, on the other hand, found that constituencies not only could help identify business slack, problems in design, shortcomings in production processes, and defects in marketing approaches, but could also help create a sense of community with business. Managers and constituencies in some areas found reasons to believe that they were more interdependent that either had realized in the past.

Regardless of whether they can continue to discover ways of dealing with each other in useful, helpful ways, neither managers nor constituencies can expect to return to the situation in which they found themselves in the decades after World War II. American business managers henceforth will have to compete as none of their predecessors ever has, and they now confront stockholders who have been transforming themselves into powerful forces. Widespread ownership of stock, through pension plans especially, and the professionalization of portfolio investment have created a constituency that possesses both legal and economic power to challenge managers more effectively than they have even been before.

The Rise of the Institutional Investors

By the end of the 1980s, financial institutions and money management firms controlled more than 40 percent of all corporate equity in the United States.[3] Their share has increased for two reasons: first, because pension funds had grown enormously since 1970, and second, because individual stockholders appeared to be putting more and more of their investments into mutual and pension funds rather than holding shares in their own accounts.

Pension funds owned assets of publicly traded stock amounting to $1.7 trillion in 1989, up from $548 billion in 1970, and the amount was rising fast toward $2.0 trillion. In 1965 pension funds held only 6 percent of all corporate equity, but by the late 1980s they owned about 25 percent of it. By the year 2000, that share should rise to 50 percent.[4] Mutual funds were increasing their share of corporate assets as well. By the end of 1986, roughly half of the forty-seven million U.S. households owning stock held it through mutual funds.[5] The institutional investors were themselves large corporations, with professional staffs whose expertise is in efficient management of diversified portfolios, that is, wide selections of various stocks. Their managers were adept at reducing *their* overall risk by continually shifting and adjusting their port-

Table 6.3. Institutional Share of Total
New York Stock Exchange Trading,
1975–1986[a]

Year	Percentage Share	Year	Percentage Share
1975	17	1981	32
1976	19	1982	41
1977	22	1983	46
1978	23	1984	50
1979	27	1985	52
1980	29	1986[b]	50
		1987	51

[a] Institutional trading refers to blocks of 10,000 or more shares.

[b] 1986 figures are through October of that year.

Source: 1975–1986, Kenneth R. Sheets et al., "How The Market Is Rigged Against You," *U.S. News and World Report*, December 1, 1986, p. 45. For 1987, The Report of the Governor's Task Force on Pension Fund Investment, *Our Money's Worth*, A Project of the New York State Industrial Cooperation Council, June, 1989, p. 2. [Reprinted with permission.]

folios, presumably to maximize total return; some of them, working with large banks and investment houses, had shown themselves eager to exploit financial opportunities that can arise through arbitrage between the current market price and the realizable underlying value of business firms. As a result, institutional investors account for an increasingly large share, of stock market activity. (See Table 6.3.)

Not only do institutional investors possess large assets, but they also have come to be larger players in the stock market, continually buying and selling their holdings. In 1976 the typical institutional investor turned over a little more than 20 percent of its portfolio annually; by 1983, the percentage had risen to around 60, and by 1985 to 80![6] Such a marked increase in turnover has created a lot of activity in the stock markets. In 1985 institutions accounted for 70 percent of all trading on the New York Stock Exchange.[8] Commanding large assets, institutional investors have, unsurprisingly, begun to exert great influence on corporate business policy; and if their exertions bring little response, then the professional investors have proved willing to go further, encouraging mergers, takeovers, divestitures, and restructuring, often at the expense of managers and despite managerial attempts to fend them off. *Business Week* presented two vignettes of recent encounters between institutional investors and corporate managers. More telling than statistics, they dramatize the changes that have taken place over the 1980s.

The setting is New York's Metropolitan Museum of Art. The date: Apr. 14. The occasion: International Paper Co.'s annual shareholder meeting. Corporate gadfly Lewis D. Gilbert stands up, his red patent-leather shoes sparkling, to remind directors that shareholders, not managers, own a company. Board members, sitting on a stage towering over the audience, smile benignly and agree wholeheartedly.

Then another shareholder rises. The smiles disappear. Richard M. Schlefer, investment manager of the $30 billion College Retirement Equities Fund (CREF), proposes a resolution condemning management for adopting a poison pill to thwart hostile takeovers. He wants it put to a vote. Chairman John A. Georges stares down at Schlefer and defends the measure. "We wanted to protect the company against opportunists," he says. Schlefer is incensed. As he later puts It: "CREF owns 800,000 shares of International Paper stock, the bulk of it for 10 years. Who's the opportunist?"[9]

Ten weeks earlier, Roger B. Smith, chairman of General Motors Corp., is on an unexpected three-state odyssey to meet with some of GM's biggest shareholders. Angered at his ousting last year of dissident director H. Ross Perot with $700 million in "hushmail," they berate Smith for GM's meager profits, its falling market share, and its poor productivity despite $40 billion in new equipment since 1979. Why, they ask, is management paying itself big bonuses when much smaller Ford Motor Co. had overtaken GM in earnings? In a booming market, GM's stock had moved mainly sideways. Unless Smith acts, some of them, members of the Council of Institutional Investors,[10] threaten to introduce a proposal at this month's [May 1987] shareholder meeting critical of management.

Within weeks, GM's management announces a series of major policy changes. The company says it will buy back stock, cut capital spending, trim production to reduce burgeoning inventories, and replace cash bonuses for managers with a stock-incentive compensation plan linked to long-term performance. The institutional shareholders withdraw their proposal.

If major earthquakes begin with minor tremors, then the events at IP and GM foretell traumatic times ahead in Corporate America. CREF's proxy proposal against the poison pill at IP didn't win, but it got more support than anyone expected—28% of the votes cast.[11]

Many managers and scholars have been surprised by the boldness with which institutional investors have challenged management policies.[12] Some managers of large business corporations have reacted with anger at the loss of their former autonomy.[13]

The appearance of a powerful and legitimate constituency—organized shareholders through institutional investors—poses a fundamental challenge to managers. Defenders of the business system have always grounded the authority of managers on responsibility to the corporate owners. Institutional investors, though, claim a more direct and immediate right to speak and act for the owners than do managers and thus can preempt the chief support of managers' authority. If the interests of managers diverge from those of institutional investors who are acting as owners, then by the logic that managers themselves have long used—indeed, have insisted upon—it is the managers, not the owners, who must give way. Managers are not able to ignore institutional investors, for law and tradition are clearly on the side of owners. Thus, the kind of managerial autonomy that characterized the Galbraithian corporations is not apt to be recovered.

DIVERGENT INTERESTS IN RISK TAKING

Managers and institutional investors tend to evaluate risk taking differently. Professor John C. Coffee, Jr., argues that each group will necessarily take a different perspective of risk taking, for each approaches it from quite a different direction. The preference for risk among institutional investors is shaped fundamentally by the concept and practice of a carefully, continually adjusted diversified portfolio, distributing investments among many different firms and a variety of assets to minimize *overall* risk. In their consideration of the risk appropriate for any particular firm (or its stock), institutional investors are either risk-neutral or at least not risk-averse: Should a firm increase its risk by increasing indebtedness, for example, the investor simply offsets the added risk with another, safer asset acquisition. Institutional investors continually assess their portfolios in the light of current economic conditions, and adjust them as their evaluation of corporate policies and programs change. They pick and choose stocks to fill specific purposes in their portfolios and by insisting upon higher debt loads, require managers of industrial business corporations to pay out rather than retain a substantial part of cash flow. They thus (1) prefer high debt, (2) do not favor diversified conglomerates unless they match specific needs in the portfolio, and (3) discourage internal financing or "satisficed," rather than maximized, profits. Consequently, institutional investors, working with or through those who are reorganizing and restructuring firms, are forcing changes in corporate characteristics that managers of the past generation found most congenial.

The congeniality of these characteristics arose from the protection they offered against managerial risk—at least less risk than that which the heavily leveraged, trimmed-down firms must confront. Whatever the relevant specific economic circumstances, industrial managers, in fact, generally do occupy a position relatively more averse to risk than that of institutional investors. First, corporate managers retain as much of their firms' cash flow as they can. In 1988, the 1,000 largest public companies, measured by sales, generated total funds of $1.6 trillion, but they distributed less than 7 percent to shareholders—$108 billion as dividends and another $51 billion through share repurchases.[14] Such high levels of retention increase managers' freedom from the discipline of capital markets, increase the size of the companies they control, and thus provide justification for higher salaries for themselves and promotions for their subordinates.

Second, managers' averseness to risk arises from the nature of both their personal portfolios of assets and their companies' portfolios (their product lines, subsidiaries, and various acquisitions), which are much less diversified than those of the institutional investors. Senior executives' most important assets are their skills employed on the job. Those skills often are quite firm-specific and not easily transferred to other firms. Even if alternative positions may be available, the rate of return is not likely to be as high.[15] Historically, mobility among top business executives has been low, though the rate of movement has increased in recent years. A large share of managers' compensation depends on the dividends received and capital gains from their companies' stock, linking their own personal fortunes to the fortunes of their firms.[16] The managers, therefore, possessing a relatively undiversified portfolio, have much more to lose and a much greater need to take care in their decisions and policies than the well-diversified institutional investors. Even if one assumes that corporate managers generally are entrepreneurial risk takers (an heroic assumption, as business history since the 1950s indicates), they would still be on average more risk-averse in evaluating policies and programs for their firm than the institutional investors would be in evaluating policies and programs *for the same firm.*

The risk-averseness of Galbraithian corporate managers has encouraged them to pursue policies that the new institutional investors often find unsatisfactorily inefficient.[17] As the investors have gained more control, they have demanded better earnings performance. Managers have responded reluctantly, offering opportunity for enterprising financiers and entrepreneurial "arbitrateurs" to move. These new business players, often invidiously labeled "raiders" and "greenmailers," have sought to restructure "undervalued" firms and turn them into more valuable properties, with corresponding gains for themselves and any stockholders allied with them.

The rise of the institutional investor has been accompanied by an increase in the number of hostile takeovers, unwilling mergers, and leveraged buyouts. Much of the public and political debate over these activities has focused upon the implications for American competition, the effects on industrial research and development, and the threat to the stability of the national economy. We will not examine the pros and cons of the cases made by various advocates, but simply note the significant shift in power away from corporate business managers. Institutional investors, as major stockholders and therefore *owners* of the firm, are a constituency whose power, legitimacy, and legal position more than matches that of managers, whatever their putative claims to prerogative, authority, and control. Even business managers who are not subject to a hostile takeover or a raid indicate by initiating restructurings that they understand the nature of the changes taking place. Managers of almost any firm, large or small, in any industry, have been challenged by outsiders, either through the initiation or with the help of institutional investors. The more traditionally they continue to manage—autonomously, in the Galbraithian manner—the more likely the challenge will continue.

The Restructuring of American Business

The restructuring of American business corporations proceeded rapidly through the 1980s, affecting more firms than ever before, and involving a volume of assets (in constant dollars) far larger than any earlier merger-and-acquisition period.[18] By the middle of the decade mergers and acquisitions were running at the rate of $180 billion in volume and over 3,000 transactions per year.[19] Some 900 divestitures took place in 1984. In 1985 alone, 23 percent of the nation's leading 850 corporations underwent an "operational restructuring," usually selling or spinning off divisions. Usually the proceeds were not reinvested in other acquisitions but were paid out to stockholders as part of a downsizing of the firms.[20]

An increasing portion of the restructuring has been accompanied by leveraged buyouts (LBOs). In 1981 LBOs accounted for less than 5 percent of the total, but by 1986 they made up more than 20 percent, and most of them involved a large increase in company debt.[21] The total value of the 16 public-company and 59 divisional buyouts in 1979 was $1.4 billion; in 1988 there were 125 public-company buyouts and 89 divisional buyouts, with a total value of $77 billion.[22] The LBOs and other forms of restructuring greatly increased the debt burden on American corporations. The rise through the mid-1980s was very rapid (See Table 6.4), and a sizable portion of it was financed by foreigners. The combined borrowings of all nonfinancial corporations in the United States rose by nearly 20 percent between 1979 and 1988, from $835 billion to almost $2 trillion, with interest charges equal to more than 20 percent of corporate cash flows. These are high by historic standards.[23]

Corporations with a heavy debt load are not as easy to manage as those that enjoy little or no debt. Managers will probably find it increasingly difficult to indulge the policies and practices that had contributed to three characteristics of the Galbraithian corporation: profit satisficing, internal financing of investment, and diversification through conglomeration. Institutional investors found these characteristics to be impermissible—they were inefficient as investors defined that term. The Galbraithian practices had become weaknesses that could no longer be tolerated. Managers had found the practices desirable, for, given their

Table 6.4. American Corporate Debt Issue,
1983–1986

Year	Amount	Share Foreign Financed
1983	$ 55 billion	10.3%
1984	$ 80 billion	23.3%
1985	$145 billion	25.7%
1986	$265 billion	14.2%

Source: *The Economist*, July 11, 1987, p. 25.

level of risk-averseness, they provided useful organizational slack and reserves of resources for bad times. The less risk-averse institutional investors found them indicators of inefficiency that had to be eliminated. The policies that created them were unnecessary.

Profit-satisficing corporations. Business managers for the most part have rejected free, competitive market price competition whenever they could. In that rejection they also discarded a basic value (and assumption) of economists: profit maximization. Managers often identified their business purpose as profit making, but the profits they sought were not those defined by economic theory. Peter Drucker maintained that for modern corporate managers profit maximization is

> not only false, it is irrelevant . . . the concept of profit maximization is, in fact, meaningless. . . . Profit and profitability are, however, crucial—for society even more than for the individual business. Yet profitability is not the purpose of but a limiting factor on business enterprise and business activity. . . . Profit is a condition of survival. It is the cost of the future, the cost of staying in business. A business that obtains enough profit to satisfy its objectives in the key areas is a business that has the means of survival. . . . [what is needed is] minimum profitability rather than . . . that meaningless shibboleth "profit maximization."[24]

Managers could not in truth claim that any of their decisions was the best in given circumstances. They suffered from what Herbert Simon called "bounded rationality."[25] Managerial decisions could not assuredly maximize profits, even if they were intended to do so, because they did not—they could not—include all relevant information. *All* information has never been available readily, already, or on a timely basis.

The best that managers can hope is that their decisions are good enough to enable them, on average, to meet their aspirations. Maximized profits cannot be identified except possibly by the economist, and that, only after the fact. They must, therefore, seek another profit goal, that of satisficing. Professor Herbert Simon argued that typically managers satisfice by seeking to attain designated target outcomes. He is not alone in describing satisficing as the usual, practical managerial goal. Over the last fifty years many scholars have investigated and confirmed the satisficing behavior of corporate managers, tough generally accepted economic analyses hardly recognize the weight of the accumulated evidence. The research studies indicate that in satisficing profits managers pursue goals much more complex and quite different from the maximizing goal specified in economic theory.[26]

Typically corporate managers own a very small part of their firms; in the 1980s, the median public-company CEO held only 0.25 percent of the firm's equity. Consequently their personal wealth increased only a very small fraction of increased shareholder wealth: Less than one-third of one percent! In contrast the personal wealth of the typical LBO business-unit managers responded twenty times more sensitively to changes in shareholder wealth.[27] It can hardly be surprising, therefore, that a number of studies have found that management-controlled firms showed a lower return on investment than firms where ownership and control were not separated.[28] The author of one study concluded that management-controlled firms have a rate of return only half that of owner-managed firms and somewhat less than the return of externally controlled firms.[29] For decades, stockholders have received only a portion of the "return on shareholders' equity" and statistical analysis shows no significant relationship between corporate performance and stockholders' realized return.[30]

Competitive market theory suggests that stockholders, through dividends or market valuation, should at least receive benefits of 100¢ for each dollar of reported net income. Research indicates that from 1970 to 1984 less than half of a sample of the largest corporations, produced even the theoretical minimal return. An investor who held General Electric or General Motors stock, for example, received only half the benefits the companies earned. Investors who held stocks in such prestigious firms as Lilly, Westinghouse, Kodak, Sears and Xerox lost money.[31]

The solution to the disappearance of "retained earnings" and the loss of stockholder enrichment appears to be that Galbraithian managers have often proved themselves to be exceedingly inefficient investors of their firm's funds.

Internal financing of corporate growth. At least since the 1920s a cardinal managerial rule for large corporations has been to finance the firm's growth internally.[32] Internal financing allowed managers to expand their firms without having to test their judgment in the marketplace; they did not have to share their control with banks or investment houses.

They did not care for the monitoring of the capital markets.[33] Research as shown that of a dozen mature industrial firms from 1969 to 1978, almost three-quarters of them generated their capital funds internally, while the remaining quarter used long-term debt; none raised capital from new equity issues.[34]

Managers' reliance upon internal financing has been significant, because careful examination of the rates of return produced have often made little or no economic sense. On average, the rates have been below market levels, and in a number of cases actually negative.[35]

During the 1980s, analysts working for institutional investors and scholars publicized the below-market rates of returns from internally financed investments. In 1984, for example, the oil industry's asset value exceeded its market value by $191 billion, with more than 80 percent of the discount accruing in the ten largest companies. The managers of the large companies found themselves with greatly increased flows of cash because of the jump in oil prices. Instead of paying out the money in dividends, many of them went on shopping sprees, buying up companies in industries with which they were unfamiliar. They also continued to drill for oil, even when it had become evident that every dollar spent was going to bring in no more than sixty cents.[36]

Such uneconomic, *irrational* policies pursued by business managers call for some explanation. Several scholars argue that these policies bought managers autonomy, organizational perquisites, and personal power. Professor Oliver Williamson found that retained earnings could be, and were, used to achieve such managerial benefits as generous bonuses, lavish offices, company limousines, planes, yachts, or hunting lodges.[37] All of these benefits could be charged off to the firm and written off corporate taxes as business expenses. Michael Jensen pointed out that top managers did not keep all the benefits of retained earnings for themselves, however. They shared them particularly with middle managers through promotion rather than year-to-year bonuses, creating a strong organizational bias toward growth of staffs. He could have added that any large flow of earnings, if retained, could be used by managers to provide benefits for employees, vendor/supplier, and other constituencies, as well as for themselves and an ever-larger organization. A number of large firms did exactly this, as will be described in the next chapter.

Diversification through conglomeration. Managers of Galbraithian corporations had molded their companies into multidivisional organizations and diversified into a variety of production areas. Professor Alfred Chandler wrote that "the years after World War II mark[ed] the triumph of [this] modern business enterprise."[38] Diversification became even more evident as managers in some industries began to acquire unrelated firms. The conglomerate firm appeared first in such diverse industries as textiles and ocean shipping, and machine tools and defense industries. The firms acquired were usually small and in competitive industries. By the early 1960s this mode of diversification had become a major corporate variation. Managers of conglomerates pursued a different investment strategy from that of managers of older type firms. Most of the latter had sought growth through internal expansion, by direct investment in an initial process, with branching out to related areas of production. The conglomerate, however, grew by buying up existing enterprises, often in completely unrelated industries and with quite different products.

In a study of mergers and acquisitions between 1950 and 1975,[39] researchers found that the typical large corporation nearly doubled the number of lines in which it operated. Firms that were relatively small in 1950 tended to diversify more aggressively than the larger firms. These increasingly conglomerate firms did not perform very well.[40] A sizable fraction of the acquisitions made during the 1960s and 1970s were subsequently resold. Apparently new lines of business were often acquired to reap the benefits that good managers had produced in building them up. However, the "acquired" managers often soon left, leaving the problems of managing the larger, more diversified company to the "acquiring managers, who did not necessarily perform at all well." Ravenscraft and Scherer conclude:

> . . . they overestimated their ability to manage a sizeable portfolio of acquisitions, large and small, related and unrelated. By the time they learned that they had erred, they had already overextended themselves and were unable to cope with the problems emerging from accumulated acquisitions. Or alternatively, they recognized their limitations but pursued a damage-limiting strategy, continuing (like Beatrice Foods) to make mergers but ruthlessly selling off acquisitions that showed signs of persistent difficulties.[41]

Given what they consider an average the mediocre records of success in earnings and profits that business firms have written for themselves through mergers and acquisitions, Ravenscraft and Scherer asked what motivation would explain the periodic surges of such activity. They suggest that it may indeed be a drive for empire building.

> It provides a plausible explanation of managers' strong desire for formal control and organizational integration, even when achieving them through merger means paying more for assets, raising bureaucratic costs and increasing the likelihood of incentive failures.[42]

Thus all three characteristics of the Galbraithian corporation considered here indirect American managers of large business corporations, well into the 1980s, as "sitting ducks" for investors who sought "efficient" management, investment, and production. T. Boone Pickens, Jr., one of the best known of the "take-over artists" warned in 1986 that stockholders increasingly refused to accept the notion that professional managers could or should continue to enjoy the "sovereign autonomy" they claimed. Lackluster performance and chronic undervaluation of the firms they controlled had to be a addressed.[43] The management consulting firm of McKinsey agreed with Pickens. Its researchers analyzed twenty-five large manufacturing companies, treating their headquarters as if they were portfolio managers. It then related the costs of headquarters to the market value of the equity that the headquarters' managers oversaw. The resulting management "fee" on average was almost 2 percent of market equity value, a fee that at a minimum is about three times larger than that charged by managements of major equity funds, and if all costs were to be included, is probably closer to six times larger. Although the analysis does not indicate that corporate headquarters are necessarily wasting money, it does warn that they had better not be wasting money when their costs are so high.[44]

MANAGING IN A NEW STYLE

It seems plausible that the earlier corporate characteristics examined in the last chapter—growth with stability, self-defined social responsibility, bureaucratic hierarchical organization, and authoritarian relationship with constituencies—founded on the "inefficiencies" of profit satisficing, internal financing, and acquisition through conglomeration. As hostile raiders and merger specialists began to press managers for more efficiency and larger returns to stockholders, and as institutional investors lent their support to that pressure, all seven characteristics of the Galbraithian corporation became increasingly obsolete. The old style of managing became endangered, an incumbrance and problem in the new emerging economy.

   Business managers of the 1990s operate in a more competitive world economy than ever before. They confront the danger of takeover by "raiders" if they do not voluntarily restructure their firms, and in either case must carry more debt and operate with less slack than was the norm in earlier decades. As a result they perform with small margins for error. The autonomy their predecessors knew in the 1950s and the decades afterward has been eroded until little remains. The exercise of managerial power henceforward will be increasingly conditional in a demanding environment and a more uncertain future than in the post–World War II decades. As managers maneuver, bargain, and negotiate with institutional investors, they may begin to appreciate the advantages of securing for themselves allies and friends. Surrounded by many constituencies of whom in the past they have taken only such notice as they chose, they may now discover among them a number whose interests match their own far better than those of the institutional investors or owners.

Two corporate managers spoke to the issue in 1986, in the midst of the restructuring of industry. Thomas H. Wyman, then-chairman of the board of CBS, stressed the need for examining old relationships and insuring new, understandable information.

> But ours is an increasingly complex world characterized by frequent and accelerated change. Business decisions affect—and are affected by—many constituencies. Successful executives will be those who understand and interpret complex relationships. To be effective demands continual reconsideration of assumptions underlying old and familiar networks . . . as well as gathering and sorting of new information.
>
> This puts a premium on managers with the ability to be flexible, critical and capable of continuous learning, managers with the skills to anticipate change . . . and not be surprised.[45]

Donald Petersen, president of a revived, reorganized, and profitable Ford Motor Company, was more specific in his recommendations for managing in the new economy:

> Managing change when the wind is shifting means increased reliance on and attention to teamwork, trust, respect, customers, people, and leadership. . . . The first step in managing change is to listen to our customers. . . . We exist to serve our customers, not to serve ourselves, and that *we* must learn from *them*. The second step in managing change is to listen to our people—all our people . . . openly and nondefensively. . . . Too often we become adversaries of the very people who helped us to our success; customers, union members, suppliers. We cannot build great products without quality work by our people and quality products and services from our suppliers. This means that *our* success must build *theirs*, rather than be taken from theirs.[46]

Petersen understood corporate managing as a task of building a *community* of effort, which will require the voluntary cooperation of all its members in productive work. If autonomous managers are replaced by "efficiency-minded" institutional investors, the economy may not gain much, and could lose a great deal. Corporate managers, after all, have usually pursued other, and sometimes larger, purposes than those of mere market efficiency. If they now find they can no longer sustain their autonomy, they might wisely consider the strengths and capabilities available in responding to their various constituencies. Both they and the constituencies need to experiment with new roles for themselves as they deal with common, as well as conflicting, problems and opportunities. Constituencies will have to concern themselves far more than they have with the requirements of efficiency and production; managers will wisely appreciate the contribution that a fair, equitable, and participative industrial society can provide.

Managers have long rationalized their autonomous role by asserting it allowed them to produce efficiently—a high value of the market ethic. As the findings reported here indicate, in fact they often followed a very different ethic, though they usually hid it from themselves. In Chapter 7 we will examine the nature of the market values they claimed and the ways in which those values clash with or even distort the values with which managers have shown themselves intuitively comfortable.

NOTES

1. *Social Responsibilities of Business Corporations,* A Statement on National Policy by the Research and Policy Committee of the Committee on Economic Development, June, 1971, pp. 11–12. [reprinted with permission.]

2. William Shepherd, "Causes of Increased Competition in The U.S. Economy, 1939–1980," *Review of Economics and Statistics* 64 (November 1982):613. [Reprinted with permission.]

3. See Michael C. Jensen, "Eclipse of the Public Corporation," *Harvard Business Review* (September–October 1989):67. Louis Lowenstein, "Pruning Deadwood in Hostile Takeovers: A Proposal for Legislation," *Columbia Law Review* 83 (1983):249, 297–298, mentioned a somewhat lower figure. He suggests that institutional investors owned more than 35 percent of shares of firms listed in the Standard & Poor's 500 stock index. Ford S. Worthy, "What's Next For The Raiders," *Fortune,* November 11, 1986, p. 23, provided an estimate in line with Jensen's and Lowenstein's. Peter Drucker gives a figure of 50 percent of all publicly traded common shares of large firms. "Corporate takeovers—what is to be done?" *The Public Interest* 12 (Winter 1986):11.

4. Anice C. Wallace, "Bill Could Shift Control Over Pension Funds," *New York Times,* August 28, 1989, and Bruce Nussbaum and Judith H. Dobrzynski, "The Battle For Corporate Control," *Business Week,* May 18, 1987, p. 103–4.

5. The number of mutual funds have increased rapidly in recent years. Through 1985 and 1986 the number of mutual fund accounts increased by 2.7 million to over 42 million. Since households sold more stock than they bought over the same period for a total of $227 billion, it is probable that many are not withdrawing from the market, but transferring their assets from individual accounts to mutual funds. George Russell, "Manic Market," *Time,* November 10, 1986, p. 67. Also see Kenneth R. Sheets et al., "How The Market Is Rigged Against You," *U.S. News and World Report,* December 1, 1986, p. 47.

6. For the earlier years see Steven Greenhouse, "The Folly of Inflating Quarterly Profits," *New York Times,* March 2, 1986, sec. III, reporting on a survey of 7,500 fund portfolios made by SEI Corporation. Louis Rukeyser provided the figure for 1985, *Wall Street Week,* January 9, 1987, Public Broadcasting System television program.

7. Taken from Sheets et al., "How The Market Is Rigged Against You," p.

8. Ibid. Of the 70 percent of the trades by institutional investors, 57 percent were made by the big, domestic firms; 15 percent by mutual funds; 10 percent by foreign firms; 5 percent by insurance companies; and 12 percent by other.

9. Schlefer has a valid point, for research shows that among the pension funds that do the most churning of stock are those owned by corporations, not the big public pension funds, such as CREF. The latter tend to hold stocks for an average of five years. Corporate pension funds turn over as much as 40 percent a year. Avon, for example had turnovers of 57 percent and 75 percent in 1986. It is worth noting that Avon's chairman believes managers have the right, indeed, duty to subordinate stockholders' interests to those of other corporate constituencies, as noted above. See Nussbaum and Dobrzynski, "The Battle For Corporate Control," p. 104.

10. The Council of Institutional Investors has more than 50 pension fund members that control $200 billion worth of investments.

11. Bruce Nussbaum and Judith H. Dobrzynski, "The Battle For Corporate Control," *Business Week*, May 18, 1987, p. 102. [Reprinted with permission.]

12. See Anthony Bianco, "American Business Has A New Kingpin: The Investment Banker," *Business Week*, November 24, 1986, p. 77. He quotes John Kenneth Galbraith, who expressed his surprise at what is happening. His earlier studies gave no hint that managers would ever have to worry about any constituency, particularly the unorganized, passive stockholders.

13. Peter C. Clapman and Richard M. Schlefer, "Recipe for a Management Autocracy," *New York Times*, December 14, 1986.

14. Michael C. Jensen, "Eclipse of the Public Corporation," *Harvard Business Review* (September–October 1989):66.

15. Oliver Williamson, "The Modern Corporation: Origins, Evolution, Attributes" *Journal of Economic Literature* 19 (1981):1548 and "Corporate Governance" *Yale Law Journal* 93 (1984):1207–17.

16. W. Lewellen, *The Ownership Income of Management* (New York: Columbia University Press, 1971), p. 150–51.

17. John C. Coffee, Jr., examines this thesis in detail. See "Shareholders Versus Managers: The Strain In The Corporate Web," *The Center for Law & Economic Studies*, Working Paper No. 17, January 1986, also published in John C. Coffee, Jr., Louis Lowenstein, and Susan Rose-Ackerman, eds., *Knights, Raiders & Targets: The Impact of the Hostile Takeover* (New York: Oxford University Press, 1988).

18. David J. Ravenscraft and F. M. Scherer, *Mergers, Sell-offs, & Economic Efficiency* (Washington, D.C.: The Brookings Institution 1987), p. 21.

19. Michael C. Jensen, "The Takeover Controversy: The Restructuring of Corporate America," *Beta Gamma Sigma, From the Podium* (September 1987):1.

20. See Coffee, "Shareholders Versus Managers," p. 44.

21. Nathaniel C. Nash, "Company Buyouts Assailed in [SEC] Study," *New York Times*, January 31, 1988.

22. Jensen, "Eclipse of the Public Corporation," p. 65. See also Leslie Wayne "Takeovers Revert to the Old Mode," *New York Times*, January 4, 1988. A large number of companies also conducted LBOs in which they bought back sizable portions of their own stock. In 1984, total corporate equity shrank by a record $85 billion, the direct result of merger activity that involved an exchange of debt and cash for equity. See Leonard Silk, "Economic Scene: Preventing Debt Disaster," *New York Times*, September 6, 1985, p. D2. In 1987 more than 1,400

companies announced plans to buy back shares, valued at more than $80 billion. That was more than double the number of 1986 transactions and nearly double the dollar value. Of course, not all of these were completed; if they had been, something like 2.4 billion shares of stock would have been removed from the market. See Kevin G. Salwen, "Share Buy-Back Plans Proliferate," *Wall Street Journal,* January 6, 1988.

23. Jensen, "Eclipse of the Public Corporation," p. 67; also see Christopher Farrell and Leah J, Nathans, "The Bills Are Coming Due," *Business Week,* September 11, 1989, p. 85. Farrell and Nathans report that U.S. corporations during the 1980s retired nearly $500 billion in equity and piled up almost $1 trillion in debt. Using Merrill Lynch & Co. estimates they indicate interest payments may take 30 percent of cash flow.

24. Peter Drucker, *Management: Tasks, Responsibilities, Practices* (New York: Harper & Row, 1974), pp. 59, 114. [Reprinted with permission.]

25. Herbert A. Simon, *Models of Man* (New York: John Wiley & Sons, 1957).

26. The reader may want to examine the writings of such economists and sociologists as William Baumol, Richard M. Cyert, John Kenneth Galbraith, James G. March, Robin L. Maris, Edith T. Penrose, or Oliver E. Williamson for fuller elaborations of these studies.

27. Jensen, "Eclipse of the Public Corporation," p. 68, forthcoming.

28. W. McEachern, *Managerial Control and Performance* (Lexington, Mass: Lexington Books, 1975), pp. 39–51. Nine such studies are reviewed.

29. Ibid.

30. Ben C. Ball, Jr., "The mysterious disappearance of retained earnings," *Harvard Business Review* (July–August 1987):56–63.

31. Ibid., p. 58.

32. "In the United States, the growth of these diversified divisionalized, managerial enterprises was largely self-financed, with new facilities being paid for from retained earnings." Alfred D. Chandler, Jr., and Herman Daem, eds., *Managerial Hierarchies* (Cambridge: Harvard University Press, 1980), p. 34. Also see Gordon Donaldson, *Corporate Debt Capacity* (Boston: Graduate School of Business Administration, Harvard University, 1961), pp. 51–56. He examined the financing of a number of large corporations over the years 1939–1958 and found great reliance upon internally generated financing.

33. "The aim [of corporate managers] is rather to reduce management's dependence on outside opinion and enable it to develop appropriate modifications of debt policy consistent with individual circumstances." Donaldson, *Corporate Debt Capacity,* p. 157. See also Michael C. Jensen, "The Takeover Controversy: The Restructuring of Corporate America," *Beta Gamma Sigma, From The Podium* (September 1987):2.

34. Gordon Donaldson, *Managing Corporate Wealth: The Operation of a Comprehensive Financial Goals System* (New York: Praeger, 1984), pp. 45–46.

35. William Baumol, Peggy Heim, Burton Malkiel, and William Quandt, "Earnings Retention, New Capital and the Growth of the Firm," *Review of Economics and Statistics* 52 (1970):345 and their "Efficiency of Corporate Investment: Reply," *Review of Economics and Statistics* 53 (1973):128. Also see, Henry G. Grabowski and Dennis C. Mueller, "Life Cycle Effects on Corporate Returns on Retentions," *Review of Economic and Statistics* 57 (1975):400.

36. Ennius Gergsma, "Do-It-Yourself Takeover Curbs," *Wall Street Journal*, February 12, 1988. Michael C. Jensen also accuses managers in the oil industry of wasting billions of dollars from the late 1970s on. They used what he called "the payout of free cash flow" on organizational inefficiencies or low-return projects.

37. "Managerial Discretion and Business Behavior," *American Economic Review* 53 (1963):1032, 1047–51.

38. Alfred Chandler, *The Visible Hand: The Managerial Revolution in American Business* (Cambridge: Harvard University Press, 1977), p. 477.

39. David J. Ravenscraft and F. M. Scherer, *Mergers, Sell-Offs, & Economic Efficiency*, Washington, D.C.: The Brookings Institution, 1987.

40. Andrei Shleifer and Robert W. Vishny, "Value Maximization and the Acquisition Process," *Journal of Economic Perspectives* 2 (Winter 1988):7–20. "The acquisition process is probably the most important vehicle by which manages enter new lines of business. In his sample of thirty-three large diversified U.S. corporations, Porter (1987) found that between 1950 and 1986 his firms entered an average of eighty new industries each and that over 70 percent of this diversification was accomplished through acquisition. Large-scale movement of U.S. manufacturing toward unrelated diversification is now thought by many observers (including Porter) to have been unsuccessful. The high level of divestiture of peripheral businesses by diversified corporations beginning in the mid-1970s is almost surely a response to that failure." [Reprinted with permission (p. 13).]

41. Ravenscraft and Scherer, *Mergers, Sell-offs, & Economic Efficiency*, p. 212. [Reprinted with permission.]

42. Ibid., p. 214.

43. T. Boone Pickens, Jr., "Professions of a short-termer," *Harvard Business Review* (May–June 1986):75.

44. Ennius Bergsma, "Do-It-Yourself Takeover Curbs," *Wall Street Journal*, February 12, 1988. Myron Magnet points out that many large companies tie up a lot of capital in investments that pay a small rate of return. Consultants have found that 80 to 90 percent of companies' values typically come from 20 percent of their investments. See "Restructuring Really Works," *Fortune*, March 2, 1987, p. 40.

45. Speech delivered at Dickinson College, quoted in *Wall Street Journal*, April 30, 1986. It should be noted that Mr. Wyman did not meet his own test; he was surprised by a takeover of CBS a few months later. He lost his position, though he was handsomely rewarded.

46. "Shifting Winds," *Perspectives*, Boston Consulting Group, 1986.

7

Market Values for Corporate Managers

Corporate constituencies regularly make demands upon managers that, at the least, require the consideration and time of staff advisers and, at the most, may preempt the attention and efforts of the chief officials. In their dealings with constituencies managers may find it necessary to change company policies or adjust major expenditures of resources. Such is the case when managers bargain in good faith with certified labor unions. Of course, by law employers must meet with such unions, though legally they do not have to agree to any particular terms. In practice, however, they may indeed have to agree to certain terms, for union leaders can at times exert considerable pressure in support of their terms. They may lawfully use concerted activities such as the strike, picketing, and in some cases, boycotts. Consequently, managers almost always pay attention to unions' approaches and give careful consideration to the requests they present.

Managers do not ordinarily feel as pressed to consider many of the presentations of other and newer constituencies. Particularly if the groups do not enjoy any special legal favor or governmentally protected position, managers' initial preference is to reject them and to treat their demands as illegitimate. A commonly expressed judgment, noted in Chapter 2, is that they are not owners and no one elected them; furthermore, "They muddy the waters of decision with foolish arguments and unrealistic objectives."[1] In fact, some constituency demands are foolish and unrealistic; some are little more than publicity gimmicks, advertisements for those involved. More often, though, they arise out of substantial and deserving issues. Through hard experience and despite their distaste, managers have learned that they had better pay respectful attention and give careful thought to their responses.

Constituencies have learned that it helps to speak as dissident stockholders. Most have discovered the excellent forum of an annual meet-

166

ing; they have discovered that ownership bestows a certain legitimacy upon them. The group may be small in members, but it can nevertheless propose actions large in impact upon the corporation. A constituency, whether a stockholder or not, may accompany its presentations to management with political lobbying, seeking change in relevant government regulations, subsidies, or tax advantages for the target company. Since constituencies may engage in the same concerted activities as unions, their demands, on occasion, may also bring the threat or reality of more public but not necessarily more effectual action—demonstration, unfavorable publicity, and protest, including even consumer boycotts.

In presenting their demands to managers, and certainly in displaying them to the public, constituency leaders appeal to as wide a range of interests as they can, supporting their claims with popular and widely held values. These values are often democratic, participative, and egalitarian. Managers will need to be able first to understand the nature, appeal, and implication of the values invoked by the groups. Second, they will find it useful to be ready to explain or defend the values upon which they base their own policies and actions. They can do so effectively and wisely only if they comprehend and understand their own values. They need an informed defense particularly when constituency proposals and the values underlying them conflict with their own corporate policies.

CAN BUSINESS MANAGERS BORROW THE ECONOMISTS' DEFENSE OF THE MARKET?

As managers are confronted more and more with constituencies critical of their actions and values, they may be tempted to seek the advice and borrow the arguments of economists, perhaps naively believing the free *market* apologia will serve their business, corporate, free enterprise interests. At first glance it appears that economists and business managers express similar values and thus could be expected to agree on major policies. Most economists in the United States are advocates of the free, competitive market. Most of those teaching in our major universities and colleges, serving in the government, and working as professionals in business firms believe that a free market is a valuable, core economic institution. They are, in Professor Amartya Sen's words, "much taken by the notion of private motivated achieving public good through the intermediary of the market mechanism."[2] If this notion is demonstrable and free market arguments can defend managerial policies, economists should be helpful to managers as they contend with their critics. Economists may then provide a powerful case that responding to constituencies in nonmarket ways is unnecessary and maybe even

168

harmful. Managers could offer counter proposals that solving constituency problems through the market is desirable and appropriate.

Surveys indicate that most economists strongly support the market.
In England 87 percent of government economists and 82 percent of
economists in private employment agreed that:

> ". . . in a free enterprise economy, the presumed harmony between individ-
> ual and public interest" is brought about by "competitive markets and pur-
> suit of self-interest by individuals" and/or "a strong desire for profit max-
> imization."[3]

It may be worth noting that the more involved respondents were in the
world of practical business affairs the less support they gave to this
statement. About three-fifths of business economists, 62 percent, accepted it, and an even smaller share of the generally pro-business Conservative members of Parliament found it acceptable. Even these smaller
proportions favoring the market are substantial, however. Most English
economists appear to believe that a free market is a useful, practical
way to solve social problems.

In the United States the proportion of economists favorably inclined
toward the market is as large as or larger than in England. More than
90 percent of a sample of economists agreed that "the free market
worked for the best" in the case of rental housing, imports and exports,
international exchange rates, and interest rates. The surveyed economists strongly supported policies that allow individuals to participate in
the market, enabling them to buy according to their own preferences
and tastes. A high proportion even favored a market procedure to help
the poor with "cash grants," rather than "goods in kind." Such grants
allow recipients to spend them in the market as they see fit. American
economists were not antigovernment, but they showed a decided preference for market allocation of resources and of goods or services over
government "interference." Such a preference probably explains why
nearly three-quarters of the American economists rejected wage-price
controls to contain inflation.[4]

In another, later poll surveying attitudes toward "government involvement with troubled basic industries," more than one-half of teaching economists chose a free market solution—"no [government] involvement and more reliance on imports." As in England, there was
displayed the apparent anomaly that the further a respondent was from
the daily workaday world of business, the stronger the support of the
market. Only 39 percent of business executives, 21 percent of government officials, and 11 percent of union leaders revealed such faith as
economists showed in the market.[5]

The strong support economists give to the market is based on their belief in the putative outcome of free, competitive exchange among individuals. Adam Smith described it more than 200 years ago.

> [An individual] neither intends to promote the public interest, nor knows how much he is promoting it. . . . he intends only his own gain, and he is in this, as in many other cases, led by an invisible hand to promote an end which was no part of his intention. Nor is it always the worse for the society that it was no part of it. By pursuing his own interest he frequently promotes it. I have never known much good done by those who affected to trade for the public good. It is an affectation, indeed, not very common among merchants, and very few words need be employed in dissuading them from it.[6]

As the surveys indicate, economists believe the market serves the public interest as well as individual interests. It reconciles the two, they maintain, providing an effective way to check greed while allowing individual freedom to flourish.

Economists offer two major arguments in favor of the competitive market: First, it provides a more efficient allocation of resources than any alternative, and second, it offers more individual freedom than nonmarket systems. Almost all economists emphasize the first argument. Efficiency is a value with which most are technically, as well as philosophically, more comfortable than with freedom; it is a value they confidently believe the market serves exquisitely well. The reader should beware, however, that economists' definition of *efficiency* is not the same as the everyday meaning of the word, an output per unit of input. Economic efficiency contemplates the allocation of market goods and services to uses most highly valued by individuals, given the structure of prices and any effective demand in accordance with a particular distribution of income. An economy like Brazil's with extremely unequal distribution can be as economically efficient as one like Sweden's with a relatively equal distribution. As long as the markets serve consumers possessing the same effective demand equally well, the economies will be equally efficient.[7] The reader should note that this definition of efficiency allows room for a contribution to, but is not the same thing as, increasing productivity or economic growth.

The Defense of Efficiency

Arthur Okun, in a popular economic essay *Equality and Efficiency: the Big Tradeoff*,[8] borrowed the popular notion of efficiency as

> . . . getting the most out of a given input. . . . If society finds a way, with the same inputs, to turn out more of some products (and no less of the others), it has scored an increase in efficiency. This concept of efficiency implies that more is better, insofar as the "more" consists of items that people want to buy. In relying on the verdicts of consumers as indications of

what they want, I, like other economists, accept people's choices as reasonably rational expressions of what makes them better off.[9]

He, like most economists in Western societies, was convinced that though the market falls short of the competitive ideal, it has long since proved itself to be the most efficient organizer of production in practice as well as in theory. The plethora of consumer goods and their ever-changing variety and improving quality, in addition to the high level of incomes in the United States, Western Europe, and Japan, are evidence of a high level of efficiency in these free-market economies. Most citizens of centrally planned economies would probably agree; their own economies have demonstrated a marked inability to foster productive efficiency.

By resting their case for the market upon efficiency, economists proceed in a "somewhat ethically evasive" way, according to Professor Thomas Schelling,[10] for they stress ways to secure the largest output, and tend to ignore how that output is distributed. Their only professional concern as economists is for allocative efficiency, which is usually focused on production. As advisers they can help both private and public policy makers to select programs that will minimize waste and maximize return from the resources used. They can thereby assure their clients that at any given cost to the rich, there will be more available, possibly to be used for the poor. Such assurance is not always persuasive to the public, however. The poor may not secure any portion of the increase, but the rich get richer, absolutely and the poor, while not worse off absolutely, discover they have lost relatively.

Furthermore, if the rich are very rich, the poor very poor, and the middle class small or absent, a market system will still produce "efficiently." Efficient markets may operate, after all, in countries where many people are starving and suffering from acute deprivation while a few families enjoy the good life, in an El Salvador, for example, or a South Africa or a Zaire. The benefits of an efficient market, under such circumstances, may not be impressive to those who realize the social tensions and political stress within any community that is burdened with a wide gap between rich and poor. Unless the poor actually improve their situation, in the market, the claim that improvement may be possible is not likely to receive much credence.

A free competitive market allocates resources efficiently in any kind of society, in accordance with effective consumer demand. In an El Salvador, for example, the few rich can afford expensive automobiles and villas and import fancy foods from abroad. The poor earn so little that they can hardly afford a basic bicycle, a tin-roofed shack, and locally grown beans and corn. With a small or nonexistent middle class, the market will furnish few goods and services in between these two kinds of demands. It will, nevertheless, provide Cadillacs for the rich and perhaps some bicycles for the poor; in so performing, it will be *efficient* as economists use that term. It will provide the goods and services for which there is an *effective* demand. But one would be surprised if many of the disadvantaged poor—the majority—in those countries or if many ordinary Americans, viewing them from afar, are impressed by such market efficiency in that setting. Market efficiency probably commends itself to Americans when it is present in societies that already enjoy an acceptable measure of equity, manifest tolerable fairness in economic affairs, and can be perceived as reasonably just. As a value in and of itself, however, efficiency is commendable only secondarily.[11]

If this conclusion is valid, economists may not be able to provide as general a defense of the free market—with convincing arguments to answer critics—as managers might hope. As noted in Chapter 4, the success of the economy in providing a cornucopia of goods and services along with a generally improving standard of living for most people in the United States also creates difficulties, generates problems, and perhaps even insures social failures. The claim of the economists for efficient production is valid only if certain social preconditions are met— if there is a general agreement on, or at least movement toward, widely shared notions of social justice. The acceptability of market efficiency is thus dependent upon the large size of the American middle class and the relative abundance of opportunity in the economy. It is the resulting American emphasis on justice that makes market efficiency a more compelling value and a better defense for business than it is in many poorer foreign lands.

Managers may take for granted that their business operations are efficient and that they operate in efficient markets. They are then almost sure to presume that their critics can only suggest changes that will interfere with efficiency, lessening or taxing it in some undesirable way. If firms already and always operate as efficiently as they can, there is no room for improvement. Critics, however, may be less impressed with the claims of efficiency than they are with their concerns for social justice. Managers may need to defend their efficient operations, but they may wisely be prepared to deal with issues of social justice as well.

Hindrances to Realizing Market Efficiency

Managers find it comfortable to assume that market efficiencies are but the sum of—the reflections of—their own and their companies' efficiencies. In protecting their policies, company programs, and industry procedures, managers convince themselves that they are thereby protecting the efficiency of the economy. The relationship between a firm's and the economy's efficiency may, of course, causally run the other way; firm efficiencies may arise to a significant degree out of public

education, public streets, and a court system that protects property and
enforces contracts. An increasingly sophisticated public, which has
learned in the past two decades to be suspicious about the pronouncements of those in authority, will not necessarily accept the managerial
assumptions. It will want to be shown in fact as well as in theory and
after-dinner talks that firms are truly efficient.

An efficient market in general is no guarantee of efficient markets in
particular; because firms do not operate at peak efficiency at all times,
especially if they are large. Corporate raiders have made the case that
the firms they target have not been operated efficiently and that new
managers, using different methods or styles, can do better. A study by
Professor Frank C. Lichtenberg supports this assertion. He found that
takeover targets exhibited significant improvements in productivity and
operating efficiency after changes in control. Most companies prior to
an ownership change showed poor performance, lower than average
and declining. After the takeovers, however, productivity increased.[12]

Managers may not keep up with technology, which continually opens
new possibilities for market efficiencies; they may maintain styles of
operating unsuited to a changing social environment. Managers obviously do not always pursue market efficiency, maintaining their efforts within the demands and requirements of free competition. As already indicated, managers continually search for shelters from the fierce,
blustery winds of competition; if they do not find them ready-made,
they are not above constructing them. A number of market failures or
noncompetition conditions help them in their search and construction.

Lack of information and externalities. First, consumers may not be well
informed about the products they choose. On the one hand, technological advance provides goods so increasingly intricate in operation and
complex in consequence that few consumers can be truly informed
buyers. On the other hand, the worldwide spread of the market has
introduced sophisticated goods to simple people, unused to modern
dangers in their use. For example, illiterate mothers, living in rural
areas and innocent of the wiles of high pressure hucksters, can be persuaded to buy an infant formula that is more costly than they can afford and that may even be dangerous to the health of their babies.

Second, spillover effects or externalities—nonmarket costs of pollution, for example, arising from the production and sale of good—may
be significant. If they are, producers will not pay the full costs of production, thus enabling them to charge a lower price and sell more than
they otherwise would. Consider the widely sold disposable diapers that
wind up in rapidly filling land dumps; because of the plastic they contain, they will deteriorate only after decades or centuries. No doubt
firms producing the diapers are profitable, and consumers find them
useful, but in providing them, the economy is probably using its re-

sources inefficiently. In a world where the spillovers of crowding and pollution have become serious problems endangering health and comfort, most people realize that narrowly measured efficiencies of polluters and the rewards they take for themselves and their customers can mock the larger promises of an efficient market.

Declining costs of large scale—Oligopolies. Third, economies of large scale can undermine competitive outcomes and deny efficient allocation in the market.[13] Professors Geoffrey Heal and D. J. Brown suggest that economies where costs decline with larger output are much more common than most economists assume. Outcomes are quite different when economies of scale are present from when costs are rising. Economies of scale can invalidate many conventional assumptions about the operations of the market.

For example, in the first third of the century as oil fields were being discovered and exploited in the contiguous United States, economies of scale led to much waste of resources. Once a driller discovered an underground reservoir, the oil belonged to whoever could pump it to the surface. The more wells drilled, the more and faster one could "capture" the oil, at least for a short time; the more oil taken out of the reservoir and sold, the lower the average costs, for the fixed costs of drilling and labor were spread over more units of production. The result was the impairment, if not destruction, of the reservoir, long before its full use. In addition, companies producing the oil pumped it out in such quantities that they depressed the market, selling for absurdly low prices. On some occasions and in certain places the glut was so vast that producers actually burned the oil or drained it into creeks, simply to dispose of it.

Another example of inefficiencies arising out of economies of scale can be found in the American automobile industry. In the three decades after the end of World War II, the industry became increasingly oligopolisitic, squeezing out the smaller, marginal producers, as the largest firms increased their market share. Given their primary market in the United States, the larger manufacturers could produce more cheaply than smaller American auto firms. Once they had achieved their dominating, oligopolistic position in the domestic market, however, the three large firms were under little competitive pressure to increase or even to maintain their efficiency. The managers no longer carefully tracked the changing demand for their products and left markets wide open for foreign entry. European and particularly Japanese manufacturers offered new, less costly, and higher quality products, highlighting the degree to which the American firms had long escaped market pressure to operate efficiently.[14]

When Japanese auto producers captured a significant portion of the domestic market, the managers of American firms had to change some of their smug noncompetitive practices that had raised their costs far above those of the Japanese. In early 1982, General Motors, for example, announced that henceforth it was going to buy steel through competitive bidding, a practice that one would have expected firms in a free-market economy to have been doing all along. Suppliers bidding competitively against each other would seem to be a more efficient means of purchasing than simply allocating shares of corporate buying among a certain set of suppliers. General Motors told its steel suppliers that it would buy from them on the basis of price, quality, and timely delivery.[15] Since GM ordinarily purchased over 7 percent of the steel industry's domestic shipments, the impact of its newfound competitive rigor was widely felt. Steven Flax, writing in *Fortune*, made clear the long-time managerial laxness that the change to bidding revealed.

> GM's shift ended an era of stunning complacency. Its steel buying had become an automatic process, conducted with the soothing reassurance of a familiar ritual. At the beginning of each year a supplier would be awarded a fixed percentage of GM's needs for particular steels at particular plants. "We did the same thing the same way every year," says Gus W. Rylander, a 27-year sales veteran at Armco. "We'd go up there [to Detroit] and get our share of the pie." . . . As Merrill Lynch steel analyst Charles A. Bradford puts it, "The steel companies' attitude was, 'We make steel. If you want it, you buy it.'" GM, by the same token, didn't want to hear from any steel maker who thought he might have a better or cheaper way of doing things . . . Neither side paid scrupulous attention to quality . . . At the same time GM continued to pay suppliers' list prices even though it had the clout to bargain. Banking on the consumer's willingness to absorb price increases, the company chose to pass costs on rather than disrupt the comforts of doing business as usual.[16]

Before the auto manufacturers turned to competitive bidding, they assured their suppliers that each would receive their accustomed share, and to meet increased costs, they would buy each succeeding year at a somewhat higher price. No one worried much about quality. Suppliers did not have to concern themselves with research and development, as long as they supplied extras as an allowance for defective parts.

With the coming of competitive bidding, however, suppliers had to seek efficiencies, rationalize their operations, sharpen their selling skills, improve the quality of their products, and cut costs. The new effort won efficiency gains though Japanese manufacturers maintained their lead, continuing to increase quality and to cut costs. American manufacturers have not found it easy to catch up to the ever-rising levels of market efficiency. It was an ironic situation that the steel, parts, and auto producers had made for themselves; they had long rewarded themselves and been hailed by industry observers as if they were efficient, deserving business leaders. Their efficiency, their quality, and their effort have been revealed as impressive only because they had little or no competition, not because in fact they had been competitive.

The example of the auto industry indicates that in the long run the market did respond efficiently, as economists measure it. When American auto producers realized that they were losing large sales to the Japanese and might well lose them permanently, they cut costs sharply, pressed for wage concessions from their employees, and improved the quality of their products significantly. In the process, they and their suppliers, including steel producers, had to close old and inefficient plants, lay off tens of thousands of employees, and adversely affect communities all over the nation, but especially in the Midwest, an area that became popularly labeled as "The Rust Belt." When the auto producers were enjoying their noncompetitive position, auto buyers paid higher prices and received lower value than a free market would have assured; but they provided many benefits to many constituencies. When they began to respond to competition, the market required actions that had widespread and destructive economic and social impact on those same, formerly favored constituents.

The largest American firm for many years, popularly assumed for many years to be an efficient producer, also turned out to be a wasteful producer. The deregulation of the telephone industry brought about the breakup of the gigantic American Telephone & Telegraph Company. When confronting market competition it had never known before, its inefficiencies became all too noticeable. Writing in the *New York Times*, Peter W. Barnes noted that: its managers had long rated manufacturing concerns higher than response to customers. They only slowly introduced new phones, offering durability, not novelty. They tolerated the highest labor costs in the telecommunications industry; for equipment installation and maintenance, AT&T paid 84 percent more than IBM and 118 percent more than MCI. In addition it tolerated a huge managerial staff, and even in 1984 had ten levels of managers between entry-level supervisors and the chairman.[17]

The High Costs of Market Uncertainty and Instability

A fourth condition that limits the public acceptability of market efficiency is that of uncertainty and instability. The market continually fluctuates. In boom times, output, employment, prices, and incomes tend to rise; in recession, they tend to decline. Popular attention focuses upon recessions' unemployment even though its incidence bears heavily on a very small portion of the total work force at any particular time. Claims for a market efficiency that appears to need or depend upon such unemployment—certainly a waste of any society's most valuable resource—are simply not well received by the public.

Economists, it would seem, offer business managers a defense that may not fit well with business practices or with public judgments. As consumers, the public at large benefits from market efficiency if it enjoys an income with which to purchase. But many among the public also benefit when firms can escape market pressures and efficiency slips. Managers and presumably stockholders, employees, suppliers, and the communities in which they live and spend their stable, certain earnings, all may share in the rewards won by a noncompetitive but secure firm. Such was the case for years with General Motors. When competition pushed its managers to pursue efficiency, as it did in the early 1980s, heavy costs were borne by those who previously benefited. In 1980 there were roughly 30,000 direct and indirect auto parts suppliers, most of them small firms, privately owned and important contributors to the communities in which they were located. Following the introduction of competitive bidding, half are expected to disappear or be forced to take up some other line of production by 1990. The pressure upon the workers involved will be greater than ever, and many will lose their jobs; those who manage to hold on will find it impossible to maintain high wage levels, let alone win increases.[18]

Despite the qualifications with which they hedge efficiency, economists strongly support a market allocation of resources. Some prefer not to rest their case entirely on the gains that supposedly accrue to customers, for they are too problematic. They bolster the case with another value of greater price: freedom. Perhaps business managers can find in *freedom* a market value to use in refuting their critics and to defend their business policies, market procedures, and organizational position.

FREEDOM AND THE MARKETPLACE

Two well known economists, F. A. Hayek and Milton Friedman, both Nobel laureates, have emphasized freedom—or liberty—as the preeminent value of the market. Other economists are not unappreciative of the freedom allowed in and encouraged through the market, but most do not find it as compelling as does Hayek; even Friedman quietly qualifies freedom as a market value. In Hayek's fullest defense of the market in *The Constitution of Liberty*,[19] he focuses continually upon freedom—"The Value of Freedom," "Freedom and The Law," and "Freedom in the Welfare State" are the three main sections of the book. Friedman entitled his shorter but probably more popularly influential book *Capitalism and Freedom*.[20] He argued in it that the competitive market is essential for economic freedom, which in turn provides the necessary basis for political freedom. Only when "economic power is kept in separate hands from political power, [can it] serve as a check and a counter to political power."[21] He asserts, without proof, demonstration, or argument, that "freedom of the individual, or perhaps the family [is] our ultimate goal in judging social arrangements." Though ultimate, he admits it is only an instrumental, and thus limited, goal.

> In a society, freedom has nothing to say about what an individual does with his freedom; it is not an all embracing ethic. Indeed, a major aim of the [classical] liberal is to leave the ethical problem for the individual to wrestle with. The "really" important ethical problems are those that face an individual in a free society—what he should do with his freedom.[22]

One might assume that Friedman perceives individual freedom in a market economy as bounded primarily by one's own ethical decisions. Obviously, the matter is hardly that simple, as Friedman implicitly recognizes. Both the laws and regulations of the state, as well as the rights and entitlement of others, impose significant limitations upon the use of one's freedom. Even in a free market, and perhaps particularly in a free market, dependent as it is upon private property rights, individual freedom is shaped and molded by the boundaries of others' freedoms. Thus, many critics of the competitive market system, and even knowledgeable supporters, doubt that its freedoms are as significant and as far-reaching as Hayek and Friedman presume. Karl Marx developed these ordinary observations into a full-blown ideology that defined personal freedom as real only in the context of just social relationships. But one does not have to be a Marxist to perceive that whether excessive or minimal, personal human freedom always has its boundaries.

Boundaries on Freedom

First, government aids and assists some citizens while restricting others. Not only does it redistribute income through various "welfare" programs, from social security to defense contracts, but it also restricts and regulates the right to produce and to sell certain products or services. It declares illegal the production and general sale of products like heroin and "crack," though it allows other dangerous or poisonous products such as cigarettes and alcohol to be widely sold at retail. It licenses a variety of trades—funeral undertaking, taxi services, medical, dental, and even such construction trades as plumbing and electrical contracting—allowing entry only to those who meet certain qualifications.

State, local, and federal governments create property out of thin air and grant use of it to some and not to others. Broadcasting licenses for radio and television are one example, and patents and corporate charters are another. It entitles some people—minority members, women, the aged, or veterans—to special consideration through a variety of laws or gives special rights through Department of Agriculture marketing orders to such groups as the growers of kiwi fruit and oranges. It also provides advantageous tax benefits for home owners but not for apartment dwellers; and it bails out whole industries like savings and loans, and some bankrupt firms, such as Chrysler, third largest auto manufacturer, or the tenth largest bank, Continental, but not the local grocery store, the nearby pizza parlor, or a failing bookshop.

Second, most people find their freedom bounded by their income and wealth; their personal household budget sets the range of many of their freedoms. Examples of free choice, often favored in economic texts, focus upon common consumer goods, so inexpensive and readily available that choice is an option for all. Typically, Friedman points out, men can choose among many different colors of ties.[23] He is correct, but surely the choice is a trivial and unimpressive test of market freedom. Critics and those unused to the market system may find such choices to be more a measure of waste than an indication of freedom. A young Mozambican, after observing the American scene commented:

> I simply do not understand America. It is the moon. The rest of the world is dealing with problems of production and distribution and you Americans seemed threatened by waste, by disposal. Tell me, do you really have twelve different kinds of toothpaste? Why?[24]

Americans can buy at least twelve and probably many more brands of toothpaste. They might reply to the young African that they presume consumers enjoy the various brands, otherwise they would not spend their money on them. In making their purchase, consumers make it worthwhile for producers to supply them. Those who buy the different kinds of toothpaste probably use them for a good purpose—at least one can assume that the use satisfies a consumer's perceived need or the consumer would not have bought the toothpaste. To an advocate of the free market there is not waste involved, only an elaboration of a product that matches consumers' expressed, and voluntary, demands. The Mozambican's question is not one that will bother most Americans, but many will agree that having many brands of toothpaste from which to choose hardly proves the existence of a significant freedom.

Budget constraints and effective choice. Not only must a variety of goods and services be available if freedom is to be a reality, but one must also enjoy an *effective* choice. With no money in one's pockets, all the variety of the market reduces to a nullity, eliminating freedom of choice. For example, the housing market offers a wide variety in the kind and quality of shelter, but without an adequate income or a good job and suitable credit rating, one may not be able to afford to choose a detached house in the suburbs. Likewise, a supermarket filled with an amazing variety of foods offers but a limited range to one whose monthly income will allow only the minimum basics. One may prefer prime steak for dinner regularly but have to be satisfied with bologna or chicken wings because the cost of steak is too great for the available take-home pay or monthly Social Security check.

Since budget constraints impose boundaries upon freedom of economic choice, one's freedom is qualified and conditioned by one's location in the distribution of income. Along with income, one's political and social position also importantly affects available freedom. Market advocates must either deny the significance of income distribution or of sociopolitical position in the enjoyment of freedom, or they must assume a particular income distribution that provides the appropriate freedom for most people. Only a few economists such as Hayek, who holds uncompromising libertarian views, are willing to rest their case for the competitive market solely on its promised freedom; even Friedman bolsters his defense by unobtrusively inserting into his argument an element of equity, a governmental, nonmarket redistribution to provide minimum incomes through a negative income tax.

Most economists probably agree that the market encourages and allows much individual freedom; unfortunately, the incidence of freedom is as varied as is the distribution of income. At best, one can point to no more than a rough and ready relationship of income and merit, and at worst, there is simply no connection.[25] It is hardly surprising, therefore, that a defense of the market based solely upon the value of freedom is less than convincing to the public, unless it had already agreed that freedom is the highest of all social values. As already noted, most economists probably agree that the market *can* encourage and even allow much individual freedom, but incomes are far from equal and so the freedom to chose in the market is exceedingly varied. All people have to live within their budgets, and poor people's choices are relatively, and sometimes seriously, limited.

THE ROLE OF FAIRNESS

Hayek strongly insists that those who examine the effects of budget limitations on freedom introduce an irrelevant value into the analysis of the competitive market. They concern themselves with merit or justice—the question of "who gets what"—which, he maintains, is logically separate, and categorically different, from freedom.[26] Justice introduces an extraneous and irrelevant value into the market system. It is an inappropriate introduction, he argues. Merit applied in the market will confuse those who work within it and can only impair effective market processes.

> . . . in a free system it is neither desirable nor practicable that material rewards should be made generally to correspond to what men recognize as

merits and that it is an essential characteristic of a free society that an individual's position should not necessarily depend on the view that his fellows hold about the merit he has acquired. . . . The fact is, of course, that we do not wish people to earn a maximum of merit but to achieve a maximum of usefulness at a minimum of pain and sacrifice, and therefore a minimum of merit. Not only would it be impossible for us to reward all merit justly, but it would not even be desirable that people should aim chiefly at earning a maximum of merit. Any attempt to induce them to do this would necessarily result in people being rewarded differently for the same service. And it is only the [market-determined] value of the result that we can judge with any degree of confidence, not the different degrees of effort and care that it has cost different people to achieve it. . . . It would probably contribute more to human happiness if, instead of trying to make remuneration correspond to merit, we make clearer how uncertain is the connection between value and merit. We are probably all much too ready to ascribe personal merit where there is, in fact, superior [economic] value.[27]

Friedman is not as definitive and sure as Hayek in excluding merit, or justice, from the market; he neither rejects nor explicitly embraces it. However, he implies and thus assumes a kind of minimal egalitarianism that requires government to redistribute some income. He recognizes that those with no income or wealth possess little freedom within the market. Insofar as they cannot buy or sell, they can cast no "vote" and play no role in the market. He was apparently bothered by the observation that large numbers of individuals enjoy little or no market input, and as a consequence the market does not reflect their preferences. He presumes an inclusive society for his economy.

Friedman illustrates his "free private enterprise economy" with a primitive example:

[the market] . . . in simplest form . . . a number of independent households—a collection of Robinson Crusoes, as it were. Each household uses the resources it controls to produce goods and services that it exchanges for goods and services produced by other households, *on terms mutually acceptable to the two parties to the bargain.* . . . *Since the household always has the alternative of producing directly for itself, it need not enter into any exchange unless it benefits from it.*[28]

His self-reliant families—his independent households—had some freedom of choice, for they possessed an ability to enter the market or to stay out. The families were not dependent for their very livelihood upon whatever terms and conditions the market supplied at the moment. Such a freedom suggests that all households in Friedman's market must possess some bargaining power. All of them, even the poorest, receive at least a minimum income and do not live hand-to-mouth. Each is protected against the very dependencies on large-scale economic organization that, as we have seen, in fact dominate the economic relations of over 90 percent of employment Americans.

Fairness as a Guarantee of Minimum Income

That Friedman carefully chose his illustration and that it reflects a social value underlying his market notion of efficiency and freedom is clear from the policy recommendations that he made in later chapters of *Capitalism and Freedom*. He proposed what many casual readers thought was a "liberal" and thus "un-Friedmanlike" scheme of income redistribution: the negative income tax (NIT). Under NIT the poor would fill out income tax forms as does everyone else, and if a family's income fell below some predetermined level, the U.S. Treasury would write it a check to bring the income up to a minimum. Other, richer people would pay their graduated taxes to the Treasury. His assumption that all members of an economy should have at least some minimum income, and thereby an ability to participate in the market, puts him and his NIT proposal in the classical liberal tradition of the free market. He wanted to help the poor use the market to serve themselves. Then they could choose with some freedom, avoiding a governmental bureaucracy that paternalistically determined the goods and services for the poor.

The NIT is not necessarily "liberal" in the contemporary political sense at all. It is a program based on Friedman's free-market assumptions. All persons (or families) can participate in the market only if, and insofar as, their incomes allow them to choose to buy and sell in it. To assure full participation, all families must have at least some income. NIT would enable all to participate at least minimally.

In advocating NIT Friedman reveals that he had secreted into the very foundation of his market defense a sense of justice. It is a concern for the poor, who might otherwise be left outside the system. He implicitly recognizes that market efficiency and freedom need the accompaniment and fundamental support of at least some sense of fairness. He appears to advocate, with some reserve it is true, a kind of democratic egalitarianism. He does *not* espouse equality of income, one must note, and he passes no judgment on the merits of any particular market return—salary, wages, and earnings—received by those in various occupations or that the different factors of production garner for themselves. Nevertheless, he clearly believes the market works best only when the poor at the bottom have some assured minimum income.[29] His NIT would provide that minimum, improving the market and serving individual freedom at the same time.

That a popular, learned American economist should temper his economics with a concern for a kind of egalitarian minimum is not surprising. Ordinary Americans have long prized fairness, though not equality, in the distribution of income. They want a nation in which all may strive for any position and all may participate at least minimally in economic as well as political affairs. They manifest their concern for fair treatment and egalitarian values whenever, as an electorate, they can exert influence upon policies and programs.

Fairness as Equal Sharing in a Time of Trial

Such concern was revealed in the program of meat rationing during World War II, for example. If asked why the government rationed meat more or less equally during the war, most people would probably reply that it must have been in short supply. That is hardly an adequate answer, though, for all economic goods are in "short supply," and must be "rationed." Usually the market carries out the rationing, allocating supplies according to those who will pay the necessary price. When rationing was introduced, the Secretary of Agriculture, Claude Wickard, implicitly endorsed the "shortage" explanation, pointing to a need for conserving supplies. The government decided to allot each person a quota of ration coupons with which to buy a certain quantity of meat. Secretary Wickard asserted that rationing would cut meat consumption by 25 percent.

In fact the productiveness of farms allowed Americans to *increase* their consumption, even in the face of greatly expanded demand elsewhere for meat. Absolute production of meat rose extraordinarily during the war; American farmers were able to supply all the meat needed by the military for its twelve million soldiers, sailors, and marines, who probably enjoyed more regular meals with meat than ever before in their lives. Farmers produced meat in such quantities that the government not only supplied its armed forces but furnished it through Lend-Lease to its allies around the world; in addition, meat was in such ample supply that the civilian population at home was able to increase its per capita consumption during the war years by an average of more than 7 percent! In 1943 civilian consumption per person rose to 153 pounds, higher than at any time in the preceding forty-four years, and far higher than any of the Depression years through the 1930s. The economy had never before produced meat in such quantities.

It was not in "short supply" compared to the past, though prosperous war workers probably wanted to buy even more meat than was available in butcher shops. In general, many wage earners found themselves unable to buy goods during the war that for the first time in their lives they could afford in the price-controlled economy. Other observers were more candid than Secretary Wickard about the purposes of meat rationing. Chester Bowles, Connecticut's rationing commissioner, who would later be director of the Office of Price Administration (OPA) for the federal government, declared in 1942 that "if uneven or unfair distribution of necessities causes bitterness, then our home morale will suffer drastically."[30] The first chief of the Ration Banking Branch, OPA, agreed, pointing out that a "maldistribution of food would react unfavorably on the war effort."[31]

Not all public commentators understood the importance of "fairness" in distributing basic foodstuffs during wartime. The editors of *Time* agreed that rationing sugar made sense, but they protested meat rationing.

> Rationing meat is a very different problem from rationing sugar. The poor normally eat almost as much sugar as the rich and so rationing everybody the same amount of sugar per week makes sense. Meat, on the other hand, is a food which the rich normally consume over three times as much as the poor.

They went on to complain that for more than 50 percent of the U.S. citizens whose income was less than $2,100 a year, the meat-rationing scheme meant eating more meat than ever before. They drew sympathetic attention to the wealthy (in that time, families with incomes of $5,000 or more a year). They would have to be satisfied with eating 40 percent less meat.[32] A major newsmagazine would not be likely to argue so nonegalitarian a position today. But even in 1942 it reflected a minority sentiment. In time of crisis and emergency, when the whole American nation needed to pull together and when most people wanted to emphasize their sense of community, the people accepted—even welcomed—shared "sacrifices," even if only symbolically, on the home front and on the battlefield. There was more meat available than ever before, but Americans found the gross inequalities of a market allocation unsatisfactory; in such a time they preferred an equitable sharing through the use of government ration coupons.

In the early 1970s Americans showed their readiness to turn to equitable sharing through government rationing when it appeared that imports of petroleum might be drastically curtailed. Congress approved and the federal government made ready for rationing civilian supplies of gasoline. It even printed millions of coupon books, which eventually were destroyed, unused and unneeded. Economists then and since pointed out that rationing was not an efficient method of allocating gasoline, but to little avail.[33]

Efficiency does not appear to be the dominant concern of those who speak and write publicly, however. One could expect the president of a United Automobile Union local to say, "It's the little guy who's going to get it. I'd rather see them ration gas than rise the price so the little guy can't buy it." One might even expect an editorialist for the *New*

York Times to write that "Rationing is fairer . . . there is a serious question of economic equity involved . . . those who do not need gasoline—whatever their income or wealth—should be required to do without." But it may be surprising to read in the conservative columns of the *Wall Street Journal* that while government rationing poses many problems, it is probably inevitable, and, provided the coupons are transferable, fairly sensible, and probably tolerable.[34]

Fairness as Egalitarian Relative Pay

Where a broad electorate can influence pay structures, an element of applied fairness is also evident. The salary scales in local, state, and federal governments show a marked egalitarian bias. Those in the lowest ranks receive higher pay, and those in the higher ranks receive considerably lower pay, than those in comparable private sector positions.[35] Almost half the public believe that elected public officials are either underpaid or worth their pay. This approval is high compared to the public's attitude toward the pay of those in other professions. Only about one-fourth believe that top corporate executives or show business and television stars are worth their pay; fewer than one of every five (19 percent) believe that professional athletes are worth their pay.[36]

Another example of the American electorate's preference for egalitarian pay is found in labor unions. The pay of union leaders, even of officers in the largest unions with up to a million members, is noticeably smaller than that of the top officers in the companies with whom they deal. In 1987 the AFL–CIO paid its president Lane Kirkland, only $150,000, and the Air Line Pilots paid their president $243,382, with the typical large union paying their top officer around $100,000.[37]

Business Week reported the average salary of 678 business executives surveyed in 1987 just short of one million dollars ($965,617), a figure that excludes long-term pay.[38] That is more than nine times the pay of the union presidents. If corporate managers were elected to office as democratically as most union leaders, one might well expect their pay to be considerably lower. Having to answer only to members of their boards of directors, whom they appoint, and confronting the electoral power of stockholders only through the less-than-democratic annual meeting procedures, they can largely ignore the widespread American sense of "fair" pay that the opinion surveys indicate. In a 1984 survey *Business Week* and Louis Harris & Associates, Inc., found that a large majority, more than three-quarters, of Americans believe that top corporate executives are overpaid.[39] Even editors of business publications like *Fortune* agree: ". . . top-level compensation doesn't make much sense."[40]

There is no evidence that the public favors equality of pay; it approves high pay and large bonuses to managers in industries that appear to be efficient, productive, and innovative. Managers in the computer industry, for example, merit their pay, according to the respondents in the polls, but those in the telephone, communications, automobile, steel, and oil industries understandably rank near the bottom. Those executives are perceived as not meriting the pay they granted themselves.

Peter Drucker warned managers that they had better pay attention to the public sense of "fairness" in pay. Since the relationship of executive compensation and company performance is erratic and uneven few Americans find much reason for executives' high pay. Japanese managers, he pointed out, receive far smaller compensation, both absolutely and relative to ordinary workers' pay. The very high pay American business executives paid themselves offended most managers as well as the public, in his opinion.[41]

Drucker's warning was not markedly differently from the arguments of business's regular fault finders, such as those associated with Ralph Nader, in the Democracy Project.[42] That the public generally, and both pro- and antibusiness critics, should find common ground in deploring the level of executive compensation suggests there is a general and prevailing American agreement on a standard of "fairness" in the market system. Some form of justice—not mere efficiency—should prevail. To convince the public that business managers deserve their compensation, the people will have to be persuaded that the executives and their companies are truly efficient contributors to economic welfare. As already noted, the managers of a number of the nation's leading and large firms have not been able to provide that conviction.

The public concern over executive pay in 1984 was stimulated by a perception that the top executives of Ford and General Motors paid themselves far beyond any notion of merit, or perhaps even in an absence of merit. Early in the year they announced very lucrative bonuses, with a number of top officers in each company receiving more than $1 million. True, 1983 profits *had* increased significantly over earlier years, but a government quota on Japanese imports significantly aided the rise in profitability. The trade restriction allowed the American firms, (1) to raise their prices, and (2) to sell more large and intermediate cars on which the markups over costs were highest. Those sales were stimulated in part because consumers could not get their fill of the cheaper, higher quality Japanese models excluded by law. There was widespread belief that the company executives were exploiting a publicly given favored position. Such exploitation, understandably, was not perceived as either fair or equitable; it was not acceptable within the terms of the implicit values Americans expect to see expressed in the market systems.

EFFICIENCY AND FREEDOM ARE NOT ENOUGH

A number of scholars have cautioned business managers that they cannot expect to find either capitalism or the business system well defended by the arguments of economists. Ernest van den Haag, a strong supporter of capitalism, has written:

> Justice is as irrelevant to the functioning of the market, to economic efficiency, and to the science of economics, as it is to a computer or to the science of meteorology. But it is not irrelevant to our attitude toward these things. People will tolerate a social or economic system, however efficient, only if they perceive it as just.[43]

He argued that there is "an uneasy and unstable relation between economic and moral valuation." In the past, religion mediated that relationship in the United States, particularly that part of religion known as the Protestant ethic. It emphasized the individual and personal values of frugality, industry, sobriety, reliability, and piety, providing a moral justification for even the harshest market outcomes—loss of job, unemployment, or serious injury. These were matter of individual responsibility, not the market. The ethic also underwrote the notion that there was a direct and causal link between these values and the distribution of power, privilege, and property. Many in the public were willing to believe that economic success might even be a sign of divine approval.

Irving Kristol, also a strong and vocal defender of the business system, has pointed out that such identification in the past allowed believers to presume the market to be just, not merely free. He writes, "Samuel Smiles or Horatio Alger would have regarded Professor Hayek's writings as slanderous of his fellow Christians, blasphemous of God, and ultimately subversive of the social order."[44] Kristol doubted that many modern business leaders and rising young managers would be as outraged by Hayek's separation of merit and the market as Smiles or Alger would have been; many of them do not possess religious convictions that allow them to find divine sanction for the market. Moreover, he did not find that they possess any source of justification for their, and their corporation's, performance. The traditional "test of the market" is unconvincing to many, for it appears in practice to be a random rewarder; *Fortune* is not a strong god to provide a modern justification for whatever success one encounters. Simply to accept whatever return the market happens to provide means accepting things as they are because they have been accepted in the past. Unfortunately for those who hope to rely upon tradition, they find it is a social asset continuously and rapidly eroded by the dynamism of capitalism.

Managers thus face a serious problem: How can they justify themselves in the American society? The values of efficiency and freedom offered by the economists are all too limited and qualified to provide an effective apologia. Although many Americans still believe in the divine justification of the market and business system, they are probably a far smaller share of the total than a century ago when Andrew Carnegie wrote his widely admired *The Gospel of Wealth*.[45] Far too many other Americans, including a large number of business managers themselves, are willing to tolerate market outcomes only if they can modify them. There exists widespread, even if tacit, understanding that market efficiency and market freedom need the leavening of other values, particularly those that generally fall under the rubric of justice.

Managers, then, will have to defend themselves not only with claims—and proofs—of efficient production, but also with the arguments of fairness and the reasons of equity. Economists may be able to help in supporting the claims, but they offer little in advancing the arguments and reasons of justice. Managers will have to seek their own understanding of and develop their own sensitivity to the demands for justice as debated, understood, and practiced in the United States.

NOTES

1. Allen R. Janger and Ronald E. Berenbeim, *External Challenges to Management Decisions* (New York: The Conference Board, 1981), p. 7.

2. Amartya Sen, "The Profit Motive," *Lloyds Bank Review* (January 1983):1.

3. Ibid., p. 2.

4. J. R. Kearl, Clayne L. Pope, Gordon C. Whiting, and Larry Wimmer, "A Confusion of Economists?" *The American Economic Review, Papers and Proceedings* 69 (May 1979):30.

5. Opinion Research Corporation Survey, reported by LTV in *Wall Street Journal*, August 31, 1983.

6. Adam Smith, *An Inquiry Into The Nature and Causes of The Wealth of Nations* (New York: The Modern Library, 1937), p. 423.

7. Martin Shubik, "Corporate Control, Efficient Markets, and the Public Good," in John C. Coffee, Jr., Louis Lowenstein, and Susan Rose-Ackerman, eds., *Knights, Raiders, & Targets: The Impact of the Hostile Takeover* (New York: Oxford University Press, 1988), p. 34, labels economists' concept of efficiency as weak. It implies fairness, equity, and justice only insofar as the distribution of property rights may provide.

8. Arthur Okun, *Equality and Efficiency: The Big Tradeoff* (Washington, D.C.: The Brookings Institution, 1975).

9. Ibid. p. 2. [Reprinted with permission].

10. Thomas C. Schelling, "Economic Reasoning and The Ethics of Policy," *The Public Interest* (Summer 1981):53.

11. For a study of an encounter between economists pursuing "efficiency" and environmentalists and industrialists seeking other and conflicting political

goals, and the mutual misperceptions of the economists and the others in the Washington scene, see Steven Kelman, "Economists and the Environmental Muddle," *The Public Interest* (Summer 1981):106–23.

12. Frank R. Lichtenberg and Donald Siegel, "Productivity and Changes in Ownership of Manufacturing Plants," in *Brookings Papers on Economic Activity, Special Issue on Microeconomics,* 3 (1987):643–73. See also, Albert O. Hirschman, *Exit, Voice and Loyalty, Responses to Decline in Firms, Organizations, and States* (Cambridge: Harvard University Press, 1970), for a discussion of how and why slack arises in organizations and may be discovered and remedied.

13. See Geoffrey Heal and D. J. Brown, "Equity, Efficiency and Increasing Returns," *Review of Economic Studies* 46 (October 1979):571–85.

14. See Graciela Chichilnisky and Geoffrey Heal, "Trade and the Evolving World Economy" *Hermes* (Winter 1986):34.

15. "G.M. to Buy Steel by Bid Rather than Contract," *New York Times,* March 24, 1982; Amal Nag, "Armco, Inland Win GM Steel Supply Contracts," *Wall Street Journal,* June 23, 1982.

16. Steven Flax, "How Detroit Is Reforming The Steelmakers," *Fortune,* May 16, 1983, p. 126 [Reprinted with permission.]

17. Peter W. Barnes, "A.T.&T.: Hot Products, High Costs," *New York Times,* August 5, 1984.

18. Amal Nag, "Auto Companies Push Parts Makers to Raise Efficiency, Cut Prices," *Wall Street Journal,* July 31, 1984.

19. F. A. Hayek, *The Constitution of Liberty* (Chicago: University of Chicago Press, 1960).

20. Milton Friedman, *Capitalism and Freedom* (Chicago: The University of Chicago Press, Phoenix Books, 1962).

21. Ibid., p. 16.

22. Ibid., p. 12. [Reprinted with permission.]

23. Ibid., p. 15.

24. *New York Times,* December 30, 1979.

25. See Lester C. Thurow, "Generating Inequality," *Mechanisms of Distribution in the U.S. Economy* (New York: Basic Books, 1975).

26. Alexis de Tocqueville made the same point. "There is good reason for distinguishing [equality from freedom]. . . . The taste which men have for liberty and that which they feel for equality are, in fact, two different things; and I am not afraid to add that among democratic nations they are two unequal things." See Tocqueville, *Democracy in America,* vol. II, (New York: Alfred A. Knopf, Vintage Books, 1945, p. 100). [Copyright 1945 renewed 1973 by Alfred A. Knopf, Inc. Reprinted with permission of the publisher.]

27. Hayek, *The Constitution of Liberty* pp. 93, 96, 98, 99. Another ardent supporter of the free market also explicitly rejected the claim that the market works with any justice or rewards merit. See Frank H. Knight, *The Ethics of Competition and Other Essays* (New York: Harper, 1935), pp. 54–58.

28. Friedman, *Capitalism and Freedom,* p. 13. [Reprinted with permission.]

29. Friedman does not explore the philosophical underpinnings of his economic thought, but he probably would agree with Michael Novak, a theologian who has written on the moral values underlying the American market-system— democratic capitalism—in *The Spirit of Democratic Capitalism* (New York: Simon & Schuster, 1962), 57–58. Novak observes that "Its moral–cultural system also

has many legitimate and indispensable roles to play in economic life, from encouraging self-restraint, hard work, discipline, and sacrifice for the future to
insisting upon generosity, compassion, integrity, and concern for the common
good. The economic activist is simultaneously a citizen of the polity and a seeker
after truth, beauty, virtue, and meaning. The differentiation of systems is intended to protect all against unitary power." [Reprinted with permission.]

30. Quoted in William J. Wilson, "The Price Control Acts of 1942," in *The Beginnings of OPA*, Part I, Office of Temporary Controls (Washington: Government Printing Office, 1947), p. 39.

31. Joseph A. Kershaw, *A History of Ration Banking*, Office of Temporary Controls (Washington: Government Printing Office, 1947), p. 1.

32. "Meat, More for the Poor," *Time*, September 14, 1942, p. 79. [Reprinted with permission.]

33. Professor Thomas Schelling found that his economics students almost always favored rationing of gasoline in case of curtailed imports. See "Economic Reasoning and the Ethics of Policy," *The Public Interest* (Summer 1981):40.

34. James W. Kuhn, "Rationing Again," *The Columbia Forum* (Winter 1974):42.

35. David Lewin and Walter Fogel, "Wage Determination In The Public Sector," *Industrial and Labor Relations Review* 27 (April 1974):410. Those in the lowest ranks receive higher pay, and those in the higher ranks receive considerably lower pay, than those in comparable private sector positions. Also see David Lewin, "Aspects of Wage Determination in Local Government Employment," *Public Administration Review* 34 (March-April 1974):149

36. "Top Executive Pay Peeves The Public," *Business Week*, June 25, 1984, p. 15.

37. "Labor Leaders' Pay in 1987," *Business Week*, May 2, 1988, p. 56. Donald Gough, a principal at Sibson & Co., a management consulting firm in Princeton, N.J., argued that the job of union president is roughly comparable to that of personnel vice-president at companies with up to $5 million in annual revenues. These executives, he says, typically earn $90,000 to $140,000, almost exactly the range paid to the heads of large unions. See "Most Union Honchos Escaped The Squeeze That Pinched The Rank and File," *Business Week*, June 25, 1984, pp. 109–12.

38. John A. Byre et al., "Who Made The Most—And Why," *Business Week*, May 2, 1988, p. 50.

39. "Top Executive Pay Peeves The Public," p. 15.

40. Graef S. Crystal, "The Wacky, Wacky World of CEO Pay," *Fortune*, June 6, 1988, p. 68.

41. "Reform Executive Pay or Congress Will," *Wall Street Journal*, April 24, 1984.

42. See Mark Green and Bonnie Tenneriello, "Executive Merit Pay, *New York Times*, April 25, 1984.

43. Ernest van den Haag, "Economics Is Not Enough—Notes on the Anticapitalist Spirit," *The Public Interest* (Fall 1976):109.

44. Irving Kristol, " 'When virtue loses all her loveliness'—some reflections on capitalism and 'the free society' " *The Public Interest* 21 (Fall 1970):10.

45. For a selection of such inspirational, religious celebration of the market system and capitalism see Moses Rischin, ed., *The American Gospel of Success* (New York: New Viewpoints, 1974).

8

The Business System and Its Values

Many corporate managers understand the need to justify their activities and those of their firms in terms other than market efficiency; they are also cognizant of the boundaries on freedom that limited incomes impose. Managers seldom explicitly express such an understanding, but their activities certainly demonstrate it. Like many of their firms' constituencies, they find market values too narrow and partial. Managers praise the economists' free competition in the abstract, but in practice they prefer *fair* competition. Free market competition brings with it uncertainties, instabilities, and insecurities that are, at the least uncomfortable, and at the worst destructive. To protect themselves, managers seek stable production arrangements and look for ways to develop control through corporate planning. They reserve the term *free* for their own corporate activities; *free* enterprise both proclaims their identification with American traditional values and argues against government "interference" in business affairs.

Though managers emphasize the importance of profits, they find as little merit in the economic concept of profit maximization as they do in the free market. They appear to be more concerned with profit satisficing. Efficiency is a major standard and value of the economists' market, but managers usually prefer to be judged by their *effectiveness*. Economists presume an independent individualism among those who participate in the marketplace, but managers find such independence at odds with the organizational requirements of corporate performance. Mainline economic theory assumed competition among individuals and units so small that none significantly affects market outcomes, though for over a century large corporations have increasingly gained economic significance. In wide areas of the market the decisions of a relatively few corporate managers can shake an industry and even tilt the whole economy. Free market advocates insist that consumers are market sovereigns, although most managers (along with their employees and those in the community whose income depends upon company livelihood) quite willingly sacrifice consumers' interests to maintain their own. Finally, managers do not easily accept within their own organizations in a number of the values that American economists often attach to or implicitly assume underlie the market—democracy, participation, and egalitarianism.

FAIR, NOT FREE COMPETITION

When business managers praise competition, they usually refer to a particular kind, quite different from the free competition of prices described by economists. More than fifty years ago Eugene G. Grace, president of Bethlehem Steel, proudly explained the unbusinesslike nature of price competition. That steel prices had not changed over eight years, 1922 to 1930, was, he insisted, good industry policy. That in 1931 he and other steel managers should raise steel prices by a dollar per ton as the economy sank into depression was also a sensible move.

> What the steel industry has done with the price structure is to put it on a basis similar to that of retail trade, with announced prices for its merchandise so that the public will know the rock-bottom figure, and will be sure that every one else is buying at the same price from any one supplier. . . . This system permits the freest sort of *legitimate* competition. . . . [T]here are penalties if any company can be shown to have sold below its published price, and further protection against the establishment of prices which are *unfair* taking into consideration cost of production and other factors.[1]

A generation later, another leading steel manager, Roger Blough of United States Steel, explained competition in much the same way, explicitly rejecting the economists' definition of free competition:

> With respect to pricing and marketing, no longer is there the competition of peddler with a pack of pots and pans on his back and a different price to every customer, nor the "perfect competition," sometimes suggested in textbooks. Today's competition is the competition of pricing policies, of quality, of consumer surveys, of mass advertising and of mass distribution devices, of research, and of production practices and conditions of employment. This kind of competition is the only kind that is workable in a society like ours, which requires large productive groups constantly moving forward to bring improved standards of living. The advantages to the consumer resulting from this naturally evolved competition are far greater than could ever be possible under the theory of "perfect competition."[2]

Later in a Senate hearing on steel pricing, Mr. Blough explained that uniform prices by all steel firms was the essence of competition, not its denial as economists assume:

> My concept is that a price that matches another is a competitive price. If you don't choose to accept that concept, then of course, you don't accept it. In the steel industry we know it is so.[3]

In his testimony he continued:

> For anyone to assume that prices are not competitive because some producers raise the price the same as other producers, I think is, as I said before, simply an erroneous assumption.[4]

The steel industry's method of fair competition, he insisted, served consumers better than free competition. If one producer were to lower the price, competition would actually be diminished; buyers would then have no option but to buy that producers' product![5]

By the late 1970s the steel industry in the United States no longer enjoyed its favored and protected position. It was in serious trouble with declining sales, a precipitous drop in employment, and marginal profits. The fair competition practiced by its managers turns out to have been in fact little more than price fixing. It allowed foreign, price-cutting competitors to increase their sales and market shares steadily. Nevertheless, steel managers continued to argue that fair competition was preferable to the economists' free competition. In 1980, Bethlehem Steel proclaimed in a full-page, nationwide advertisement that:

> America's system of free enterprise is based on a concept of *fair* competition. That's what free enterprise is all about. Based on fair competition and independent action, it affords any group of investors the opportunity to succeed, or to fail. . . . [The government should provide] firm assurances that imported products, either by excessive volume or unfair pricing, will not disrupt the marketplace, particularly during this nation's period of revitalization.[6]

Bethlehem's managers and those of other American steel firms continued to insist that markets should be fair, *not free*. In a letter sent to all economic professors in the country, Bethlehem Steel in February 1984 complained that "Unfairly traded imported steel is our number one problem." United States Steel, in its 1984 annual report, announced with pride that the Steel Caucus of the House of Representatives had introduced a Fair Trade in Steel Act to impose a 15 percent quota on imported steel for five years. Further, it noted with approval federal and state fair trade legislation requiring government agencies to "Buy American" as a way of bucking up domestic production.

Managers of firms in one section of the steel industry, the minimills, have not been concerned with fair competition, however. They have operated in free-market fashion, pricing their product low enough to induce consumers to buy it from them. Since 1960 they have expanded and prospered in their free market, as economic theory predicts they should. Then there were only ten or twelve minimills that shared about 2 percent of the steel market. By 1984 there were fifty minimills, accounting for between 15 and 18 percent of the market, an amount equal to four-fifths of total imports![7] Despite the increased tonnage of sales lost to minimills, managers of large steel firms have in their public statements carefully ignored the free market success of the minimills. They have concentrated on the "unfairness" of competition from abroad.

Fair, Not Efficient, Allocation among Airlines

Through the 1980s after deregulation, airline managers have insisted that the Federal Aviation Agency (FAA) *fairly* allocate the limited landing slots at airports. The air controllers' strike in 1981 provoked the federal government to fire the strikers, about two-thirds of the then-employed controllers. Consequently, the FAA had to restrict the number of landings at many major airports. In subsequent months, new controllers were hired and airlines sought more landing permits to accommodate the increased business that came with deregulation. The FAA tried one method after another of allocating landing slots among the rival airlines. First, it apportioned slots among the carriers in a way that managers of existing carriers thought fair in proportion to their business. Existing carriers were favored, of course, but newcomers and carriers like People Express who were expanding their business found nothing fair in the proportional allocation. In late 1981 and early 1982, the FAA attempted another "fair" method, this time granting preference to new entrants but randomly allocating 40 percent of the new slots among all carriers. After each allocation, it allowed a limited amount of bartering among carriers, for winning carriers often received slots that were of little use to them. Managers of several airlines, particularly the major, well-established carriers, then attacked this system as lacking any logic and as *unfair*.[8]

Professor Alfred Kahn, a defender of market competition and an advocate of deregulation, suggested that the FAA forget about a fair allocation; he recommended that the allotment be carried out efficiently by auctioning the slots; free-market competition would efficiently distribute slots to those carriers that could make best use of them. Carrier managers ignored Kahn's "academic" solution and the opportunity to handle the problem efficiently. A spokesperson for Pan American World Airways, for example, insisted that the allocation had to be carried out in a way that did not put his company, and in general any company, at a competitive disadvantage.[9] The FAA assured the industry that it would make sure that all carriers got "their fair share of flights."[10]

It is clear that managers in many industries besides steel and air travel regularly prefer nonmarket rationing to free-market allocation. In 1979 electronic firms discovered that two critical semiconductor products, the low power Shottky and the 16K RAMS, were in short supply. Fairchild, Texas Instruments, and Motorola were among the firms that instituted rationing rather than allow free-market bidding to ration the scarce parts. A Fairchild manager explained, "We want to stick with our loyal customers . . . [and not] dilute our efforts by taking in some potential flake who will disappear when things improve."[11]

Fair, Not Efficient, Allocation
Among Auto Suppliers

We have already examined how General Motors for years "fairly" allocated its steel purchases and buying from auto parts suppliers. Only in 1982 did foreign competition become acute enough that its managers decided that fairness would have to give way to free-market, price, and quality competition among suppliers. Many other business representatives have also indicated that fairness is a higher value than the efficiency of free, competitive market allocation. Professor Steven Kelman discovered that trade association lobbyists in Washington generally did not like free-market approaches to pollution control. They opposed the economically efficient use of market price incentives as a way to reduce pollution. Significantly, they and the environmentalists interviewed offered similar reasons: Both were concerned with the matter of fairness. First, a quarter of the industry representatives and nearly a third of the environmentalists argued that using market incentives was bad because they would be licenses to pollute. Second, a fifth of the industry representatives and over a tenth of the environmentalists also saw a market approach as "Inequitable . . . [because it] would hit the poor companies too hard."[12]

That corporate managers may prefer the fairness of nonprice competition to the freedom of price competition is understandable. Price competition imposes severe constraints upon the seller, often with few benefits to the producer, employer, workers, or anyone—except, of course, consumers. It usually tends to cut revenues and eventually profits.[13] Thus it is not surprising that managers in many industries besides steel have found fair competition to be the business—and the "American"—way. At least since the latter decades of the nineteenth century, business managers have made clear their dislike of the economists' free competition.

A Century of Business Fairness

John D. Rockefeller, Sr., early in his business career denounced free (price) competition both for its "idiotic senseless destruction" and "the wasteful conditions it produced."[14] The head of the U. S. Envelope Company, James Logan, argued in 1901 that:

> Competition is industrial war. Ignorant, unrestricted competition, carried to its logical conclusion, means death to some of the combatants and injury for all. Even the victor does not soon recover from the wounds received in the conflict.[15]

Shortly afterward an officer of the American Tobacco Company suggested that the Supreme Court did not understand the necessity of fair, and the dangers of free, competition when it upheld, in 1911, the dissolution of his company.

> Unrestricted [free] competition had been tried out to a conclusion with the result that the industrial fabric of the nation was confronted with an almost tragic condition of impending bankruptcy. Unrestricted competition had proven a deceptive mirage, and its victims were struggling on every hand to find some means of escape from the perils of their environment. In this trying situation, it was perfectly natural that the idea of rational cooperation in lieu of cut-throat competition should suggest itself.[16]

It was not only corporate managers who found free competition to be dangerous and destructive; Edward Bellamy and Henry Demarest Lloyd found in industrial competition an explanation of vicious behavior among people. In *Looking Backward* (1988), Bellamy suggested that free competition destroyed opportunities to develop fraternal cooperation, contributing to the social evils of inequality and corruption. He argued that only when cooperation replaced competition would Americans be able to enjoy a truly just society.[17] Even so noted a trust-buster as Theodore Roosevelt, in accepting the nomination of the Progressive (Bull Moose) Party in 1912, warned of the dangers of competition.

> [U]nrestricted competition as an economic principle has become too destructive to be permitted to exist, and . . . the small men must be allowed to co-operate under penalty of succumbing before their big competitors. . . . Through co-operation we may limit the wastes of the competitive system. . . . Concentration and co-operation in industry in order to secure efficiency are a world-wide movement. . . . [They] are conditions imperatively essential for industrial advance.[18]

Small business managers strongly agreed with Roosevelt. They found free competition to be anathema. It undermined their own economic well-being and threatened to destroy their status and role in society. They sought relief from free market pressures through protective legislation. In 1936 Congress passed the Robinson–Patman Act exempting price maintenance agreements from antitrust regulations. Fourteen states soon followed suit with their own "fair trade" laws. The next year at the instigation of the National Association of Retail Druggists and other small-business associations, Congress passed the Miller–Tydings Resale Price Maintenance Act, legalizing the states' fair trade statutes. Within a few years, almost all states extended such protection to retail stores; by 1941 only the states of Missouri, Texas, and Vermont, and the District of Columbia had failed to enact fair trade laws.

In 1952 Congress extended the reach of the Miller–Tydings Act, permitting the use of interstate contracts to fix resale prices within those states that allowed intrastate price-fixing contracts. In the early 1960s fair-trade advocates made a determined effort to get more comprehensive protection. They sought federal legislation that would extend fair trade over the whole economy. A number of liberal legislators joined with business leaders to secure passage of the proposed bills. Senator Hubert Humphrey (the son of a drugstore manager) argued that fair trade promoted competition and helped prevent monopoly.

> . . . fair-trade laws curb certain types of unfair competition in order to promote fair competition, and thus prevent the growth of monopoly power. . . . Our free enterprise system cannot long endure if economic power is concentrated in relatively few hands. We must have large numbers of businesses, and we must make it possible for men to have a fair chance of success in business enterprise, if we are to avoid the statism about which there is much fear. Accordingly, the restoration of effective fair trade, as a bulwark for the preservation of our competitive small business economy, should have the support of every American citizen who believes in free enterprise.[19]

Despite the votes of Humphrey and another noted liberal, Senator William Proxmire, the bill was tabled. Nonetheless, for nearly forty years retail trade took place in an environment of fair, not free, competition, protected by a variety of state statutes.

Meanwhile consumer groups were growing in strength and effectiveness. They attacked fair-trade laws as nothing more than price-fixing that gouged consumers. In 1975, their lobbying secured a repeal of Miller–Tydings through the Consumer Goods Pricing Act, to the dismay of, and in opposition to, many business managers—not only those in small businesses, but also those among the giant corporate producers. The Reagan administration, ever sympathetic to business interests, proposed to reinstate fair trade. William F. Baxter, the U.S. Department of Justice antitrust chief, asked the Supreme Court late in 1983 to reinterpret its earlier decisions to allow producers and dealers to work out arrangements that would guarantee prices and high enough profit margins to enable the offering of extra services that consumers often want and need.

It is evident that managers in many firms, large and small, and in quite different industries seek to shelter their business from free-market competition as much as possible. Almost all managers attempt to soften price competition by trying to distinguish their own product from similar products offered by rival firms. They may tout their brand name or emphasize an advertised attribute, whether physically present or not, in an attempt to distract consumers' attention away from price. Some managers confronting foreign competition like those in steel, automobiles, and textiles seek legal quotas and other governmental trade restrictions to limit imports; others may seek legislative regulation, such as product or work safety codes, pollution restraints, or quality standards that impose barriers to entry. A variety of occupational groups—barbers, beauticians, funeral undertakers, architects, egg graders, electricians, medical doctors, airplane pilots, electricians, pest controllers, social workers, well drillers, abstracters, and hoisting engineers—have sought governmentally instituted or approved licenses with the intention of restricting entry to fair competitors, though they invariably argue that they are only seeking to assure high quality service and increased safety for consumers.

Business managers, as well as other Americans, frequently choose the adjective "fair" in describing preferred economic relationships and desirable economic outcomes. Most Americans know and probably approve the folk saying, "A fair day's wage for a fair day's work"; an efficient daily wage for a day of efficient work does not carry an ethically compelling message. Stockholders are also entitled to a fair return on their investment, and profits above a fair level are not proper.[20] The fixed prices normal in most American retailing imply a fair price —the best price a business firm can afford to offer as it did for Eugene Grace of Bethlehem. Such a sense of fairness is not limited to small businesses, however. Managers of large businesses also imply that haggling over price is an invidious activity, indicative of a lack of faith in the fairness of the seller.[21] In their concern for "fairness," business managers prove themselves to be thoroughly American. They resemble the corporate constituencies in proclaiming fairness as their preferred goal. A cynic may suspect that both may be merely covering their own self-interest with an acceptable gloss, but nevertheless one can wonder that both should choose the same term—and value—to espouse publicly. It is enough to make one doubt that the old medieval church ethic of the "just price" ever really died the death that market economists have proclaimed for the past 200 years.

MANAGING EFFECTIVELY, NOT MANAGING EFFICIENTLY

A term closely related to fairness, with the same aura of ambiguity, is frequently found in managerial speeches, company communications, and business studies. It is *effective*—effective management, an effective program, an effective policy, or an effective record. For economists it is a meaningless term unless it is tied to some market measure.[22] Such a tie is not easily discerned, however, and if managers do perceive it, they may find the two incompatible. Chester I. Barnard, an influential management theorist and one-time president of the New Jersey Bell Telephone Company, pointed out more than half a century ago the vital difference between managerial effectiveness and a firm's efficiency.[23] Managers are effective, he asserted, when they attain a specific desired end, and a firm is efficient when it is growing or at least not shrinking.[24] Barnard's definition of managerial effectiveness was but an early attempt. There is no agreement on what effectiveness is or how to measure it. Both scholars and managers have offered many definitions. In recent years, though, those who seek to define effectiveness have focused on the achievement of managerial goals[25] through the integration of personal and organizational purposes.[26]

Effective Inside, Efficient Outside

Effective management became more important than efficient management as corporate organizations grew large, bureaucratic, and hierarchically layered. Job assignment, selection, hiring, promotion, and performance evaluation, all had to be made by fiat, far removed from any market indicators of success or failure. The very size of a firm often insulated it from minor swings and changes in the market in any case. With little or no guidance from any market signals, how were managers to be evaluated? The general answer has been to establish criteria against which they could be evaluated—personality traits, behavior, achievement of goals, comparisons with others in similar situations. Performance appraisal, with the intent to measure effective managing, has become an important, specialized function in business corporations.

Since appraisals are made on the basis of performance goals established by managers, usually those in superior positions and not by outside—market—standards, they may be merely circular judgments: A performance is only as good as the established goal by which it is evaluated, but the goal may well be determined by the kind of performance desired. As a consequence of such circularity, companies often measure and reward managerial performance on the basis of behavior that is not in the firm's long–term interest.[27] It is a system well suited for fitting goals to profit-satisficing performance. If one's performance is measured against an internally generated profit goal, members of the organization will probably exert great pressures to keep the goal both reasonable and attainable. This is one of the weaknesses of goal-oriented management, including goal-oriented ethics. Either we set for ourselves unreachable ideal goals or we frame our goals in our capacities, substituting means for ends. We deal further with this problem in later chapters. Are the less abstract ethics of corporate social responsibility—corporate effectiveness—the more fitting ethic for managers in a complex, changing society? Studies of the relationship of performance and goal setting suggest that neither may have much to do with objectively measured productivity or economic efficiency. Managers look to their basic social and personal values to guide them, not to economics.[28]

The distinction between the efficiency of a firm rewarded in the market and the effectiveness of managers as they choose to define that effectiveness in terms of their own goals leads to much confusion about managerial success—and probably to great uneasiness among managers. However effective managers may be in meeting the internally established goals of the firm, the goals themselves may be inappropriate, given the changed and changeable responses of the market. However effective managers may judge *themselves* to be, outsiders may find them inadequate. Investors, for example, usually prefer to evaluate managers by objective, impersonal standards—return to stockholders' equity, return on total capital, growth in sales and earnings, stock-market performance, debt/equity ratios, and net profit margins—that may be quite different from a firm's internally generated standards. But investors may need to be called to account for the narrowing of these very criteria. In a just, caring, and responsible society, no participant may have the luxury of living by a single-value criterion.

The Unfairness of an Uncertain Market

Attempts to evaluate managers' performance by market standards are bound to be problematic. Market success often depends upon luck and good fortune. However able and competent managers may be, the unforeseen and the unforeseeable may undo their best efforts. The Task Force on Corporate Social Performance, in a 1980 report to the U.S. Department of Commerce, pointed out this fact of business life.

> The genius of American business has been and remains its ability to adapt to change. This ability has been greatly tested during the past two decades. It will be tested anew in the decades ahead. The price and supply of energy are likely to remain uncertain. The world economy is sure to become more interdependent. Economic competition will increase, in both domestic and international markets (which are becoming harder to separate). New tech-

nologies will threaten established industries and also create whole new industries. The dangers of inflation and recession will remain. Undoubtedly, these developments will make the economic environment more complex and more uncertain. . . . Unanticipated change has frequently had adverse consequences for American industry. Shifts in consumer preferences, rising energy costs, equal employment opportunity for women and minorities, and environmental protection are just several developments that caught many companies unaware and unprepared in the 1960's and 1970's.[29]

Business managers indeed operate in an uncertain world. Even those who command large corporations and dominate whole industries possess little power to influence the course of major economic fluctuations that may easily upset the best laid plans and most carefully devised strategies. Managers are well aware of these fluctuations. Since 1854 the economy has experienced thirty business cycles; even since 1933 it has ridden ten cycles up and down.[30] Not only does the economy suffer wide swings but also technology and consumer tastes continually threaten to upset the plans of corporate managers. Consider the sad experience of Coca Cola's managers in 1985 when they attempted to change the formula for their soft drink's syrup. They made the change only after a long and complicated series of market tests and extensive consumer surveys. Yet within weeks, adverse consumer reaction forced them to admit that their research had failed. Rational management, geared to supposedly scientific market research, was overruled by unanticipated human behavior.

The uncertainties of the market and the fluctuations of the economy produce continual business turnover. Business firms die in great numbers, and many of even the largest and most successful companies of one era are the forgotten names of another. The great railroad companies that dominated American business in the late nineteenth century are hardly known to ordinary Americans today; the profitable, powerful meat packers at the turn of the century—Swift, Armour, and Wilson—sell little meat now, their names lingering on as minor subsidiaries of new conglomerates. Steel and auto firms that once commanded wide recognition—Youngstown Sheet and Tube or Republic, and Studebaker or Willys—are hardly even remembered. The turnover of firms among the Fortune 500 or any other listing of the largest companies is considerable.[31]

One can hardly be surprised that managers wish to be evaluated by some other measure than market performance. If boom times provide a buoyant lift to the whole economy, even relatively poor managers may do well. If depression tugs at earnings, not even good managers may be able to turn a profit. Furthermore, the work and performance of any manager in the vast bureaucratic structures of modern large corporations may be so far removed from the market and his or her influence so diffused that a market evaluation is impossible and irrelevant. Thus managers seek an alternate, complementary method of evaluating their activities. Business consultants and management scholars have attempted to provide such a method. Richard E. Boyatzis, chief executive officer of McBer and Company, has identified the essential characteristics of the competent manager. Grounding his conclusions on a study of over 2,000 managers and the work of many scholars, he argues that managerial competence requires both efficiency and effectiveness above and beyond any efficiency as measured by the market.[32]

> *Effective performance of a job is the attainment of specific results (i.e., outcomes) required by the job through specific actions while maintaining or being consistent with policies, procedures, and conditions of the organizational environment.* . . . At the very core of every manager's job is the requirement to make things happen toward a goal or consistent with a plan. Whether in terms of making profits, staying within budget, or producing to deadline, there are many forms of the "bottom line" to which managers attempt to move their organization.[33]

Market indices, he believes, measure only a portion of the more important characteristics of performance effectiveness.[34]

In recognizing organizationally defined effectiveness as important as or perhaps more important than economic efficiency, business managers afford themselves a norm only indirectly and loosely connected to the market; with it they can justify their efforts to plan, stabilize, and administer within the large organizations subject to their command, apart from any market measure of efficiency. Even if market indicators provide little or no basis for considering them efficient, they can nevertheless reward themselves for their effectiveness.

For more than a generation General Motors Corporation, for example, was considered by the public and by its own managers to be a superbly managed firm; its managers were effective for they met or surpassed their own performance goals, which were all too flexible. Such performance, many concluded, also must indicate that they were efficient. It is now apparent that however effective they were, they were not at all efficient by the standards that Japanese managers were then developing. A significant portion of their profitability was due to their protected markets, the lack of competition. GM managers allowed many a market inefficiency to develop and they tolerated them over long periods. Evidently they could have cut costs, reduced prices, and improved profits far more than they were willing to contemplate. The same conclusion can be drawn for any firm that enjoys a protected market position. Its managers prefer to evaluate themselves by their effectiveness rather than the potential return that profit maximization in an efficient free market can offer.

Not only do managers prefer fair to free competition, profit satificing to profit maximizing, and effective to efficient performance, but they have also discarded or reinterpreted many of the other characteristics and values that economists find in the market system. They apply the adjective "free" to their enterprises and an absence of government regulation rather than to the market. They are ambivalent about a free market unregulated by government, for "proper" probusiness policies can help them greatly. They do not gainsay subsidies, protective legislation, or tax advantages. Fair competition is better than free competition because a more planned, controlled, and manageable market is preferable to the harsh uncertainties and instabilities of the market. Likewise managerial effectiveness—the ability to meet self-defined goals and to serve self-proclaimed purposes—is more certain of achievement than is the pursuit of market efficiency.

The individualism assumed by advocates of the competitive market system does not easily fit the preferences and needs of business managers any more than did the other market values already mentioned. Individualism and the values it identifies are so deeply embedded in the American ethos, however, that managers have not rejected but reinterpreted it. Early corporate leaders attacked individualism as out of date and harmful in an industrial society. Later, managers discovered that it was more prudent to maintain the term, but define it in such a way that it could serve the possibilities of the large corporation.

INDIVIDUALISM: A THREAT TO CORPORATE EFFORT

Managers did not easily effect the reconciliation of traditional American individualism with the requirements of the emerging large business corporation. Some, like John D. Rockefeller, Sr., bluntly rejected individualism as wasteful and incompatible with the operation of the new business organizations.

> [The combination movement] had to come, though all we saw at that moment was the need to save ourselves from wasteful conditions. . . . The day of combination is here to stay. Individualism has gone never to return.[35]

Chauncey M. Depew, in an 1897 address at the unveiling of a statue in honor of Vanderbilt University's benefactor, Cornelius Vanderbilt, disagreed with Rockefeller. He asserted that "the American Commonwealth is built upon the individual," of which Commodore Vanderbilt was a conspicuous example.[36] He may have reached his conclusion by looking backward rather than forward, however. In Depew's day Andrew Carnegie exemplified, even better than Vanderbilt, the promise of and the opportunities for individualism. He reveled in his own "intense individualism," mixing with it a thoroughgoing paternalistic concern for his workers and poorer neighbors. Even as Carnegie preached and wrote about his individualism, the giant company that he had helped create, Carnegie Steel, became in 1901 the core of the largest corporation the world had ever seen, United States Steel. Depew's and Carnegie's praise of individualism accorded well with the traditional American ethos, but their own business activities flagrantly contradicted it.

Some business spokespersons confronted directly the problems that the individualistic tradition created in an economy of large-scale organizations. The *American Bankers' Magazine,* whose editors were always ready to downgrade traditional individualism and praise the new business combinations, suggested in 1902 that the decline of individualism and the rise of the business corporations would help, not hurt, ordinary Americans:

> The general character of the charges against modern business methods is that they injure the individual in his opportunities to acquire wealth. . . . It is probable that the new business order may reduce the number of great prizes, but that it will increase the opportunity for a larger number to arrive at a competence. . . . It is too soon to assert that in the long run, after the combination system has been fully developed and perfected in all its details, it will bring about a greater equality in the distribution of wealth than could be hoped for under the competitive system, but there are many signs that lead to this conclusion. It is highly probable from existing indications that the intense individuality that characterized the average citizen of the United States during the first century of the republic is gradually softening.[37]

To attack individualism as Rockefeller did, or to point to the benefits available in giving it up as the editorialist of the *American Banker's Magazine* did, was not convincing to many Americans. Tocqueville had first recognized and named the American social philosophy of individualism.[38] He recognized that it was a powerful and appealing social ideal. It arose from the condition of social equality common throughout the nation. The praise and defense of individualism were woven tightly into almost every strand of American thought. Ralph Waldo Emerson, characteristic of his time, provided an individualistic philosophy; Andrew Jackson offered a political expression for it; and Frederick Jackson Turner and his followers constructed a history around the notion of individuals moving from society—from others—to a new beginning.[39] As late as the 1930s, in the midst of the worst depression the economy had experienced, Herbert Hoover extolled individualism in extravagant terms:

> While I can make no claim for having introduced the term "rugged individualism," I should be proud to have invented it. It has been used by American leaders for over a half a century in eulogy of those God-fearing men and

women of honesty whose stamina and character and fearless assertion of
rights led them to make their own way of life.[40]

Redefining Corporate Togetherness as Individualism

Rather than reject so deeply rooted a value as individualism, business
managers and their defenders discovered that a redefinition was a wiser
course to follow. Advancing technology and the growth of the large
corporation wove the threads of economic and social interdependence
ever more tightly. They made the older concept and practice of individualism anachronistic. Individualism, as Tocqueville perceived it and
as managers feared it, was a detachment of the individual from the
group—the cutting off of persons from each other. Such action was
not desirable in the large corporation. The threads of interdependence
could not be unraveled without destroying economic progress and the
American level of well-being. The editors of *Fortune* in 1951 announced their solution—a reconciliation of traditional individualism with
the needs of the interdependent corporation:

> The concept that appears to be emerging, as the answer of the modern in-
> dividual to this challenge, is the concept of the *team*. It is an old concept but
> it is being put to new uses. As a member of a team an individual can find
> full opportunity for self-expression and still retain a dynamic relationship to
> other individuals . . . the concept of the team has the power to challenge
> the individual to seek his self-expression, not along purely egotistic chan-
> nels. . . . A community—big or little—is created, and through it the indi-
> vidual finds a higher expression of himself.[41]

Employees and even middle level managers may confuse the old individualism with the new. Workers who inventively bargain with supervisors for special benefits for themselves and the members of their work
groups may realize a self-expression at the place of work that helps
themselves, but their side agreements create precedents that can cause
trouble for the personnel office and the managers of industrial relations. The subordinate manager who refuses on orders from his superiors to sell dangerous, below-standard oil-well casing for use under
conditions of high pressure, can lose his job.[42] A bank manager who
protests his superior's illegal transfers of funds to avoid paying taxes
can be fired.[43] Such expressions of individualism are not often appreciated by managers in many firms. Employees are too often expected
to follow the injunction of Speaker Sam Rayburn to incoming members
of the House of Representatives: To get along, *go* along.

Through verbal magic, modern managers have inverted Tocqueville's individualism, or the freedom to detach oneself from others. It

has now become in the redefinition offered by *Fortune* and now widely accepted in business—the freedom to participate in society or at least a corporate community. Whereas individualism was originally freedom *from* society, it now becomes freedom *within* society. Self-reliance has been transformed into other-directed sensitivity; independence has been changed into team play; autonomy has became initiative within organizational bounds. Professor John W. Ward points out that insofar as the *Fortune* redefinition of individualism has become the accepted norm, there are

> . . . two different systems of values both going under the [same] name . . .
> that verge toward opposites—one toward the isolated person, the other toward
> society—and [yet both are] . . . descriptively correct in saying that at differ-
> ent moments in our history Americans have moved toward one and then the
> other.[44]

The old individualism, however, rested upon a moral foundation that the new may not. That foundation was rooted in the Puritan tradition that gave religious sanction to both individualism and work as a calling. Prudence, sobriety, propriety, and dignity in work easily attached themselves to the worker inbued with the old values of individualism; whether they are accompanying values of the new individualism is a question we will explore later. It is a question of significance to managers who must lead hundreds, thousands, and even tens of thousands of people, all employees within our vast corporate organizations. As those who sit at the apex of many-layered hierarchies, they should have a careful regard for the values generally held by those beneath them. It is through respect for—even affirmation of—those values that managers can win the needed cooperation that makes productive, effective, and efficient coordination of their corporate enterprises possible.

THE DIFFERENCES BETWEEN THE MARKET SYSTEM AND THE BUSINESS SYSTEM: A SUMMARY

The business system, as managers practice and interpret it, is quite different from the market system that economists tout. In sum, whereas economists eschew any sense of justice, arguing that justice is best left to judges, legislators, philosophers, and theologians, managers apply a variety of norms of justice to all their relationships, though they are often implied rather than recognized. In meeting consumers through the market, managers argue for *fair*, not free, competition. In dealing with stockholders they seek *satisfactory*, not efficient (or maximum), returns. In evaluating themselves and their subordinates, they aim for *effective* and (but sometimes rather than) efficient performance. In relating to employees within the organization, managers insist upon an *hierarchical authority of command* to which all should submit with unquestioning, cooperative willingness. The market envisions only such managerial right as may be won from the voluntary agreement of individuals, free to accept or reject terms at any time. In addition, managers have redefined *individualism* in a way that helps contain employee initiative and deflect any challenge to managerial control. The individualism of the market presumes an independence and self-direction that is not easily possible within corporate employment.

Market Rhetoric as a Cloak for Corporate Power

With such wide differences between the managers' business system and the market system, why should managers have adopted the rhetoric and symbols of the free, competitive market for their own and pretend that their business system and its values are but minor variations of those in the market? One attractive reason, we believe, was that market theory provided camouflage for power.[45] Since the theory takes little note of managers, explaining economic outcome as the result of impersonal influences of prices, costs, demand, and supply, it presumes that only the market, not managers, exercises power. A generation ago Francis X. Sutton, Seymour E. Harris, Carl Kaysen, and James Tobin explained:

> The classical [economic] creed . . . assures the businessman that he really has no choice in the matter [of seeking profits and economic efficiency] anyway, since he is merely the agent of inexorable market forces . . . he has no choice, that he simply does what he has to do and what any other man in his place would do. The classical creed enables him to say this by telling him he is simply the agent and victim of impersonal market forces. His power and discretion, according to the classical creed, are severely limited by the competition of actual or potential rivals for the "votes" of consumers. . . . The power attributed to the consumer in the doctrine of "consumer sovereignty" is a principal manifestation in the ideology of the sensitivity of businessmen about the amount and legitimacy of their power. One way of shedding awkward responsibility is to believe that the consumer is the real boss, that the businessman merely carries out his (or perhaps more frequently in the ideology, her) orders.[46]

In the 1950s and through the 1960s many managers of large corporate enterprises that dominated whole communities tried to deny their power. They maintained that it was inappropriate for them to express any but personal views about racial discrimination, women's rights. local schooling, or other contentious public issues. We have already noted earlier the refusal of the chief officers of United States Steel to denounce racial segregation in Birmingham, Alabama, during the travail over civil rights in the early 1960s. Though the company was by far the largest employer in the city and its managers were leading citizens in the community, they insisted that they exercised power only over the quality, costs, and prices of their products; to attempt to change social practices and to intervene in local politics was beyond the ability of officers of a mere *private* corporation. One may doubt that such intervention was actually beyond their ability, though they may have wisely decided to exercise that ability with restraint. Business managers do not always possess either the knowledge or the wisdom to intervene in local politics.

Many politicians, community leaders, and public commentators attacked such a position, deriding the notion that corporate influence was bounded by its own internal business interests. They pointed out that corporate leaders did not enjoy the option of *not* exerting community influence; their firms were so large, and often so dominated the locality in which they were situated, that they spoke loudly, even when they kept silent. Like the silence of Sir Thomas More about Henry VIII's usurpation of papal authority, the silence of corporate managers conveys messages and influences the public.

Some managers of large business corporations are still tempted to deny that they wield power, but by the early 1970s more farseeing and responsive business leaders admitted that they could no longer pretend that they were simple pawns of the market.

> The self-interest of the modern corporation and the way it is pursued have diverged a great deal from the classic *laissez-faire* model. There is broad recognition today that corporate self-interest is inexorably involved in the well-being of the society of which business is an integral part, and from which it draws the basic requirements needed for it to function at all . . . and it has become clear that the essential resources and goodwill of society are not naturally forthcoming to corporations whenever needed, but must be worked for and developed. This body of understanding is the basis for the doctrine that it is in the "enlightened self-interest" of corporations to promote the public welfare in a positive way.[47]

These progressive business leaders, however, emphasized the need for managers to act on their own initiative so that they could preserve a necessary flexibility and avoid the restrictions that government might impose or to ward off the constraints that constituencies might demand.

Such outside participation in managerial policy would reduce business efficiency and adaptiveness, they maintained. They were more willing than business leaders of an earlier time to recognize the social power they possessed, but they were unwilling to admit that the power could or should be shared in any way. They insisted that they must be allowed to exercise power as they saw fit. Social responsibility (or in the language of the competitive market that they still felt compelled to use, "enlightened self-interest"), rather than the market itself, had become the cloak under which they hid from public view such power as they enjoyed.

It was a cloak of privilege, but one that has become threadbare. The privilege it allowed rested upon a protected position in a market economy. Only to the degree that the managers and their firms were sheltered from the "gales of creative destruction," to use Schumpeter's phrase, or shielded from market competition were they able to talk grandly about social responsibility, and sometimes to follow talk with action.

The Erosion of Corporate Power

As we observed in Chapter 5, in the late 1960s when the CED was examining "social responsibility" and in the early 1970s when it published its findings, business was still basking in a world economy where American industry was unrivaled and hardly challenged. That condition did not long continue and was disappearing even as the CED report was issued.

Managers in the 1990s confront an unprecedented situation. Many are officers of very large organizations upon whom most Americans are dependent for the goods and services that enable them to enjoy their present standard of living. Those same managers also find themselves to be the employers of a sizable portion of the labor force, with tens of thousands of people dependent upon them for their livelihood. The most perceptive managers have recognized in recent years the power they exercise and the influence they command, not only over these dependents, but also throughout the society. Ironically, just as that recognition came, unexpected market competition has imposed more stringent limits upon the exercise of that power than most managers have ever known. The changed economic environment within which they operate—more complex, competitive, and uncertain—however, has made adept use of that power more vital and important to business success and managerial effectiveness than ever before. The economic environment and the social climate are challenging managers as they have not been challenged since the rise of the large corporation over a century ago. They will have to contemplate new, different, and nontraditional ways of managing.

A New Style of Managing for a New Environment

It is not antibusiness commentators, managerial critics, or labor leaders who have taken the lead in condemning the traditional business system;

rather, business scholars, industry consultants, and those most sympathetic to continued corporate success have expressed their concern about the outmoded ideas, anachronistic values, and unexamined practices of present day business managers. Many have spoken out forthrightly and sharply, particularly attacking the prevailing emphasis on managerial prerogatives and privilege. Harry Levinson, a management consultant and teacher at the Harvard Medical School and Graduate School of Education, has commented:

> . . . an executive these days . . . can no longer be an executive in the classical sense. Traditional ways of wielding managerial power have become obsolete. . . . With less outright power to face greater responsibility, the executive can no longer function effectively by control and command. The Imperial German General Staff Plan—the most commonly used organization chart, with its underlying assumption of control from the top—is obsolete, for the more rigidly an organization is controlled the less its adaptability. The executive role has changed, too. Current executive practice is as radically different from nineteenth-century procedure as is driving a stagecoach from piloting a supersonic airliner.[48]

With Stuart Rosenthal, Levinson conducted a detailed study of the managerial style and practices of the chief executive officers of six large, successful corporations and concluded that managers need to understand well the difference between authoritative and authoritarian styles. Authoritativeness is useful and even necessary, but authoritarianism is more likely to cause problems than to help corporate managers today. The managers studied by the two researchers recognized the need to hold the members of their firm together by more than shared ambitions or goals, for such sharing produced only expedient relationships.

> Meaning [within the organization], is derived from purpose beyond goals. . . . One of the characteristics of each of these organizations is a shared morality. We heard repeatedly not only of business morality, but also of concern with and fairness to competitors, to the government, to the community and of actions demonstrating concern for employees.[49]

We will examine this motion of morality later.

Peter Drucker points out that managers have to deal with a new kind of worker, especially the highly educated young people who are challenging the traditional economic and power relationships within firms. He argues that the old managerial style will not work much longer:

> For the weapon of fear—fear of economic suffering, fear of job security, physical fear of company guards or of the state's police power—which for so long substituted for managing manual work and the manual worker, is simply not operative at all in the context of knowledge work and knowledge

worker. The knowledge worker, except on the very lowest levels of knowledge work, is not productive under the spur of fear; only self-motivation and self-direction can make him productive. He has to be achieving in order to produce at all.[50]

David Ewing, managing editor of the *Harvard Business Review*, has examined managerial prerogatives and employee rights in two recent books.[51] He seconds Drucker. He makes the case that corporate managers have indeed trampled on the rights of employees, that employees have learned to fight back, and that managers must abandon autocratic styles of management if they are to take advantage of the unique skills available in today's work force.

The critics of traditional managerial values and practices find that too many corporate managers in the past ignored or failed to celebrate the most obvious and valuable creation of business: The voluntary cooperative production community. Levinson and Rosenthal have found that successful managers of large corporations often appeal to the standard of "fairness." Our examination of the business system confirms managerial concern about and use of fairness. Too often, though, it has been applied partially and certainly undemocratically to defend managers' own status and privileges. Managers have yet to demonstrate that they show as lively a concern for fairness to all their constituencies as they do for their own special claims and entitlement.

THE CORPORATION AS A COMMUNITY—
UNDER PRESSURE

Managers might wisely consider the specific community that forms their business corporations and the broader communities within which corporations are able to exist. They need to give careful attention to both sustaining and maintaining those communities. Those communities are the precondition for continued business success, however it may be measured. We can agree with Professor Robert A. Nisbet, who pointed out that business and the economic system can be sustained only within a larger community.

> There is indeed a sense in which the so-called free market never existed at all save in the imaginations of the rationalists. What has so often been called the natural economic order of the nineteenth century turns out to be, when carefully examined, a special set of political controls and immunities existing on the foundations of institutions, most notably the family and local community, which had nothing whatsoever to do with the essence of capitalism. Freedom of contract, the fluidity of capital, the mobility of labor, and the whole factory system were able to thrive and to give the appearance of internal stability only because of the continued existence of institutional and cul-

tural allegiances which were, in every sense, precapitalist. Despite the rationalist faith in natural economic harmonies, the real roots of economic stability lay in groups and association that were not essentially economic at all.[52]

Nisbet insisted that the business system as managers have operated it does not call forth moral allegiances—too few people will really care to stand up for it. The freedom that it promotes cannot ultimately rest upon moral atomism or upon large-scale impersonalities. Those involved in the large corporation and those whose lives are significantly affected by its activities need to participate in its governance and discover in its foundations a morally legitimate rationale if they are to provide it any lasting support and approval.

In the next chapters we will examine the institutional, cultural, and moral allegiances that provide the foundations of both the market and the business systems. We will examine the problems that have developed as many of these foundations have eroded. We will consider the possibilities for developing new institutions and reformulating old and honored values to serve the needs of both corporate managers and their many and new constituencies.

NOTES

1. Eugene G. Grace, "Industry and the Recovery Act," *Scribner's*, February, 1934, p. 99. Italics added.

2. Roger Blough, *Free Man and the Corporation* (New York: McGraw-Hill, 1959), p. 11. [Reprinted with permission (italics added).]

3. Quoted by Estes Kefauver, *In A Few Hands, Monopoly Power In America* (Baltimore: Penguin Books, 1965), p. 131.

4. U.S. Senate Committee on the Judiciary, Subcommittee on Antitrust and Monopoly, 86th Congress, 1st Session, *Hearings On Administered Prices*, Part II, (1957), p. 314.

5. See Paul A. Tiffany, *The Decline of American Steel: How Management, Labor, and Government Went Wrong* (New York: Oxford University Press, 1988). He points out that after 1901, when United States Steel was formed, managers rather than entrepreneurs took over the industry. They were interested more in cooperation and conciliation among producers than Schumpeterian creative destruction. Until recent times, the industry has been dominated by managers who took few risks and attempted to keep competition under strict control.

6. *Wall Street Journal*, August 27, 1980.

7. Jack Robert Miller, "Steel Minimills," *Scientific American* 250 (May 1984):33.

8. "FAA Plans to Alter Method of Allotting Landing Slots," *Wall Street Journal*, June 10, 1982. In 1988 the Massachusetts Port Authority (MPA), which operated Logan International Airport, began experimenting with differential landing fees as a way of reducing airport congestion. After their imposition in June, 1988, small plane operators appealed. The federal district court found the fees legal, but an administrative law judge in the Department of Transpor-

tation found the fees "not fair and reasonable . . . unjustly discriminatory and unlawfully [in] conflict with Federal authority." (Italics added) See John H. Cushman, Jr., "F.A.A. Proposing New Rules On Airline Access to Airports," *New York Times,* December 20, 1988.

9. Christopher Conte, "Bill is Proposed To Give Airlines Trust Immunity," *Wall Street Journal,* August 9, 1984, and Andy Pasztor, "Upgrading Airports Won't Solve Delays, But Revised Scheduling May, Study Says," *Wall Street Journal,* August 27, 1984.

10. Reginald Stuart, "Airlines Accept La Guardia Plan," *New York Times,* September 10, 1984. Later in April of 1986, the FAA authorized the purchase and sale of slots at four "slot-constrained" airports, O'Hare, La Guardia, Kennedy, and National. All airlines were free to participate in the slot market, but the FAA retained the right to withdraw the slot privilege and reallocate them at its discretion. From the time of the April authorization until the end of 1987 the airlines bought and sold more than 1,000 slots at the four airports. Industry uneasiness about a slot market and FAA reluctance to support wholeheartedly tradable slots impairs the effectiveness of efficient allocations. Given the tentativeness with which the industry approaches a free, competitive market for slots, the airlines at the four constrained airports may make short-term leasing arrangements rather than outright purchases. See *The Economic Report of the President,* 1988 (Washington, D.C.: Government Printing Office, 1988), pp. 218–19.

11. Marilyn Chase, "Industry Faces a Semiconductor Shortage as Makers Begin to Ration Their Supplies," *Wall Street Journal,* May 29, 1979.

12. Steven Kelman, *What Price Incentives? Economists and the Environment* (Boston: Auburn House Publishing Company, 1981), pp. 107–22.

13. Econometric studies indicate low price elasticities of demand for many products when an entire structure of demand is examined. One study found that only about 20 percent of a sample of eighty-three products had an elasticity greater than one (a 1 percent cut in price increased sales by more than 1 percent). Nearly 70 percent had a zero elasticity—a decline in price had no effect on sales, and almost 10 percent showed elasticities of less than 0.5 percent. See H. S. Houthakker and Lester C. Taylor, *Consumer Demand in the United States* 2nd ed. (Cambridge: Harvard University Press, 1970), 265–80.

14. John Chamberlain, *The Enterprising Americans: A Business History of the United States* (New York: Harper & Row, 1963), pp. 140–61.

15. Quoted by Gabriel Kolko, *The Triumph of Conservatism* (Chicago: Quadrangle Paperbacks, 1967), p. 13. See Logan's article in *North American Review* clxxii (1901):647, 689.

16. Kolko, *The Triumph of Conservatism,* pp. 13–14.

17. Robert H. Wiebe, *The Search For Order,* 1877–1920 (New York: Hill and Wang, 1969), p. 138.

18. Reprinted in Richard Hofstadter, ed., *Great Issues in American History,* vol. 2, (New York: Vintage Books, 1958), pp. 287–88. Roosevelt was summarizing the arguments of Dr. Charles R. Van Hise, president of the University of Wisconsin.

19. Senator Hubert H. Humphrey, "The Case of Fair Trade," *Congressional Record,* Proceedings and Debates of the 82nd Congress, 2nd Session 98, June 12, 1952, A3668.

20. Francis X. Sutton, Seymour E. Harris, Carl Kaysen, and James Tobin, *The American Business Creed* (Cambridge: Harvard University Press, 1956), pp. 65, 357.

21. Ibid., p. 142. George W. Perkins, a former president of the New York Life Insurance Company and a Morgan partner, in 1908 stressed the fairness inherent in a corporation's ability to stabilize prices. "The highly developed competitive system gave ruinously low prices at one time and unwarrantedly high prices at another time. When the low prices prevailed labor was cruelly hurt; when the high prices prevailed the public paid the bills." "The Modern Corporation," in Moses Rischin, ed., *The American Gospel of Success: Individualism and Beyond,* (New York: New Viewpoints, 1974), p. 117.

22. Note the emphasis given in this statement: *"Producer efficiency in the absence of consumer utility is without economic meaning."* Vincent Ostrom, "The Design of Public Organizational Arrangements," in Gene Wunderlich and W. L. Gibson, Jr., eds., *Perspectives of Property* (Institute for Research on Land and Water Resources, The Pennsylvania State University, 1972), p. 168.

23. Chester I. Barnard, *The Functions of the Executive* (Cambridge: Harvard University Press, 1982), p. 19.

24. Ibid., p. 251.

25. Harold Koontz, *Appraising Managers as Managers,* (New York: McGraw-Hill, 1971).

26. See R. Likert, *The Human Organization* (New York: McGraw-Hill, 1967), and Douglas McGregor, *The Professional Manager* (New York: McGraw-Hill, 1967).

27. Paul J. Stonich, "The Performance Measurement and Reward System: Critical to Strategic Management," *Organizational Dynamics* 12 (Winter 1984):45–57.

28. Jack N. Kondrasuk, "Studies in MBO Effectiveness," *Academy of Management Review* 6 (July 1981):419–30. The author finds that the less sophisticated the research approach, the more likely the study is to show MBO effectiveness on productivity and/or job satisfactions.

29. Report of the Task Force on Corporate Social Performance, *Business and Society: Strategies for the 1980's* (Washington: U.S. Department of Commerce, December 1980), pp. 3, 5.

30. Victor Zarnowitz, "Recent Work on Business Cycles in Historical Perspective: A Review of Theories and Evidence," *Journal of Economic Literature* xxiii (June 1985):526.

31. Dun & Bradstreet, Inc., *The Failure Record through 1969* (New York: Dun & Bradstreet, 1970), p. 2; and E. D. Hollander et al., *The Future of Small Business* (New York: Frederick A. Praeger, 1967), pp. 222–23.

32. Richard E. Boyatzis, *The Competent Manager: A Model for Effective Performance* (New York: Wiley & Sons, 1982).

33. Ibid., pp. 12, 60. [Reprinted with permission.]

34. Peter Drucker agrees. (His terminology is somewhat different from ours, however.) He argues that managerial effectiveness means keeping up with market trends, while managerial efficiency is turning out the same good with the fewest resources: "But even the most efficient business cannot survive, let alone succeed, if it is efficient in doing the wrong things, that is, if it lacks effectiveness. No amount of efficiency would have enabled the manufacturer of buggy whips to survive. Effectiveness is the foundation of success—efficiency is a min-

imum condition for survival *after* success has been achieved. Efficiency is concerned with doing things right. Effectiveness is doing the right thing." See Peter F. Drucker, *Management, Tasks, Responsibilities, Practices* (New York: Harper & Row, 1974), p. 45.

35. From *Study in Power: John D. Rockefeller, Industrialist and Philanthropist,* Volume I, by Allan Nevins. Copyright 1953 Charles Scribner's Sons; copyright renewed © 1981 Mary R. Nevins [Reprinted with permission of Charles Scribner's Sons, and imprint of Macmillan Publishing Company.]

36. Sigmund Diamond, *The Reputation of the American Businessman* (Cambridge: Harvard University Press, 1955), p. 53.

37. Alfred L. Thimm, *Business Ideologies in the Reform-Progressive Era, 1880–1914* (Tuskaloosa, Ala.: The University of Alabama Press, 1976), p. 166–67.

38. John William Ward, "The Ideal of Individualism and the Reality of Organization," Earl F. Cheit, ed., *The Business Establishment,* (New York: John Wiley & Sons, 1964), p. 43. The ideas of this section are based on Ward's arguments.

39. See Henry Nash Smith, *Virgin Land: The American West as Symbol and Myth* (Cambridge: Harvard University Press, 1950).

40. Herbert Hoover, *The Challenge to Liberty* (New York: Charles Scribner's Sons 1934), pp. 54–55.

41. John William Ward, "The Ideal of Individualism and the Reality of Organization," in Earl F. Cheit, ed., *The Business Establishment,* pp. 39–40.

42. *Geary* v. *U.S. Steel Corporation,* 319A. 2d 174, Supreme Court of Pennsylvania, 1974. For other such cases see, David W. Ewing, *"Do It My Way or You're Fired": Employee Rights and the Changing Role of Management Prerogatives* (New York: John Wiley & Sons, 1983).

43. See Roy Rowan, "The Maverick Who Yelled Foul at Citibank," *Fortune,* January 10, 1983, for a whistle-blowing employee in Citicorp's, Paris office.

44. John William Ward, in Cheit, ed., *The Business Establishment,* p. 73.

45. Paul Weaver, "After Social Responsibility," *The U.S. Business Corporation An Institution in Transition* (Cambridge: Ballinger Publishing Company, 1988) pp. 141–43, makes a similar point. "The promarket, antigovernment posture [of business leaders in the 1930s] was disingenuous, of course. Businessmen remained as certain as ever of the need for cartels and more government spending . . . in the 1960s and 1970s, when an explosion in social policy and regulatory intervention led to a vast buildup of the business lobby in Washington, business's professions of corporate social responsibility were displaced and somewhat undercut by an increasingly reactive and opportunistic approach to policy. Business still sought entitlements in the name of social responsibility, [however]."

46. Sutton et al., *The American Business Creed* (Cambridge: Harvard University Press, 1956), pp. 356, 359, 361.

47. *Social Responsibilities of Business Corporations,* Research and Policy Committee of the Committee For Economic Development, June 1971, pp. 26–27. [Reprinted with permission.]

48. Harry Levinson, with Cynthia Lang, *Executive* (Cambridge: Harvard University Press, 1981), pp. 4–5. [Reprinted with permission.]

49. Harry Levinson and Stuart Rosenthal, *CEO: Corporate Leadership in Action* (New York: Basic Books, 1984), p. 286. [Reprinted with permission.]

50. Peter F. Drucker, *Management, Tasks, Responsibilities, Practices* (New York: Harper & Row, 1974), p. 176. [Reprinted with permission.]

51. David Ewing, *Freedom Inside the Organization: Bringing Civil Liberties to the Workplace* (New York: E. P. Dutton, 1977), and *"Do It My Way or You're Fired!" Employee Rights and the Changing Role of Management Prerogatives.*

52. Robert A. Nisbet, *Community and Power* (New York: Oxford University Press, (A Galaxy Book), 1962), p. 237. [Reprinted with permission.]

9

A Minimal Ethic of Market-Oriented Responsibility: The Nestlé Case

A widely publicized encounter over a decade ago between the managers of a large multinational business corporation and a disparate group of self-declared constituencies illustrates the problems business managers create for themselves and their companies when they become blind to, or ignore, the implicit assumptions of the democratic free market. Though economic theorists seldom consider the institutional and moral foundations that Nisbet argues underlie an effective, successful market system, business managers ignore those foundations at their peril, especially in the modern world of quick and wide communications and professionally led constituencies.

Economic theory commends market outcomes as efficient only if most of a citizenry participates, at least with enough income or resources to register some effective market demand, in choosing jobs, goods, and services in a voluntary, informed way.[1] If a business sells in a market where those conditions do not hold and profits thereby, not only does the company exploit those who should be served, it also fails to further its own or general economic efficiency, a basic, prime value of the market system.

The well-publicized encounter, which illustrates the problems that beset—and the opportunities that present themselves to—managers who disregard market assumption, was between the managers of Nestlé S. A., Vevey, Switzerland and a variety of well-organized, inventive, and persistent critics in the United States and Europe. What is most remarkable about the conflict of the parties is its duration, covering a span of more than a decade, from the early 1960s to 1984. Seldom have encounters of business managers and constituencies continued for so long; as noted later, the parties usually settle their differences within a period of weeks or months. If the clashes continue for a long time, it is almost surely a sign that (1) the managers are not grappling effectively with the underlying problems, (2) the problem itself is basically a violation of free-market assumptions and values, (3) the constituencies seek other and larger social or political objectives, or (4) all or any combination of the conditions exists.

The Nestlé encounter took place in many different countries on two continents, at a variety of levels, in private and in public. Members of churches, college students, stockholder groups, news columnists, and government officials became involved, some taking sides with the company and others passionately opposing its stand. Newspapers, business magazines, television specials, congressional hearings, and scholarly journals chronicled the charges, countercharges, and long, involved negotiations, court hearings, agency rulings, and boycott.

THE INFANT FORMULA CASE

The encounter began quietly enough in the early 1960s among international agencies of the United Nations and gradually grew to receive wide attention from the public in the United States and Western Europe. Until 1970 the company managers were able to recognize only one story about infant formula: their own. It was surprisingly simple—traditional, limited, and self-contained—for a giant corporation. That the managers continued to reiterate and defend the story long after the public found it obsolete and certainly incomplete suggests that they deluded themselves and the company in ways that increasingly diverged from the realism of the constituencies and outsiders.

The Nestlé Story

Infant formula had long been a profitable Nestlé product that was milk-white in color, pure, and useful. It was simple to manufacture, made mostly of powdered cow's milk with various additives to make it a near-substitute for mother's milk. For mothers who could not supply sufficient milk for their infants, or because of work, lifestyle, or preference did not want to breast-feed, Nestlé provided a needed and much appreciated product. Nestlé first produced infant formula for mothers in industrial nations, where breast-feeding of babies increasingly had become an "old-fashioned" custom. As the post–World War II baby boom in the United States and Europe began to dwindle, Nestlé and other producers increasingly promoted the use of infant formula in developing and Third World nations, where the birth rate continued at a high level.[2] In redirecting its capacity to serve new consumer, Nestlé managers believed they were meeting a genuine need. The chairman of Nestlé-Brazil explained:

> Doctors had been concerned by the use of inadequate foods when breast milk alone was insufficient to meet the needs of babies. . . . bottle-feeding was already practiced and was not introduced by the infant food companies . . . [Because many mothers fed their babies milk substitutes with low nutritional content] doctors therefore welcomed the possibility of prescribing a safe and nutritious product for babies whose mothers had to supplement or replace breast milk.[3]

Nestlé managers were convinced that they were profitably doing good in the world by producing a healthful product for those who needed it. In addition they were contributing to the common good of the countries in which they operated. In a company promotional pamphlet they argued:

> While Nestlé is not a philanthropic society, facts and figures clearly prove that the nature of its activities in developing countries is self-evident as a factor that contributes to economic development. The company's constant need for local raw materials, processing and staff, and the particular contribution it brings to local industry, support the fact that Nestlé's presence in the Third World is based on common interests in which the progress of one is always to the benefit of the other.[4]

This company story was one of mutual benefit for the consumers of its product and its owners and producers; it implied a clear purposeful ethic, pragmatic and result-oriented. It hardly allowed room for criticism or questions. That it should be attacked for providing so good a product for people who both needed and desired it, when the company efforts also promoted economic development, could only indicate either ignorance or intentional misrepresentation of the company and a lack of understanding of the managers' intentions and their accomplishments.

Not only was the Nestlé managers' story simple, but it also excluded the possibility that "business" could possibly be properly defined or efficiently practiced by any persons outside the conventional form. Most seriously, perhaps, it failed to allow for the possibility that economic conditions and the misinformed practices of people in another culture might lead to disastrous uses of the product in spite of the managers' good intentions. That such an undeserved *outcome* might occur never entered the calculations of Nestlé managers. They tenaciously resisted evidence that such a consequence not only could but did on occasion result.

The story of the new constituencies. The evidence that Nestlé's infant formula did lead to disastrous outcomes was slowly accumulated by doctors and nutrition scientists, assisted by professionals working on the
scene where the formula was sold to Third World consumers. Several
developments brought to the attention of world health officials a concern for infant nutrition: first, the increasing presence of international
and United Nations agencies; second, a rising popular American and
European interest in Third World economic development; and third,
shifting markets for infant formula from industrial to poorer nations.
As they observed the effects of commercial promotion of infant formula in those countries, they found unfolding a story quite different
from the one the managers told themselves.

Professionals serving in international organizations were seeking ways
of improving nutritional standards around the world. Through their
interest and work they unwittingly began to create a new political and
social environment within which the producers of infant formula were
going to be selling their products. Managers in the various producing
firms apparently viewed such professionals as only bothersome bureaucrats whose interests were to be deflected from any effect upon company policies and certainly any influence on corporate business strategies. In no way did Nestlé's managers think of them as a corporate
constituency; they were outside, and "the other."

Among the most active professionals were those associated with the
United Nations Protein-Calorie Advisory Group (PAG). They had been
charged with the responsibility of coordinating nutrition research and
aid programs carried out by the World Health Organization (WHO),
the Food and Agriculture Organization (FAO), and the United Nations
Children's Emergency Fund (UNICEF). In December of 1969 a PAG
working group began a study of nutrition of unborn children and of
babies during the first months after birth. It outlined a preliminary
action program addressed to the governments of states belonging to
the UN, health professionals, and managers of the infant food industry. In November of 1970 representatives from these groups met in a
conference in Bogota, Colombia. Among various topics discussed were
two especially pertinent to Nestlé—the importance of prolonged breast-
feeding and guidelines for the marketing of breast-milk substitutes.

Some health specialists expressed concern that aggressive marketing
of breast-milk substitutes was exacerbating the worldwide trend away
from breast-feeding. It was a disturbing trend, for breast milk, when
available, was safe, healthy, nutritious, and inexpensive. Yet most of the
experts acknowledged that the overall effect of the infant-formula industry was positive: It provided a good, high-quality food in Third World
countries and it probably saved the lives of many babies who could not
be breast-fed. One participant, Dr. Derrick B. Jelliffe, a PAG consul-

tant with long experience as a medical doctor in the Caribbean, strongly disagreed with this majority view. He argued that:

> . . . industry marketing practices were the *major factor* contributing to the decline in breast-feeding and the parallel growth in the consumption of products that given the poor hygienic conditions prevalent in most developing countries, contributed to infant disease, malnutrition and death.[5]

Few of the participants found convincing evidence to support Dr. Jelliffe's thesis, but the conference ended with a call for more research into the causes and consequences of the decline in breast-feeding—a typical scientific response to uncertainty. Undeterred by uncertainty and despite Jelliffe's arguments, industry participants had won conference endorsement of their basic claims that infant formula was a good product serving good purposes. " 'Modern processed infant foods' were to be made 'widely available.' " They were not troubled by the call for "improving the information provided by labels and educational materials [provided with infant formulas]."[6] After all, their companies were always upgrading educational materials and changing labels. At the most they needed simply to add warnings that breast-feeding was the most desirable way of feeding babies.

Jelliffe discerned a context more inclusive and involved than that perceived by the company's managers. He insisted that the producing firms had to take responsibility for all aspects of their marketing, even if it might be carried out by distant subsidiaries. He raised questions that the company managers had never seriously pondered: Who bought infant formula? Did they have the ability to use it without impairing their health and their babies? Jelliffe was concerned, for example, that the use of uninformed health workers in hospitals to offer free samples of formula to new mothers too easily persuaded poor, uninformed mothers to give up breast-feeding. Once having become dependent on formula, they then had little choice but to continue its use, even if they could not afford sufficient supply and did not have access to clean water with which to mix it or proper facilities to store it. One might say in-defense of the Nestlé managers in Vevey that they could hardly be expected to know what it is like to be a mother in a rural village in Zambia. Jelliffe, however, believed that the Nestlé managers' business intimately connected them to that mother. They had a responsibility to understand her life situation if they were to assume their right to influence it. Unfortunately, the notion that managers might travel to Zambia or Kenya to acquaint themselves with the nutritional situations of mothers in those countries did not occur to anyone in Vevey this early in the story.

The free market: Ideas as well as products. Jelliffe strongly urged conversations between industry leaders and those like himself. He called for a dialogue about what he imaginatively labeled "commerciogenic malnutrition," caused by the "thoughtless promotion" of infant formula.[7] It was a label at once scary and catchy. Furthermore, many people could find his bluntly simple argument compelling. In his initiative, search for supporters (customers), and use of imaginative publicity, he was acting something like a Schumpeterian business entrepreneur: He was beginning to create a new market for an idea, a value, and a service. Even if he were not a true innovator, he prepared the way for those who could and would use his notions to fashion an innovative service: a means of influencing the decision making of a large business firm and the practices of a whole industry to serve humanitarian ends.

Nestlé managers, however, perceived Jelliffe only as an overwrought critic. They continued to maintain that the cause-and-effect relationships were anything but simple, a truth that if adequately documented and followed might have led managers sooner to acknowledge the truth of Jelliffe's claims. Instead they and other industry representatives argued in later PAG meetings, in Paris in 1972 and in New York in November of 1973, that "it is essential [i.e., good] to make formulas, foods and instructions available to those mothers who do not breast-feed for various reasons."[8] They recognized that malnutrition and hunger were serious problems in many Third World nations, and thus endorsed a call to governments—not producing companies—to make available at low or no cost breast-milk substitutes and weaning foods to those in great need. In the process of negotiating policy language, the industry participants accepted statements that they should avoid aggressive sales and promotional practices that could possibly encourage unnecessary consumption of breast-milk substitutes. They also agreed to undertake programs of education in feeding methods, including instructions on the proper and safe use of commercial products. Such agreement suggested that company learning and the needed responsibility were emerging. The statements were mere exhortation, however, for the agencies had not insisted upon, and the firms had not agreed to, any monitoring or enforcement procedures.

In fact, little came of the agreement that had been worked out with such effort. Later observers in a company report pointed out that not only had bureaucratic inertia stymied any implementation but also "the industry was also at fault in its slowness to capitalize on the effective stamp of legitimization that this UN group had just given its role in Third World infant nutrition." Though Nestlé managers agreed with the policy, they did not command an organization that could easily translate policy into action. Although they effectively controlled the quantity and quality of production, they apparently exercised only loose, indirect, and distant authority over marketing, as events were to prove. Marketing had been, and continued to be, an activity not included in the top managers' main concerns and sense of responsibility.

Critics and constituency groups had no difficulty insisting that promotional practices were part of the Nestlé story; for them, consumers were company constituents and Nestlé marketing practices were salient activities of the firm. In the August 1973 issue of *The New Internationalist*, a British publication produced under the sponsorship of three charity organizations (Oxfam, Christian Aid, and Third World First), the editors featured a question-and-answer interview with two child-health specialists experienced in Third World nutrition issues. The specialists' criticism was much sharper than that of Dr. Jelliffe, and Nestlé managers charged that, in singling them out, the critics seriously misrepresented the company's marketing practices. In a letter to the editor, they maintained that:

> . . . it would be impossible to demonstrate in the space of a letter the enormous efforts made by the Nestlé organization to ensure the correct usage of their infant food products, and the way in which the PAG guidelines have been applied by the Nestlé subsidiaries.[9]

The editors declined to come to Vevey to see firsthand the evidence of the enormous effort, but a free-lance journalist, Mike Muller, on assignment for War on Want, another British charity organization, asked for an interview, which the managers welcomed. They wanted to explain the complexities of infant nutrition, not only the acknowledged hazards in using commercial formula but also their continuing efforts to reduce the risks.

Dr. H. Mueller, director of Nestlé's Dietetics Division, explained to Muller the way in which company ads and instructions attempted to help consumers use infant formula in a healthy and safe way. He admitted, however, that the company had not researched ways in which illiterate mothers' understanding of the ads and line drawings could be improved. He declared, "We can't undertake work of that nature."[10] It was a disputable assertion of limits on responsibility for a corporation with Nestlé's financial resources. But in any case such expensive and difficult studies were not necessary, he maintained, for company promotion and ads were not aimed at persuading mothers to abandon breast-feeding or at expanding sales. Their primary purpose was in maintaining existing consumer confidence in the product.

Mueller admitted that abuses had unfortunately developed but maintained that they were outside the competence or control of the company. What doctors in hospitals did and how consumers used infant formula was not a Nestlé responsibility.[11] More than five years later, ten months after the Infant Formula Action Coalition (INFACT) had begun its American boycott, Nestlé officials still insisted that the company did not have—could not have—the responsibility to find out the extent of consumer misuse of the product in Third World countries.[12] Such a claim could easily be translated by Nestlé critics into the old slogan, "Let the buyer beware!"

Such claims delimited Nestlé's role in a curious way: they produced a good product for good purposes, and delivered it to a competitive market in which they were largely passive providers, and to willing consumers beyond their reach. As kindly and concerned people, they recognized the adverse effects of competitive promotion, consumer abuses of their product, and the terrible, complicated problems of poverty, disease, and ignorance that contributed to malnutrition among Third World people. But these effects were not part of *their* assumed responsibility, though they charitably wanted to cooperate with others in mitigating them. The irony and tragedy of this stance did not seem to penetrate corporate ethical awareness. They had retreated to an ethic of intention, apparently unaware that "The good they meant to do could be an occasion for evil."

As the months passed, the story told by critics of Nestlé (and of the infant formula industry) differed markedly from that which the managers told themselves. The critics granted the quality of the product but were disturbed by aggressive promotion that increased sales and enhanced profits. They knew that many consumers could not afford the relatively expensive breast-milk substitute and feared that illiterate mothers did not understand the consequences of switching to formula. Uninformed consumers could not evaluate the dangers of formula and thus could not make rational, free-market choices. And more ironic and subtle was the cultural gap between consumers and producers. What *is* rationality for an African village mother who may be persuaded to regard the white-coated purveyor of Western science as the possessor of superior knowledge?

The critics' story recognized that Nestlé and the other firms selling infant formula were not meeting a fundamental free-market condition.[13] If consumers cannot make informed choices, they cannot act rationally to serve their individual interests, and the market will not maximize their utility; they may in fact be subject to exploitation. Paradoxically, the critics entertained a deeper, more profound appreciation of the free market and its special assumptions than did Nestlé managers. The company's managers, claiming to be defenders of the capitalist system,[14] refused to admit what was obvious to many people in Europe and in the United States, that their marketing and promotion did not meet fundamental free-market "rules." They were unwilling to admit that using sophisticated, high-pressure, mass-media sales efforts among certain people, many of whom could not read and were unused to modern advertising techniques, was as inadmissible in a free market as using such means to induce children to buy sophisticated, high-tech toys whose use might injure the unwary.

Confrontation: Charge and Countercharge

Although Nestlé may not have been the greatest offender in its promotion, the critics knew it was the leading producer of infant formula,
annually selling about $300 million worth and accounting for some-
where between 35 and 40 percent of the total foreign market.[15] It was
a conspicuous, readily identifiable company and its managers' inability
to comprehend, let alone recognize, any story but their own, made it a
perfect symbol of the critics' concern—heedless business profit seeking.

In the years of controversy that followed Jelliffe's initial public charges
against the infant formula producers, other critics focused on Nestlé
and added other complaints and accusations. Some denounced Nestlé
simply because it was a large multinational corporation and others joined
in attacking what they saw as a prominent symbol of Western economic
imperialism that had replaced the pre–World War II, old-fashioned
colonial system.[16] That this Marxist interpretation caused its proponents to overlook some truths in the history of the episode did not
cancel the partial validity of such an interpretation. Had not Nestlé
turned to the Third World as a promising new market, and did it not
promote profit under the ideological (colonialistlike) guise of doing good?
Some of the critics incorrectly blamed Nestlé and other infant formula
firms for the generally low standards of living, inadequate nutrition,
unsafe drinking water, and high death rate of babies in many of the
Third World nations. These additions to Jelliffe's free-market story were
clearly excesses, misleading and untrue; the Nestlé managers resented
them deeply and were angered that their company should be so badly
portrayed. But focusing their attention upon these excesses and bend-
ing every effort to rebut them amounted to self-deception. It allowed
them to excuse a basic flaw in their company policy and practices by
downplaying or ignoring the core of the critics' story: Nestlé's failure
to abide by free-market standards of *informed* consumers.

After Mike Muller had published his version of the interview with
company managers, *The Baby Killer,* in March, 1974, Nestlé managers
realized that at least they had to reexamine their promotional activities.
In a later review of the lengthy controversy, company writers noted
that although *The Baby Killer* had directly misrepresented some of [Nes-
tlé's] policies . . . it had underscored some very real problems of which
the industry was aware."[17] The managers must have had some notion
that the company salesforce in Third World countries was not always
following headquarter policies or carrying out its directives. Slowly and
painfully Vevey managers seemed to be learning new truths about their
possible responsibility for how African mothers used the company
product.

The officials in Vevey decided they must attempt to enforce PAG's
guidelines against aggressive promotion "even more strictly" than be-

fore, suggesting that enforcement had not been particularly strict before 1974. In addition Nestlé top managers conducted an internal review of their marketing practices in developing countries and ordered more stringent control over the distribution of product samples, elimination of direct contact between company representatives and new mothers, and suspension of advertising that did not meet the full approval of public authorities.[18] Belatedly and privately, they recognized that they could experience more control over the marketing of their product than they publicly had been willing to admit or privately to consider.

Nestlé critics did not know about this internal review. If they had known, it is unlikely they would have been impressed. Two activist groups, the Arbeitsgruppe Dritte Welt (Third World Working Group) and the Schweizerische Arbeitgruppen fuer Entwicklungspolitik (Swiss Working Group for Development Politics), in June, 1974, published a translated version of Muller's pamphlet *The Baby Killer* under the attention-getting title *Nestlé tötet Babys* (Nestlé Kills Babies). Readers were told that infant formula was the direct cause of infant deaths, and the title clearly suggested that Nestlé was the responsible party, a sobering illustration of a simplistic, irresponsible account of "the enemy." Nestlé managers were outraged and decided that they had to take legal action against the two groups. They brought suit in the Bern District Court, charging criminal libel. The case dragged on for nearly two years into the middle of 1976 and provided much publicity for Nestlé critics, catching the attention of activitists around the world. The judge finally concluded that Nestlé, indeed, had been libeled, but he admonished the company:

> . . . the need ensues for the Nestlé company to fundamentally rethink its advertising practices in developing countries as concerns bottle feeding, for its advertising practice up to now can transform a life-saving product into one that is dangerous and life-destroying. If [Nestlé S.A.] in the future wants to be spared the accusations of immoral and unethical conduct, it will have to change its advertising practices.[19]

The judge, presumably a neutral and an outside observer, himself a native of Switzerland, had weighed the substance of the critics' claims in the balance and found them, at the core, responsible. Even before the judge had censured the company, its managers realized that they could no longer ignore their promotion of infant formula and the consumer response to their marketing. In late 1975 they and seven other large producers of infant formula created the International Council of Infant Food Industries (ICIFI) to develop a code of conduct that would include the hortatory principles PAG had urged. Under the code, Nestlé for the first time placed a moratorium on mass-media advertising.[20] It was not clear how effectively the policy would be—or could be—enforced. Critics still doubted that the industry was acting in good faith and included Nestlé in their doubts. They found the new council and its code weak and inadequate.

Others also found the new code unacceptable, including PAG and Abbott Laboratories, an American firm whose Ross Laboratories subsidiary made an infant formula. Abbott managers had encouraged the founding of ICIFI but decided not to join when they found the newly created code to be weaker than their own. Abbott had already given up advertising in medical publications in late 1975 or early 1975; it did not market to health professionals or to governments in Third World countries. Its managers explained to their stockholders:

> The most important area is to reduce the impact of advertising on the low-income, poorly-educated populations where the risk is the greatest. The ICIFI code does not address this very important issue.

> Our company decided not to join ICIFI because the organization is not prepared to go far enough in answering this legitimate criticism of our industry. We feel that for Abbot/Ross to identify with this organization and its code would limit our ability to speak on the important issues.[21]

Of course American firms selling infant formula could give up mass-media advertising and high-pressured marketing at small cost because, unlike Nestlé, they sold largely through doctors and druggists, not through groceries and regular food shops. The objectionable sales techniques were hardly vital to them and their sales. Nonetheless, as fellow managers their opinion about the correctness of the risks and responsibilities inherent in the industry carried weight and put them inevitably on the side of Nestlé critics.

The Controversy Comes to the United States

The publicity about the trial in Switzerland, the refusal of PAG to endorse the ICIFI code, and the response to American producers of infant formula brought the controversy over "aggressive promotion" of the product by Nestlé to the attention of a number of groups in the United States. On July 4, 1977, Douglas Johnson, head of INFACT, announced the beginning of a nationwide consumer boycott of Nestlé products, even though its American subsidiary neither made nor sold infant formula. When the American managers complained that the action was misdirected, INFACT insisted that Nestlé in the U.S. was merely the "hand" of the transnational "body" and thus had vital links to the corporate "head" in Switzerland.

Though Nestlé managers met with the leaders of the boycott in October, 1977, and later in February, 1978, nothing came of the meetings. The issue of infant-formula marketing soon became a matter of wide public attention and political calculation. Senator Edward Kennedy conducted a hearing into the matter in May, 1978, and the National Council of Churches and a number of other smaller church groups supported INFACT's call for a boycott. With public disapproval spreading ever wider, Nestlé managers decided that they should end all mass-media advertising of infant formula, even though they had already modified it to contain more educational content than ever before. They also limited the role of medical representatives and qualified mothercraft personnel in hospitals as salespersons. The critics and boycotters hardly noticed, for Nestlé proved to be remarkably incapable of convincing anyone that it finally was changing its role in the story—that is managers had finally recognized that marketing was a company responsibility.

WHO and UNICEF convened a conference on infant feeding in October, 1979, at which the boycott leaders and other supporting groups were active. In a "consensus" report the conference members recommended a variety of marketing reforms that Nestlé managers found compatible with many of the changes they had already effected. Nevertheless, the boycott and criticism continued to expand.

After the conference, the Americans joined several other groups to form an international coordinating organization more effectively to conduct their activities in Europe, Asia, and the United States. Industry managers now realized that their response to their various critics had to become more convincing. ICIFI drew up yet another "new" marketing code in 1980. The successive "new," improved and strengthened codes suggested to critics that Nestlé and other firms in the industry had been exceedingly slow to take any effective action or to show a sense of real responsibility for marketing abuses. "Real" responsibility would have included not only a readiness to listen to critics once, but in a concept now forced upon the company to keep on learning from them.

The World Health Assembly, meeting in May of 1981, approved a redrafted version of the earlier, 1979 "consensus" WHO code. Much to the satisfaction of industry representatives, it included an endorsement of their claim for a good product for good purposes. "Proper use" of breast-milk substitutes, accompanied by "adequate information" and "appropriate marketing and distribution," was approved. At the 1980 meeting of the World Health Assembly, however, an incident occurred that suggested that there was another chapter in the Nestlé story to which the managers had not given much attention, and that was then, and has been since, ignored by the critics. It was a story of managerial ineptitude and poor company performance.

At the conference in Geneva, Nestlé and the other firms selling infant formula used the occasion to publicize the new, more demanding ICIFI code of self-regulation. They were also able to point to comparable codes adopted by the governments of several developing nations. The publicity was undercut, however, when the worldwide organization of boycotters held a press conference immediately after industry spokespersons had made their announcement. The boycott leaders cited some 200 instances of alleged marketing abuses from around the world. Press attention focused on the dramatic revelations rather than on the new code, particularly since Nestlé and the other firms were caught unaware. They had no reply to the allegations. They made no response because none of the firms had any quick way to check out the stories or even closely supervise what was going on in the countries where the abuses were supposed to have occurred. They did not have enough information available either to deny or to confirm the allegations. ICIFI and its member companies began an investigation of the charges, a process that took several weeks. They found that of the 200 incidents only 15 could be substantiated and traced to activities of the member companies, who accounted for 85 percent of total formula sales in developing countries. Many of the abuses either were not confirmed or involved nonmember companies.[22]

The International Baby Food Action Network (IBFAN), which coordinated the worldwide boycott, however, had established a wide-ranging observation and reporting system to monitor the marketing of infant formula in most developing countries. Its monitoring activities were far more comprehensive than, though not as accurate as, those of the companies themselves. Only slowly and with extra effort were the companies, and especially Nestlé, able to find out what was going on among their own sale forces in the Third World nations. Organizations that refuse to trust the eyes of their critics can end up blind; and discovering truth submerged in error is an intellectual capability much to be coveted by those who seek to manage.

The internal Nestlé story—An inefficient, bumbling firm in need of help. Helmut Maucher became Nestlé's new chief executive officer in 1984. He quickly reported to the public that the company had not been well managed in the preceding years, including those of the controversy over the marketing of infant formula. Shortly after assuming his office, Maucher described his job as "getting this somewhat sleepy giant moving again." In his first two years he closed 35 of the company's 317 factories and slashed 15,000 people from a payroll of 155,000. Profits rose 14.8 percent, to $492 million, though sales were almost flat, at $10.9 billion.[23] He set about improving the company's productivity, not only by closing unprofitable operations but also by combating the inefficiencies of the company's decentralization. The Vevey offices employed only 1,700 out of a total work force of 140,000. There was no central purchasing, and much financing was carried out by subsidiaries locally. Obviously marketing had been exceedingly decentralized.[24] He set about unclogging the corporate lines of communications, insisting that company executives get closer to their markets to find out when and where consumers purchased Nestlé products. He urged managers to speed decisions in the fifty or so countries where Nestlé operated. According to José Daniel, another member of the top management team, they installed a system to insure "there's less paper, more talk; there are more contacts, and you travel more."[25]

Maucher soon understood that the company had to monitor its sales force much more effectively than it had done in the past, assuring its top managers that the company's code of marketing was followed by all, particularly in the developing countries. To convince its critics, to make clear to its own employees its serious commitment to the code, and to improve communications with its various, scattered subsidiaries, Nestlé had already established the Nestlé Coordination Center for Nutrition, Inc., in Washington, D.C., under an able manager, Rafael D. Pagan. His apparent mission was to coordinate the company's efforts to deal with the critics and to end the boycott. For some time after his appointment Mr. Pagan was convinced he was dealing with groups of antibusiness activists. He concluded that they had

> . . . done a better job in the political arena [than Nestlé, having] . . . allied themselves to some of the world's resentments, and . . . called these resentments into battle against us. We cannot match the activists in creating resentments, nor is it our functions as creators of wealth to do so. What we can do—what indeed we must do—is to learn to think and act politically; to establish political goals; to develop political techniques and expertise. . . .[26]

He oversaw the appointment of an audit commission with members so distinguished that their periodic reports on Nestlé's enforcement of its marketing code would convince even skeptics that Nestlé had reformed its marketing practices. The members came from a variety of nonbusiness organizations and experiences—government, churches, and universities. Pagan's public comments in 1982 suggested that he and Nestlé managers saw their problem as one of political activism. If they could convince their constituents that they offered a reasonable and responsible program, they could win the day. In fact, and as experience demonstrated to them, their problem turned out to be more than that. The commission helped them understand that the company itself confronted serious internal problems.

The audit commission's collection of information pointed to the need for improved operational efficiency in implementing its marketing code. Data and reports from around the world revealed there were indeed marketing violations that needed *investigation* and attention. The next year, 1983, Pagan admitted that:

It is a difficult, hard process. You can't be a Pollyanna. It takes a lot of give and take, but the results pay for the effort. We feel this exercise will help us in developing markets in coming years.[27]

He was echoing the view of the audit commission itself when in its *Third Quarterly Report,* March 31, 1983, it reported:

. . . the complaints [from critics and boycotters who reported alleged code violations around the world] have shown themselves to be a valuable tool for Nestlé to implement its provisions, making them company policies in the field. Valid complaints have shown Nestlé where its personnel are simply not following Nestlé policy, whether intentionally or accidentally. They have shown Nestlé where it needs to focus its own auditing and implementation efforts. They have identified issues in which the Nestlé instructions and the Nestlé policies have needed clarification and refinement.

Such conclusions come down to the assertion that their critics were helping Nestlé managers learn how to manage *responsibility,* that their critics were usefully providing helpful eyes and ears in monitoring the company marketing code. Rather than destroying the company, sabotaging its sales efforts, or undermining capitalism, they were helping the managers fulfill their own promises and carry out company sales policies. Responding quickly and honestly now to the critics' reports by enforcing the code, Nestlé was soon able to convince the boycotters that the company was ready to negotiate an end to the six-and-a-half-year boycott.

WHY THE PROBLEM CONTINUED SO LONG: FAILURE TO RECOGNIZE THE MARKET ETHIC

Why did the boycott and the controversy between Nestlé and the various constituencies continue for so long? American firms producing infant formula settled with their critics in a matter of weeks or, at the most, months. They differed from Nestlé in that as drug companies they did not sell through grocery and general stores and seldom promoted their products through mass media. For the most part they relied upon doctors to prescribe their infant formula, directing their advertising at doctors, nurses, and health workers, all schooled and well trained. In agreeing to the critics' terms, American firms had to change their marketing methods only slightly. Even if they had used Nestlé's sales methods, however, it is doubtful that American managers as sophisticated and experienced as those in the drug companies would have allowed the dispute to linger on year after year. When pressed by constituencies or subjected to public criticism, American managers usually

respond with alacrity and work out a settlement. The example of Johnson & Johnson will be discussed in Chapter 13.

Nestlé's managers hardly appeared to understand that they operated
in a market; they concentrated their attention and their defense of "a
good product for good purposes," focusing their attention on production and ignoring their responsibility in selling infant formula to consumers. The description of the product was accurate, but irrelevant to
both the constituencies' criticism and to the assumption of proper market conditions. As long as sizable numbers of consumers were illiterate
mothers who did not understand the effects and costs of becoming dependent on infant formula and consequently foregoing breast-feeding,
a basic market assumption was not present. The consumers were not
informed and thus could not be presumed to be acting rationally or
even necessarily voluntarily. Thus, they were not choosing Nestlé's infant formula to satisfy their knowledgeable wants, but more likely were
being manipulated by Nestlé's sales promotion to serve the interests of
the company. The outcome could not be defended as efficient (in economic terminology) or as socially desirable.

Had Nestlé managers examined their role in marketing infant formula as a free-market activity, they would have realized that their practices were inappropriate and out of keeping with the fundamental values of a free market. In a sad alienation of themselves from free-market
doctrine, they placed themselves above and beyond it, denying themselves the defense of the good business practice of effectively informing
their customers. Ironically, they thought of themselves as the defenders
of capitalism against a crusade of anticapitalist ideologues. They saw
themselves not only defending their company but also protecting the
capitalist system itself. Convincing themselves that they had identified
a dangerous conspiracy against the system, they blinded themselves to
the possibility that the critics understood the assumptions and values of
the market better than they did. When the critics appealed to American
consumers to support the boycott of all Nestlé products, those who responded positively probably did so with an almost intuitive understanding that hard-sell advertising to illiterates is tinged with exploitation.
Many Americans are already sensitive to and concerned about the way
in which companies "pitch" their incessant television messages for toys
and candy to young children. In their inability to comprehend their
own disregard for market principles, Nestlé managers allowed IN-
FACT and the other constituencies to preempt what should at the very
least be a basic defense of any business firm, that it is operating in
accordance with free-market doctrine.

Not all, and maybe only a few, of Nestlé critics were concerned about
the free market, of course, and some of them may have been opposed
to the very notion of a free market. Those who saw in any large multinational corporation a threat to the well-being of world economy were

quite happy to join the boycott and to criticize Nestlé; supporters of
the Third World nations' New Economic Order saw in the attack upon
Nestlé an opportunity to discredit existing international economic arrangements; people who disapproved of the modern search for and
use of technological fixes joined the coalition, for they believed that
breast-feeding, as all things natural, were superior practices; liberal
church members who felt a sense of obligation to help the poor, disadvantaged, and dispossessed of the world lent their support to the
critics because they believed Nestlé was taking advantage of people who
were already helpless; and they were joined by others, internationalists,
who believed that the rich industrial nations could contribute to world
political and social stability by helping Third World nations improve
their lot.

There should have been little surprise that INFACT persuaded so
many people to support its boycott; its leaders ably took advantage of
Nestlé's mistakes and expertly stitched together a broad coalition of
groups with parallel interests. Consequently, INFACT confronted Nestlé managers with a far broader constituency than that of any one of
the coalition members. If INFACT did not express the public interest,
it clearly represented the interests of several publics, whose combined
numbers were impressive, finally, even to Nestlé managers.

The Nestlé case, with its various and changing stories—both told and
enacted—suggest that business firms are not the clearly defined, distinct organizations that managers and the public have traditionally
imagined. It also suggests that even though the increased fuzziness of
organizational boundaries certainly makes managers' essential core-
coordinating function more difficult, it offers opportunities for generating information about a swiftly changing world and for gathering
new, innovative ideas about policies, programs, and products or services.

As those affected by company's policies and programs—corporate
constituencies—have learned to respond both through the market in
an organized way and through mobilizing public attention or shareholder concern, managers are discovering that the range of their en-
gagement must be far wider than ever imagined in the past. The increased and expanded involvement of constituencies in the affairs of
business firms is currently accompanied in many industrial countries by
the quickened pace of global economic competitiveness, a general lessening of governmental "protective controls," and the removal of anti-
competitive regulations. More competition increases managers' vulnerability, reducing their margin for error, but the rise of constituencies
and their demand for a role to play in the affairs of the firm offers
possibilities for increasing managers' effectiveness and even efficiency
in meeting the new competition.

As we have noted in earlier chapters, the rise of large institutional
investors and those ready to "raid," "takeover," or "merge" any com-

pany deemed inefficient multiplies the new competitive pressures and makes more urgent than ever before the need for managers who can make the very best of the situation in which they and their firms find themselves. They do not need to borrow additional trouble from other constituencies; indeed, continuing and long-standing trouble from constituencies may be an indicator, as with Nestlé, that managers are not handling their firm's other activities any more successfully than they are those of the constituencies. Bad management in one area of business is often a sign that management throughout the firm may be slack.

Neither managers nor readers will be persuaded that constituencies and other interest groups directly and seriously concern themselves with the efficiency—and thus profitability—of the firms with which they deal. As argued in Chapter 4, constituencies are most interested in what they define as justice for themselves. We may presume that usually they pursue most avidly those purposes and goals that benefit themselves primarily. In two kinds of situations, though, even narrowly focused constituencies may serve larger—even business—interests. First, constituencies may find frequently that they are too small or lack sufficient economic force to press managers effectively. They may broaden their purposes expediently to include those of others whose interests complement, partially overlap, or reinforce their own, and so form coalitions as INFACT did in its confrontations with Nestlé. Second, constituencies find that they can win wide public support or even discover legal assistance when they identify management mistakes, shortcomings, and inefficiencies, particularly those that arise from disregard of basic market values and assumption. Nestlé's critics, we argue, found themsleves within both situations and expertly availed themselves of the opportunities.

BEYOND THE MARKET ETHIC: NESTLÉ MANAGERS DISCOVER RESPONSIBILITY

Eventually Nestlé's managers learned what General Motors' managers had discovered some years before, that critics are an opportunity as well as a problem. A wise management will take critics seriously even if they appear to be more concerned with their own agenda than in helping the company to continue in business.[28]

In 1983, fourteen years after the first international criticisms of infant-formula marketing, Mr. Pagan summarized the lessons that Nestlé managers had learned from the long, often bitter, and certainly unpleasant encounter with their critics and the boycott:

> The politicization of institutions has brought forces into the business environment with which business is ill-equipped to deal. The church, for ex-

ample, from individual denominations to the National Council of Churches and the World Council of Churches, has an agenda which specifically addresses the changes it seeks in the way we do business. How many corporations have a church-relations program or are sensitive to the churches' social and economic concerns? Consumer organizations no longer are simply concerned with the quality of our products or the verity of our advertisement, they seek changes in the decision-making process within business and question the "appropriateness" of some of our products. Yet how many corporate consumers affairs officers are prepared to deal with the political strategies of these groups? Very few.

. . . there are a number of steps we can take to adapt to these new forces and to make positive contributions to the betterment of man and society. First, we must develop better corporate information systems. We must know what the churches are thinking, we must know what the International Consumer Groups are thinking, we must be sensitive to the political demands of important groups, such as environmental, women's and hunger coalitions. Once we develop solid sources of information, we can reach an understanding of the objectives of others and perceive accurately their reactions to our decisions. We can then also develop practical strategies to deal with the issues and the forces involved in a constructive way.

Once that understanding is achieved, we must share it throughout our companies. We must make the most far-flung field personnel aware of the implications of their every action. . . . Once we understand the objectives and strategies of these new forces which affect us, we must find a way to deal with them. An obvious first step is to meet with the leadership of these groups and try to find common ground or affect their understanding of the issues. . . . An industry organization determined to work closely with the churches would head off many of the problems which we only find out about when the stockholder resolution is filed or a church leader testifies before some Congressional Committee about our mendacity.

Similarly, companies and industries must reach out to form alliances with their closest publics: employee, customers, suppliers and neighboring communities . . . I don't think it helps business to wage a negative rearguard action against workers, customers and communities that have legitimate concerns about our operations and practices. It is better to listen, then work to create guidelines that address their concerns, and that we can work with.

. . . I believe that, if business prospers in the 1980s, it will be in large measure because business changes its attitudes, opens its spirit, admits its humanity, and acknowledges that its presence in and acceptance by a community implies acceptance of a broad array of responsibilities to that community and to the larger world outside our offices over and above the privilege to create wealth. . . . It would involve genuine changes in attitude and some changes in practices. But they would be changes well-worth making, because they ensure our survival as part of a world of free people who would have the opportunity to make something of their lives.[29]

Pagan's statement is a remarkable summary of the elements of "responsibility" that we will explore in the coming chapters. Against the context of Nestlé's long struggle with its critics, as a shift of policy, ethics, and public stance, the statement is riveting. Pagan acknowledges, at the start, that the "politicization" of the company's market found it "ill-equipped." The new forces were unprecedented; they constituted an unanticipated, unanticipatable event. The policy of ignoring or merely resisting this event is no longer tenable for Nestlé, he goes on to concede. Company response there must be—it must "deal with" these new forces. But in what interpretation? "The betterment of humankind and society"—a very broad and general umbrella—will do for a tag. He then specifies the vital, positive role which extra-company constituencies will have to play en route to that betterment. Especially astonishing is the wide net of confidence spread around to perspectives and data as diverse as those of church people, environmentalists, consumers, hunger action organizations, and groups devoted to women's rights. There is already here an anticipation that those other groups will have "reactions to our decisions," and the company must listen accordingly, in an ongoing process of dialogue, in which "we . . . find a way to deal with" these new partners in our enterprise.

The last paragraphs are replete with the language of solidarity and exploration: finding, trying to find, listening, searching for common ground. This new *style* of company constituency relations is substantive, not mere "public relations." It entails ongoing vulnerability to what may yet happen in those relations to shape the future of this corporation.

Most startling of all is the final paragraph of the text. The multiplication of terms from the realm of humanism and even religion implies a radical break with the closeted, specialized, elitist stance of a decade before. Pagan indicates a concern for "spirit," "humanity," and a broad array of responsibilities beyond the creation of wealth in a world of free people who want "to make something of their lives."

Does a business executive have any right to expect that citizens of a modern, pluralistic (not to say relativistic), culture will accept this language as serious, tenable, and emblematic of things actual and possible? Is this language rather an occasion for smirks and cynicism? One's reaction will signal much more about one's own position on the philosophical issues posed in these pages. Pagan's speech can be characterized as idealistic or realistic, a vital or a phony speech, depending on the reader's point of view.

Even if the speech was a feat of skillful window dressing, however, a phenomenal change had occurred in the company story that one Nestlé official was publicly telling.[29a] A new *theory* henceforth infused their public pronouncements: hospitality to a multiplicity of views and interests now belonged to their new definition of good business practice. No longer did they define their business concern and their organizational responsibilities as narrowly as they had tried to do through the 1970s and earlier. The Nestlé story, they discovered, involved their marketing as well as their production practices; the organization had to be far more permeable and open-ended than their original conception of "company" allowed them to contemplate. Customers and many other groups and interests had to be accepted as legitimate members of the organization—constituent parts of their firm—and could serve it well or badly, depending in part upon managers' response to them.

Another lesson can be drawn from the Nestlé experience. Since the controversy continued so long and the boycott covered so many years, Nestlé inadvertently educated its critics in business strategies and trained them in the intricacies of policy making, corporate implementation, and day-to-day, bottom-line operations. Many people were involved in organizing the boycott, testifying before Congress, publicizing the anti-Nestlé campaign on television and in the press, and coordinating a variety of affiliated groups. This experience provided a trained cadre of activists and constituency leaders who have been able to apply their skills in challenging other companies and confronting managers in other industries. The Nestlé case thus becomes one of the great "work-up" battles of the corporate responsibility movement, as crucial for its history as were the great Detroit and Cleveland sit-down strikes conducted by the auto workers in 1937[30] and the Birmingham campaign in 1962 for the civil rights movement. As long as managers react adamantly to constituency proposals and confront them as enemies and hostile influences, they run the risk of protracted battles and will almost surely educate their constituencies to react adversarially and combatively.[31]

APPENDIX: PLANT CLOSINGS
AS A FREE-MARKET ISSUE

With the same disregard for free-market doctrine that Nestlé managers showed throughout the controversy with their critics, many business managers, leading business-lobbying organizations (including the Chamber of Commerce and the National Association of Manufacturers), and the Reagan administration took the non-free-market side of this debate. The majority vote of the Congress for the plant-closing provision, the public opinion surveys that showed decided preferences for plant-closing notification, and the common practice among large corporations of giving ample notification of closings, indicated a widespread appreciation that such notice was both expected and "fair." It is fair, not only in accord with some sense of justice, but also in its accord with free-market assumptions that those in the market cannot act rationally and thus efficiently if they are poorly informed.

For managers first to decide to close and then actually to close abruptly a facility where workers believe they will remain employed has drastic effects on a community. Not only does the closing dash the employees' expectations of continued work and income, it also dashes the hopes and expectations of other community members—retailers, service operations, and government—who depend on income derived from workers spending. With little nor no advanced notice, none of those affected by a closing can begin the process of adjustment, seeking the most efficient and effective ways of looking for new work, cutting expenditures, and accumulating savings for the hard times almost sure to follow. They are denied the economic, free-market right of acting as informed, rational participants in the economic system.

There can be little doubt that some employers, particularly those who operate small firms, may find the costs of early notice difficult and expensive. The solution, however, is not to deny employees and community members the information they need to make sensible, efficient adjustments, but for the state or federal governments to provide some kind of minimal, but reasonable, assistance for the affected firms. Unexpected unemployment is almost sure to impose heavier costs upon the unemployment insurance system, and thus the industry at large, and upon the community than expected job loss. Continued, rigid insistence by business leaders and lobbyists that plant-closing legislation is inefficient and therefore unacceptable is almost sure to create even more urgent political pressure for governmentally imposed, uniform, inflexible restrictions on plant closings. Moreover, the leaders betray their own narrow interpretation of inefficiency, limiting it to their particular situation and ignoring the wider losses to other community members. Managers' opposition to prior notification of plant closings is based on a misreading of business interest and a misunderstanding or ignorance of free-market imperatives. Americans support, by overwhelming majorities in public opinion surveys, early notification of plant closings, and some local government officials have found it wise to respond to their electorates by suing firms that close factories with little regard for their communities. Of course, legislation prescribing a particular kind of notification may not be efficient. We do not argue for a law on notification, but for company policy of notification, something that both accords with free-market assumptions and receives warm public endorsement.[32]

NOTES

1. See Milton Friedman, *Capitalism and Freedom* (Chicago: University of Chicago Press, 1962), p. 13.

2. In its 1973 annual report, Nestlé managers wrote: ". . . the continual decline in birth rates, particularly in countries with a high standard of living, retarded the growth of the market. . . . In the developing countries our own products continue to sell well thanks to the growth of population and improved living standards."

3. Testimony by Oswaldo Ballarin, Chair, Nestlé-Brazil, before the Senate Subcommittee on Health and Scientific Research, May 23, 1978.

4. From the cover of the publication, *Nestlé in the developing countries* (Vevey, Switzerland: Nestlé Alimentana S.A., 1975). (A richly illustrated 228-page survey of Nestlé S.A. operations in the developing countries.)

5. Maggie McComas, Geoffrey Fookes, and George Taucher, *The Dilemma of Third World Nutrition: Nestlé and the Role of Infant Formula,* n.d. 1984? p. 4— published by Nestlé after the boycott settlement.

6. Ibid., p. 4.

7. D. B. Jelliffe, "Commerciogenic Malnutrition? Time for a Dialogue," *Food Technology* 25 (February 1971):56.

8. Introduction to PAG Statement 23, "Promotion of Special Foods (Infant Formula and Processed Protein Foods) for Vulnerable Groups," November 28, 1973.

9. "Milk and Murder," (editorial) *New Internationalist,* October 1973.

10. Mike Muller, *The Baby Killer* (London: War on Want, 1974), p. 10.

11. Dr. Mueller explained that "We do not deny that there is a trend to use that milk for feeding babies but we can say there are good reasons for doing so in some cases. . . . We make a product available on the market but the onus cannot be on us to stop mothers from taking such a product into consideration. . . . So we would like to be able to control the consumption to those people who can afford our product. It's good business to do so. Our problem is how do you stop that product getting to people who shouldn't be using it?" He indicated that governments should be setting up infant health centers, following the policy worked out with PAG. See Muller, *The Baby Killer,* p. 13.

12. Dr. Oswaldo Ballarin, speaking for Nestlé, S.A., was questioned by Senator Edward Kennedy at a hearing in May, 1978.

SENATOR KENNEDY: . . . It seemed to me that they [other witnesses] were expressing a very deep compassion and concern about the well-being of infants, the most vulnerable in this . . . face of the world. Would you agree with me that your product should not be used where there is impure water? Yes or no?

DR. BALLARIN: Uh, we give all the instructions . . .

SENATOR KENNEDY: Just . . . just answer. What would you . . . what is your position?

DR. BALLARIN: Of course not. But we cannot cope with that.

SENATOR KENNEDY: Well, as I understand what you say, is where there's impure water, it should not be used.

DR. BALLARIN: Yes.

SENATOR KENNEDY: Where the people are so poor that they're not gonna realistically be able to continue to purchase it, and which is gonna . . . that they're going to dilute it to a point, which is going to endanger the health, that it should not be used.

DR. BALLARIN: Yes. I believe . . .

SENATOR KENNEDY: Alright now . . . then my final question is . . . is what do you . . . or what do you feel is your corporate responsibility to find out the extent of the use of your product in those circumstances in the developing part of the world? Do you feel that you have any responsibility?

DR. BALLARIN: We can't have that responsibility, sir. May I make a reference to . . .

SENATOR KENNEDY: You can't have that responsibility?

DR. BALLARIN: No.

From *Transcript*, "Into the Mouth of Babes," CBS Reports, Edition VI, as broadcast over the CBS Television Network, Wednesday, July 5, 1978, produced by Janet Roach.

13. Milton Friedman, Nobel laureate economist, and well-known popularizer of free market values and strong advocate of the competitive economic system, emphasized the assumption that parties to market transactions should act voluntarily and be fully informed. In *Capitalism and Freedom*, he wrote: ". . . both parties to an economic transaction benefit from it, *provided the transaction is bilaterally voluntary and informed.*" (italics in original), p. 13.

When he held hearings on the Nestlé issue in 1978, Senator Kennedy asked the question about how well the assumptions of the free market were being met in the marketing of infant formula: "Can a product which requires clean water, good sanitation, adequate family income, and a literate parent to follow printed instructions be properly and safely used in areas where water is contaminated, sewage runs in the streets, poverty is severe and illiteracy is high?" Quoted by McComas, Fookes, and Taucher, *The Dilemma of Third World Nutrition*, p. 13.

14. In testimony before the Kennedy subcommittee, Dr. Oswaldo Ballarin began his testimony by declaring, "United Nestlé Company has advised me that their research indicates this is actually an indirect attack on the free world's economic system: a worldwide church organization with its stated purpose of undermining the free enterprise system is at the forefront of this activity." *Transcript*, "Into the Mouth of Babes." After the boycott, a retired vice-president of marketing, Nestlé S.A., visited a number of business schools in the United States, presenting the same argument that the church groups, under the leadership of the Communists and World Council of Churches, probably directed from Moscow, had at the very least displayed parallel interests and followed similar strategies in attacking major capitalist, business corporations.

15. Ann Crittenden, "Baby Formula Sales in Third World Criticized," the *New York Times*, September 11, 1975.

16. Some of the critics argued that the basic issue was one of class conflict. Mark Ritchie, an INFACT coordinator, proclaimed before a church group, "It's not just babies, it's not just multinational corporations, it's class conflict and class struggle. . . . I think ultimately what we're trying to do is take an issue-specific focus campaign and move in conjunction with other issue-specific campaigns into a larger, very wide, very class-conscious campaign and reassert our power in this country, our power in this world." Remarks before symposium, "The Human Costs of Corporate Power," organized by Clergy and Laity Concerned (CALC), August 16–17, 1978. Quoted by McComas, Fookes, and Taucher, *The Dilemma of Third World Nutrition*, p.15.

17. McComas, Fookes, and Taucher, *The Dilemma of Third World Nutrition*, p. 8.

18. Ibid., pp. 6, 7.

19. *INFACT* (Winter 1978), p. 5.

20. *PAG Bulletin* (September–December, 1977), p. 79.

21. Abbott Laboratories, *Commitment* (Spring 1976), p. 12.

22. McComas, Fookes, and Taucher, *The Dilemma of Third World Nutrition*, p. 15.

23. Maile Hulihan, "Nestlé Plots Aggressive Acquisition Program," the *Wall Street Journal*, May 21, 1984.

24. John Tagliabue, "Pruning Nestlé's Operations," the *New York Times*, January 3, 1983.

25. Ibid.

26. Rafael D. Pagan, Jr., "Carrying the Fight to the Critics of Multinational Capitalism," remarks made before the Public Affairs Council, New York, April 22, 1982, p. 2.

27. Kevin Higgins, "Infant Formula Protest Teaches Nestlé a Tactical Lesson," *Marketing News*, June 10, 1983.

28. Michael Kinsley, in a *Fortune* review of a book by Naderites Mark Green and John F. Berry, *The Challenge of Hidden Profits: Reducing Corporate Bureaucracy and Waste* (New York: Morrow, 1985), p. 193, pointed out that "most of the left-wing critique of American business is perfectly compatible with concern for profits and economic growth. Indeed there's a remarkable overlap between some of the things Nader & Co. have been saying all these years and the more recent criticisms by business consultants like Tom Peters, academics like the late William Abernathy of Harvard Business School, and the business publications like *Fortune*." *Business Week* writers some time ago pointed out that "General Motors Corp. and other auto makers paid dearly for failing to recognize early enough that Ralph Nader's objection to the Corvair model was a forerunner of a broad-based consumer movement for safer products and tougher liability standards. Similarly, by ignoring early warnings from environmentalists, hundreds of manufacturers were forced to retro-fit plants with pollution-control gear that could have been incorporated more cheaply in the original plant design." See "Capitalizing on Social Change," *Business Week*, October 29, 1979, p. 105.

29. "The Politicization of Institutions: The Responsibilities of Multinational Corporations," *Vital Speeches* (October 15, 1983):16–27.

29a. There is reason to conclude that Pagan's speech was, if not window dressing, something less than the conclusions of Nestlé's top managers. October 4, 1989, the Minneapolis-based Action for Corporate Accountability (ACA) announced a renewal of the Nestlé boycott and initiated a boycott of American Home Products. ACA charged that both companies were marketing their infant formula products in poor communities around the world in violation of the World Health Organization and UNICEF marketing code. The companies had refused to end their practices of providing free supplies of formula to hospitals, and ACA charged that the purpose of the donations was to induce sales by "hooking" uneducated, unsuspecting mothers on infant formula. See Carol-Linnea Salmon, "Milking Deadly Dollars From the Third World," *Business & Society Review* 68 (Winter 1989):43–51.

30. During these strikes in the major plants of General Motors and Chrysler "management was utterly shaken, for although there had been some minor sit-downs in rubber the previous summer, the extension of this anarcho-syndicalist weapon in the heart of American industry was entirely unexpected. . . . It is difficult even in retrospect to appreciate the revolutionary character of the sit-

down strike, a labor weapon that was used with great effectiveness during the first six months of 1937, and then disappeared." See Walter Galenson, "1937: The Turning Point for American Labor," in Ray Marshall and Richard Perlman, eds., *An Anthology of Labor Economics: Readings and Commentary,* (New York: John Wiley & Sons, 1972), p. 79.

31. The leaders of INFACT, for example, have made use of their expertise, honed and sharpened in the Nestlé campaign, by mounting efforts to persuade major defense contractors to give up the production of nuclear armaments. Their immediate goal was to drive General Electric out of the nuclear weapons business. They chose the goal and target company after studying likely public causes and conducting hundreds of street surveys, narrowing their possibilities to two projects—drug companies that may be "dumping" unsafe products abroad and makers of nuclear weapons. They decided that the latter was the greater threat, and selected General Electric as a target because it is the country's largest nuclear contractor. While their campaign did not quickly attract much attention, the leaders and supporters were not discouraged. After all, the campaign against Nestlé gathered momentum only slowly.

32. For examples, see Joseph B. White, "Factory Towns Start to Fight Back Angrily When Firms Pull Out," the *Wall Street Journal,* March 8, 1988.

10

The Content of
Human Responsibility:
Values and Principles

Suppose we treat action as a way of creating interesting goals at the same time as we treat goals as a way of justifying action. . . . Individuals and organizations sometimes need ways of doing things for which they have no good reasons. They need to act before they think. . . . In an organization that wants to continue to develop new objectives, a manager needs to be tolerant of the idea that he will discover the meaning of yesterday's action in the experiences and interpretations of today.[1]

A character in a novel by E. M. Forster remarks, "How do I know what I think until I hear what I say?" In the last chapter we urged American business leaders to consider the wisdom of an alternate version of this principle: "How will we know what we really think until we acknowledge what we really do?"

Business managers enact a more complex set of beliefs, principles, and values than they tend to acknowledge in their public rhetoric, perhaps like all who aspire to the title *professional*. Managers' calling requires them first of all to *do* something believed to be of value or service to other people. Those who enjoy the esteem of their peers and the public realize that they earn professional respect from a certain consistency of performance judged successful, effective, competent, and excellent. Peer judgment alone is seldom enough to win the highest professional accolade. Professionals, among whom business managers may number themselves, must deal with various nonprofessional "constituents," variously called clients, patients, or customers. True professionals cannot, and do not, ignore their judgments.

We have assumed in this study that human behavior is a good index of what human beings really value. The apparent discrepancies between popularly acclaimed business "values" and the *real* values evident in the behavior of many American business managers in the 1980s can be explained partly in terms of actual, multiple accountabilities now built into corporate managerial roles. The statement by the Committee for Economic Development (CED) in 1971 describes those multiple accountabilities as like those of "a trustee balancing the interest of many diverse participants" in the business system.[2] Interest balancing, of course, requires talk about fairness, and underneath fairness is usually the more comprehensive value *justice*. The business managers who prepared the CED report recognized that whether or not in theory the competitive "free" enterprise system recognizes justice, in their managerial role they are continually embroiled in questions of justice. By 1971 unions, civil rights groups, and members of the corporate responsibility movement had forced the managers of large business corporations to address explicitly their demands for justice. By not acknowledging the force of the corporate responsibility movement, those who wrote the CED statement gave it a decidedly paternatlistic flavor.

Questions of justice complicate many a managerial decision, even in situations little buffeted by constituency pressure. One of the authors saw a convincing example in a discussion among a dozen top managers in a southern textile-dominated community. At the time of the discussion, the managers' firms were already under strong competitive pressure to modernize their production processes. New capital investments had to be made, old, inefficient plants closed, and workers laid off. Near the beginning of the meeting one manager, in a mood to instruct the discussion leaders about the "reality" of business, remarked, "You know, everyone around this table competes with everyone else. We're in business to put each other out of business." The discussion then turned to two policy questions that were currently the concern of the managers of several of the firms represented. First, "should a management give 'regular customers' preference when supplies are limited?" Second, "what should be done about superannuated, inefficient employees?"

The customer in the first question was another textile company in need of locally produced combed yarn. As the discussion developed, two principles emerged as the consensus of the group. The first was that regular customers ought to be protected and given preference in the long-run interests of one's own company, even if greater short-run profit could be earned from selling to another company with a known reputation for being "inconstant." The second was that even aside from profits, there is something inherently admirable about constancy, which should be honored and rewarded in business exchanges. In the managers' conclusions was an issue familiar to students of ethics: Is an act to be valued in itself or for the sake of another value, in this case either long- or short-run profits?

The superannuated employee who had occasioned the second question worked for the company whose manager had opened the discussion with the warning that each firm was in bitter competition with every other firm. The lack of pension plans for most textile workers, someone noted, made laying off older, inefficient employees a difficult decision. The discussion leader asked if keeping such employees on the payroll cut into profits and lowered overall efficiency. The managers agreed that it did. "Then, why have you not yet fired him?" the leader asked the manager. He hesitantly replied, "We didn't have the heart to fire him."[3]

Twenty years later, when for the very survival of their companies in an even more competitive, international market, companies are laying off thousands of employees, managers may react to the story in several ways. Our aim here is to examine, in a theoretical way, which of the responses makes the best economic, managerial, social, and ethical sense. What is the most sensible response to issues of customers constancy and employee inefficiency?

FOUR OPTIONS FOR "THE JUST RESPONSE"

1. Managers should dismiss a sense of loyalty either to constant customers or to aged employees as so much economic nonsense.
2. Managers should try to translate "human relations" into profit terms, acknowledging that a reputation for sensitivity to employee need will help a company recruit more capable workers, who in turn will be more efficient and productive.
3. Managers should adopt the CED's argument that those in charge of business firms be left to balance the claims and rights of constituencies, much in the way legislators resolve the conflicts among members of their electorate.
4. Managers should recognize and accept the tensions among the various demands made upon them and their firms and seek to resolve the resulting conflicts, contradictions, and dissidence with the aid, assistance, and participation of those whose lives they affect.

To these four options we will add a fifth one not so familiar to most American managers and deal with it in following chapters. We might call this proposal a way "beyond participation towards responsibility," but it will take some spelling out.

Regarding the first response, Milton Friedman and other advocates of the free, competitive market would argue that although it makes certain *extra*-economic sense for society through government to show compassion for the disadvantaged—by providing pensions for the elderly, for example it is hardly profitable, efficient, or appropriate for a business firm to do so. Compassion, or "heart," is impossible to justify as a virtue for a business firm that owes legal responsibility to those who own the firm and others like suppliers, retailers, and employees, with whom it has made legally binding contracts.

The second response is one recommended by Elton Mayo. It allows managers to introduce justice or compassion into their decisions as values contributing to the overriding virtue of profitability. Any number of other conditions and benefits for workers or other constituents might be justified in this fashion. One may wonder, however, just how credible this response is for managers confronting the competitive pressures of the day. For example, Japanese and Korean textile firms typically do not have pension plans for their workers. When American textile firms provide pensions for their employees, they incur added costs that impair their ability to compete with foreign imports. Pension plans do cost money. Only in times of acute shortage of workers will a pension plan be a competitive advantage in recruiting workers. In times of high unemployment when firms need to lay off workers, pension expenses hurt American businesses vis à vis lower cost, foreign producers. The attempt to justify a human relations approach in service to profits fails all too quickly when applied to difficult problems.

The third response is one that General Motors long made and finally had to give up in 1982. Its managers had traditionally dealt with constant suppliers in a noncompetitive way. As foreign sales of automobiles increased, the cost-saving advantages of competitive bidding became too great for them to continue their preferred, traditional relationships. During the many years that the managers had dealt with their suppliers, they had been able to rationalize their failure to maximize profits. Not only were American consumers ready and able to pay the increased costs (absent effective competition for automobiles), but suppliers were able to provide more stable employment for their workers than otherwise would have been the case. In addition, General Motors gained stability for its own planning and production efforts. Traditionally, deals with suppliers were based on friendly personal relations among GM's sales personnel and those with whom they dealt in the supplying companies. Even national prosperity could be used to justify the company's relationship with its suppliers, for the giant firm was large enough to make plausible the claim that its good and the good of the country were the same!

Those who have argued for socialist economies have long urged similar justifications. They have argued that employment, stable planning, interpersonal loyalty, and service to country are purposes ethically prior to those of profit, which serves only a particular economic class. Indeed, at the level of the values and interests they presume to promote, such business managers are at least superficially vulnerable to the criticism that they advocate "corporate socialism," wherein the business system combines economic efficiency with comprehensive justice and compassion. At its most inflated, this reasoning amounts to a claim that business firms can be model societies—and models of justice—for the rest of society.

Critics of American business often react with amusement or skepticism at the notion of the CED's socially responsible firm, even though the description is not far from that of the 1952 corporation quite accurately described by John Kenneth Galbraith in *The New Industrial State*. But something is missing here, without which the description becomes a distortion. That something is a context of democratic social structure and tradition, both of which are distinctly American.

The fourth response follows a more complex, credible argument than the other three. It justifies not only profit but also justice and compassion, as purposes pulling in different directions, subjecting managers to stress and strain. The resulting tension almost requires, and its relief surely recommends, that managers make use of more than their own judgments. To seek relief they will have to seek help in resolving the conflicts and contradictions of the various purposes to which they want to respond.

In this fourth philosophical context, the textile manager who gives preference to regular customers and cannot dismiss aged employees need not be embarrassed at his apparent inconsistency in proclaiming his enacted values. He need not strain to promote his feelings about these matters under the cover of "hard" profit priorities, nor should he be embarrassed even at the *real* inconsistency, or tension, among the various demands made upon him by his managerial role and his position as a community citizen. The human relations approach to management, after all, may rightly begin with managers' own acknowledgment that their lives do not consist only of a single relationship, either inside the firm or outside of it.

For example, from somewhere (perhaps the church) the textile manager had acquired "heart," and in his use of the collective "we" he ascribed heart to his fellow managers as well. From somewhere (perhaps in public school) he had learned that fair play in human relations consists of doing unto others at least as kindly as they have done to you. Such a modest sense of justice could suffice to call for a certain constancy in relation to both constant customers and constant employees. From somewhere (perhaps his family) he learned the importance of the ties that bind human beings together, even in disagreement and competition. He is not really single-minded in his readiness to "put everyone else around this table out of business." Some around the table are his friends, and in the name of friendship he might compromise quite a range of competitive behavior. (GM's salespeople did so for half a century.) Furthermore, from somewhere (most certainly his acquaintance with and consent to the institutional and political traditions of

American society) he learned to accept limits on his personal and his company behavior imposed by law, government, and a vague reality called popular opinion. Like most of his peers, this manager had a high sense of obligation to obey the law. In this he shares a characteristic notable in most American business leaders. Again we recognize an element in American democratic political traditions.

Criticize governments as they may and strive as they do for changes of various laws, managers do not honor disobedience to law as virtuous behavior. In part their stance is that of other interest groups in a democracy. They have helped write the laws that touch on their interests and thus have a stake in them. Beyond that obedience to law is a trait of character that Tocqueville observed in Americans, a trait observable by foreigners to this day: The declaration, "It's the law," is the end of many an argument about what is to be done. This response is in great tension with the individualistic statement, "No law or government tells me what to do."

The battle between the two assertions is one of the great value conflicts of American history. We see it in the mythical contests of almost any Western movie. These dramas present the great American epic as a struggle between personal initiative and collective solidarity. The plots of Westerns usually resolve the struggle by depicting some heroic individual forcefully dominant over lawbreakers. The hero tames lawbreaking and prepares the way for others to settle down to a government of "law, not of men." Moral individualism, in the myth, serves social order by controlling immoral individualism.

Obviously we have made an incomplete account of the values, the commitments, the beliefs, and the community memberships that Americans, including business managers, are likely to bring to their work in this society. Americans brought those relationships with them from the beginning of European settlement of this country, clashing at once with those of Native Americans, whose relationships are similarly numerous but radically different. But our account does aid us in making a simple empirical observation, easy to ignore in economic and managerial theory, that human beings bring their whole selves to work every day. In American society, a whole self is likely to consist of connections with a great diversity of other selves, systems, loyalties, traditions, memories, and intentions. Whole selves do not originate in the workplace. They are very much present in and around it.

A NEW MIX OF TRADITION AND CHANGE FOR AMERICANS

What are managers to make of this commonsense observation about them*selves* as they confront uncommon, new experiences of change?

We recommend that they, along with everyone else in our society, pause more regularly to ponder what we do in contrast to what we say. In Chapter 8 we explored a number of the important shifts in values and the diverse forms in which they now appear. There we offered a framework for understanding what has happened. Managers continually violate the classical values of freedom and efficiency, as we saw in Chapter 3. They may look like hypocrites, but are they not merely expressing the internal and external realities in the lives of working Americans, including themselves?

The longer one assesses value commitments as managers actually fulfill them, the more the distinctions between internal and external business values in fact look artificial and unreal. The notion that compassion in addition to profit might govern the work of people and organizations may seem to fit some subjective–objective split in human affairs. That division may even appear philosophically sophisticated. Compassion is subjective, profit is objective. The managerial maxim is that business realism must work with the objective and general, not with the subjective and personal. The division is more apparent than practical, however, and the maxim is questionable philosophically. Compassion can be a response of people or organizations to objective human need. Profit's place in a hierarchy of values is a subjective judgment by the persons and organizations. Even the human relations movement of the 1950s and its implicit managerial philosophy rested on narrow theories of human nature.

After their country's defeat in World War II, Japanese managers recognized that all involved in the corporation are functioning parts of each other's lives, both individually and socially. Not even the realities that economic theory likes to dub the "externalities of exchange" (such as environmental pollution) are, viewed concretely, external to human lives. Occasionally the pollution of some industrial plant will dramatically damage the lives of workers in the plant and of many other people also. In the Chernobyl and Three Mile Island accidents, for example, the real distinction between events internal and external to nuclear power plants was illusory.

A less philosophical and more socioscientific observation seems to follow from these admissions. If business managers or any other leaders in American society are perplexed about their role in the economy and politics of the oncoming world of the 1990s; if some of the old certainties of economic theory and managerial strategy seem no longer to make logical or practical sense; if the very meaning of terms like leadership, efffectiveness, satisfactory profit, and even business integrity seem clouded with ambiguity and strange disagreement inside and outside of corporate culture; if discussion about corporate responsibility hangs, even—or especially—in a book like this, under a cloud of ambiguity, then *our problem may be that we are all on the way toward some new, unprec-*

edented experiential contexts for doing business in the modern world, though we
may not yet be far enough into the experience to know exactly what it will mean.[4]

Since books are often supposed in our society to furnish answers to
problems, it may be disconcerting at this point to read that, at certain
junctures in human history, perplexity is wiser than certainty and that
a sense of journeying is nearer reality than an assuredness of having
arrived. Those who have arrived in their profession may interpret arrival as permission from others to tell the members of their organizations what to do. They may become the eager constructors of hierarchical systems believed to reflect their own competence and authority;
but if really profound changes are taking place in and around them
and their organizations, then authoritarian leadership may be as thoroughly impractical in a business economy as it has been in the national
history of various peoples.

One observer of the Nazi era in Germany said that Hitler so centralized governmental administration in Germany that the only mistakes
he *could* make were major ones. The very success of German armies in
World War II often depended on the flexibility that the Prussian tradition granted commanders in the field. Practical leadership in a time
of great social change may, in fact, consist of blends of perplexity, cu-
riosity, tentativeness, consultation, imagination, firmness, and courage.
Confident leadership often merely implements old rules in a new and
inappropriate situation. It may also ignore old-fashioned rules whose
time has come again. A modern style of leadership for our times is the
one recommended earlier in the words of the former French president,
Valéry Giscard d'Estaing: listening before deciding, followed by yet more
listening. As he readily conceded, that kind of leadership is "in itself
complex, diversified, frustrating and evolutionary." But for our time, it
may well be the most effective, efficient, and *realistic* leadership of all.

A notable example of that new, realistic leadership was offered in
1970 by Louis B. Lundborg, then CEO of the Bank of America. In the
late spring of that year a group of leftist demonstrators raging against
American involvement in the Vietnam War burned down a local bank
office at Isla Vista, California. A few weeks later Lundborg delivered a
speech to the Rotary Club of Seattle, Washington. His title was "The
Lessons of Isla Vista."[5] To the undoubted surprise of his audience,
Lundborg declared that for him the lesson of the bank burning dis-
tilled to one bit of practical advice to business managers:

> I would suggest . . . that each of us find a college-age youth—student or
> not—and spend some time with him, to find out what's going on in that
> world that is crowding in on our heels. And if you do—remember that God
> gave you two ears and only one tongue—use them in that proportion.[6]

One is tempted to speculate on the sort of education, experience,
family upbringing, and other inculturation processes, including reli-

gion, that enabled Louis Lundborg to interpret an assault on the assets of his own firm not only with equanimity but also with the confidence that it had for him and other business persons in 1970 an important lesson. Such openness to a positive possibility in a negative event is as rare among people generally as it rare among managers particularly.

The situation and Lundborg's response to it had much in common with an incident in the South in the midst of the civil rights demonstrations of the early 1960s. A black leader of the movement sat in the office of a white business manager and said:

> When World War II came, Americans white and black were drafted into the service of their country to defend it. We fought the Germans in Europe and the Japanese in the Pacific. . . . Today a German or a Japanese can eat in any restaurant [in Fayetteville, N.C.], sit anywhere he desires in any theater, find employment in places where I, an American citizen, cannot go. Is that fair?

Later in a conversation with his pastor, the white manager expressed great anger at the blunt-spoken black visitor, but he then conceded, "That man had enough truth on his side to keep me from throwing him out."[7]

How did he come to this concession? Why did Lundborg think that the bank burners were worth listening to? The answer cannot be particular to only two individuals. Human beings behave, justify their behavior, change their ways, and justify the change for reasons that may not be readily discerned on the surface. Nor do those reasons often originate in one institution, in one segment of society, or at a single time. People live, move, and change in very complicated but nonetheless real *contexts* spread out in time and space. To go behind the public behavior and speech of the two men discussed here would require the study of their biographies.

In these final chapters, we offer our own summary answers to the questions, "What is corporate responsibility?" and "What are ethics in business?" Less important than our specific answers, however, may well be the contexts we propose for understanding the questions themselves. Some of our answers, already implicit, may so far seem partially or wholly inadequate to the questions to which American managers are seeking answers. We dare not attempt to tell them exactly what to do.

Rather we offer an analysis and discussion that may coach managers toward more careful intellectual and practical connections between some old concepts—efficiency, effectiveness, public good, justice, and compassion—and the new business realities of our time—unprecedented world competition, new constituency pressures, unavoidable government regulation, the vagaries of popular cultural change, and the unanticipated consequences of each of these realities along with the complicated outcomes of the interactions among all of them. Authors like us who propose business policies, who urge ethical standards for business, and who criticize the usefulness of authoritarian solutions to so huge an agenda of problems must take their own advice. The most constructive part of our examination is at hand, but it is the construction of an exploration, not another set of "best" answers.

THE CASE OF JUSTICE:
ROBERT NOZICK AND JOHN RAWLS

No word is more frequently on the lips of social protesters than the word "justice." Ancient in the literature of philosophical and religious ethics, the word's very regularity in public utterance suggests that it must have a relatively stable meaning. After reading innumerable stockholder resolutions and listening to their presentation in annual meetings, business managers are entitled to doubt that justice means anything more than what the speakers want it to mean. But as we have noted, business managers themselves use the concept usually to mean fairness, a word that introduces but does not cover the subject of justice as we use the word here. The problem of getting a fix on justice afflicts business managers as well as their critics. Here we mean to translate claims about fair prices, fair competition, and fair wages for a fair day's work into claims about justice. Are managers to use the word with both consistency and richness of meaning? Can they and their critics agree on a context in which the word finds its right limit, scope, and use?

Philosophers argue at length over how to answer the question. None of them escapes the problem of bounded rationality, which Professor H. A. Simon finds in business decision making and which, in fact, characterizes human thought and actions generally. Boundaries in this phrase are simply those facts, truths, and presumptions that people believe *must* be accepted if they are to think or decide about some matter at hand. One vital difference between infants and adults is that infants literally do not know what they are doing and must be protected from what they might do, for example, falling down stairs. Adults, too, can fall down stairs; but they must *count on* some stairways as safe before trusting their weight to them. Even so, when they step into yet-undetermined futures or some other already-determined contexts, they count on the stability of truth. Stairs may collapse and truths may prove illusory, but trust some of them we must.

Thus there is no exit from the necessity of trusting the stability of some conceptual or contextual boundaries to make possible our most sophisticated rationality. That trust can be demonstrated by brief summaries of two contemporary approaches to the definition of justice, those of Robert Nozick and of John Rawls, both professors at Harvard University. The basic thinking of these two astute philosophers has claimed the attention of journalists and business educators in remarkable ways over the past decade, so hungry have many been to gain clarity in the meaning of justice. Even a cursory examination of the book of Nozick and Rawls offers the reader two clearly different ways of defining the term. A closer examination suggests that differences turn largely on the choice, made by the two thinkers, of the boundary-context in terms of which they carry on their explorations. Our description of their two intellectual styles is introductory only. We move later in this chapter and in the next to examine three other thinkers whose ways of *beginning* ethical reflection, we believe, suit the realities of business managers and constituency members more nearly than do the theories of Nozick and Rawls. For a first exercise of imagination and logic in anyone's formation of an ethical theory for economic life, however, the famous Nozick–Rawls debate can be very useful. We turn here to describe that debate in brief.

Nozick: Fair Contract as the Beginning of Justice

When people lay the charge of injustice against some social arrangement, they often mean that some benefit—resources, money, or goods—is wrongly distributed between two or among a number of persons. Their concern is centered on the matter of distributive justice.

What then, asks Nozick, does it mean to say that a certain distribution of benefits is just? One usually means, he answers, that the distribution occurred by a *just procedure* or that it results in a balance of benefits *inherently right or good*. Nozick prefers the first, procedural context for defining justice. "Things come into the world already attached to people having entitlements over them."[8] A just distribution of wealth is one in which an acquisition was made by a proper procedure. Wilt Chamberlain, the great basketball professional, earned a just share of wealth, according to Nozick, when he played for millions of spectators, each of whom voluntarily paid twenty-five cents extra for the privilege of watching him rather than the second-best player. The example displays a typical mass-market transaction in which all those involved are free contractors. Chamberlain garnered a lot of wealth because each fan had given up a small bit of wealth. *Free* choice for individuals, groups, and all parties in economic exchanges make much sense to Nozick because he views such choice as a fundamental feature or boundary of the society in which he wants to live—a free or democratic society.

Another theory of justice comes into play when a government proposes to tax Wilt Chamberlain's annual income at a rate higher or in an amount different from that applied to persons of lower income. If one approves the procedure by which wealth is acquired, one may accept the resulting economic inequality as justified. But one may also believe that inequality must be limited in a just society. Is it right that some should feast while others starve? Is it right to allow others to starve even when they seem to choose to starve? Laws against suicide assume that the answer is no. Whether voluntarily or coercively, should not those who benefit most in the community share some of their wealth with those who benefit least?

Nozick poses the choice between these two definitions of distributive justice with eminent clarity:

> No doubt people will not long accept a distribution which they believe is *unjust*. People want their society to be and to look just. But must the look of justice reside in a resulting pattern rather than in the underlying generating [procedural] principles?[9]

The practical consequences of Nozick's preference for procedural justice are easy to discern in the debates about fair prices and fair competition in the American business system. A fair price becomes one that producers and consumers freely consent to pay, a principle that probably spells doom for all "fair trade" laws. Nozick's vision of economic justice here is profoundly attuned to classical capitalist theory. He would have no more patience than Adam Smith with business managers "meeting together for some public good." He, too, sees that managers from various firms agreeing on fair market shares or fair uniform prices are a threat to freedom of contract in a free society.

But more fundamental than capitalist theory for Nozick is his resolute preference for historical principles of justice over end-result principles.

> In contrast to end-result principles of justice, *historical principles* of justice hold that past circumstances or actions of people can create differential entitlements or differential deserts to things.[10]

On the one hand, if the white settlers of America took over the lands of the Native Americans by military force, one may charge the transaction with procedural injustice. On the other hand, if they bought the land for a price acceptable to both parties at the time, they cannot be charged with injustice 200 years later when the land has become expensive urban real estate. In the former case, Nozick's principles might require the rectification of an injustice—it might justify large compensations to the descendants of the originally pillaged landowners. But it would not require compensation for the exchange of beads and hunting rights into which the original inhabitants of Manhattan Island entered with the new Dutch settlers. (Nozick leaves ambiguous, for purpose of illustration, how one should deal with the practical consequences of the misunderstanding that afflicts most such exchanges in the early colonial history of this country. Native American culture had little or no concept of "selling" land or "developing" it.)

However historically oriented the Nozick view of justice; however focused on the virtues of free decisions by all parties to economic exchanges; and however perennially appealing may be certain procedures for acquiring wealth in any conceivable society, his view of justice escapes neither ambiguity nor conflict with other values that many a modern American espouses. The view is conservative in a strict sense of the term. What happened once upon a time takes ordinary precedence over what may be happening now. In one sense or another, the present distribution of society's benefits has been tied to the past. Even if present and future events—or more accurate knowledge of the past— cause us to revise our notions of a just procedure, we still have only a past event generating a measure of present justice. Moreover, however faithful and apparently unarbitrary Nozick may be in his devotion to historical principles, his concept of freedom is at once culture-bound and, oddly, as abstract as many of the end-principles he rejects. No specific past in history is free of ambiguity, of course. No historical justice is or can be pure. The freedom that most of us see in the human past or experience in our own present is *always* a bounded freedom. The boundaries are many and complex: limits of time, circumstance, knowledge, intention, understanding, and opposing human interests. Every economic transaction exhibits such boundaries. Surely the sale of Manhattan Island exhibited a mix of trickery, fear, greed, need, and honesty on both sides.

A more contemporary problem illustrates the way such boundaries appear in corporate business transactions. In late 1988 RJR Nabisco agreed to a $25 billion leverage buyout (LBO) of its shareholders. It distributed to its shareholders substantially all its equity, and replaced it with $19 billion in new debt obligations. The Metropolitan Life Insurance Company (MetLife) and Jefferson-Pilot Life Insurance Company held about $350 million worth of the company's low-coupon senior bonds, purchased between July, 1975 and July, 1988. The day after the LBO was announced, RJR's bond yields jumped from just under 10 percent to nearly 12 percent.[11] Metlife estimated that as a result its bonds declined in value by more than 25 percent.[12] MetLife charged RJR with a breach of the implied covenant of good faith and fair dealing and an assault upon its equitable rights.

Federal judges in both district and circuit courts rejected MetLife's claim. They maintained that so large a company as MetLife, with professional staffs investing its funds, could only be considered sophisticated and informed; those who made the decision to buy the bonds must have been and were, as their internal records showed, well aware of dangers of possible LBOs. Nevertheless, MetLife officials and many observers criticized the transfer of wealth from bondholders to share-holders. The editors of *Forbes* called the action a "rape";[13] *Time* writers described the transfer as "a game of greed";[14] and the *Financial Times* referred to it as "A Bid Too Far."[15] Though the courts judged MetLife officials to have voluntarily, freely, and with information invested in the bonds, many Americans believed that fairness was lacking.

As we move back and forth across the fuzzy boundaries among what we do, what we believe to be of value, and what our "primary" values are, we may not always pursue consistency. The attempt to distinguish and relate these levels of human action is the object of many a philosophical and sociological analysis. But clarity in the matter is an elusive goal, especially for realists with good memories. A former automobile dealer has observed in the negotiations preceding a car sale, "The sale comes at the moment when each side believes that it is getting the better of the other." Such an observation may be one definition of a bargain, but its shrewdness stems from a rich, complex assessment of human nature at work in market transactions. On both sides, greed complicates freedom, just as does ignorance. Modern law suits for product deficiencies often turn on an assessment of the ignorance that one or both parties freely accepted in the bargaining process as one of the limits in the situation. Freedom-in-exchange is not thereby a fiction, but it is only one of the virtues ascribable to a just bargain. A careful analysis of anyone's concrete experience of "freedom" or "justice," in fact, will often acknowledge the presence of other value-claims that parties to the experience will all assume as important: truth telling and assumptions about trustworthiness, for example. Thus, the word "free" no more summarizes the full range of primary goods we want for ourselves and our society than does the word "just." Those who speak only in terms of a free society or a just society considerably impoverish the lexicon from which we can describe a society worth calling truly human. But in advancing these judgments, our discourse runs ahead to writers whose thought we pursue below. Let us turn to *A Theory of Justice* by John Rawls.

Rawls: Justice as Freedom of Opportunity for All, Especially the Least Favored

One can accuse Nozick of being insufficiently rich and empirical in his descriptions of exchange relationships in human history, but one cannot fault him for believing that the concrete human past has some just claim as precedent for the behavior of humans in the present. We are indeed bound by some events of the past. But John Rawls is aware, along with many philosophers and not a few politicians, that people use other measurements in assessing their goals and purposes and rights and duties. When American colonists revolted against British imperialism, they had recourse through the Declaration of Independence to "self-evident truths." They were self-evident certainly to the lawyers and merchants who voted for the Declaration with full confidence in the "natural law" of the eighteenth-century rationalists. Why should people have to accept George III's justice, given its particular distribution of benefits provided by the accidents of history? Are there not other *ideas* of justice accessible to the human mind and human experience in every age?

Without confessing himself to be an advocate of natural law, John Rawls proposes that there are such ideas. One such idea is the "original position" in which everyone of us could imagine ourselves. Political philosophers from Plato to Thomas Hobbes and John Locke preceded Rawls with such a notion. They, too, tried to imagine what kind of political order people would set up if they could start afresh.

In an adroit "let's suppose" mood, Rawls asks his readers to imagine that they are at the start of society but, with infantlike ignorance, they have no way of knowing what their future position (status, fortune, looks, situation of social benefit or loss) will be. Before the slings and arrows of outrageous fortune descend upon them (or for some the happy converse), what kind of society would they want to live in, especially in view of the uncertainty of their future? They would want, he answers, personal opportunity to achieve good fortune and personal protection against disaster. They would logically want this same double-insurance for *everyone* in the society to come. They would want it, for as yet they have no reason to argue as Nozick might that their good fortune in the past must be preserved; nor do they possess any knowledge to calculate as Karl Marx's proletarian might that their bad fortune can be and must be reversed. In short, they must logically desire a society of equal opportunity, enabling all members to strive for social goods and equal protection against gross failure to achieve those goods.

In one respect, the basic context of Rawlsian justice contrasts starkly with that of Nozick. Rawls imagines a beginning idealistically, while Nozick imagines a beginning historically. Both philosophers, however, have a high regard for the value of individual freedom and socially assured liberty that allows persons to achieve a wide range of social goods. Rawls's formulation of his "first principle" of justice rings with the tonality of both Jeffersonian liberalism and Nozickian conservatism:

> Each person is to have an equal right to the most extensive total system of equal basic liberties compatible with a similar system of liberty for all.[16]

In expanding on the principle, Rawls insists on the rule that "liberty can be restricted only for the sake of liberty,"[17] thus reminding us of the familiar democratic maxim that "your liberty ends where mine begins." His definition of justice, therefore, begins with the ideas and interests of individuals. Its first principle concerns the liberty that the society will ensure for all people *individually*.

For Rawls there is a vital second principle, and it is for this addition that his view of justice has become famous. Except for the idealism of his original position, his first principle of justice is relatively compatible with Nozick's free society. But Rawls knows that the great modern debate over justice swirls around the phenomenon of what, if anything, people of good fortune owe to people of bad fortune. Gross inequality of health, wealth, education, position, and honor *do* afflict most human societies. Is that inequality incidental to any notion of justice? Doubtless, as Nozick writes, "people want their society to be and to look just." Would a society that, through just procedures, ended up with inequalities very like those of ancient China, medieval Europe, or modern America, still be just? Nozick briskly answers, "Yes!" Rawls replies, "No!"

Even in the extreme case, writes Rawls, where people misuse their liberties and end up in allegedly deserved poverty, we cannot say that it is just for their society to write them off and consign them to the social trash bin. Social inequality is a profound problem for Rawls. Indeed, for him a society is just if its very inequalities are so arranged as to be of continuing help to those people in the bottom stratum. A society may provide many explanations for inequality and offer many justifications of it, but the only Rawlsian defense is that the inequalities must help *lessen* inequality by offering opportunities and real benefits to the least favored in society. Only if inequality can be harnessed to the cause of lessening inequality, can it be reconciled with his first principle: liberty for all.

Thus Rawls also might acknowledge the justice of allowing a starring athlete like Wilt Chamberlain to make a fortune from basketball, but his justifying argument would be quite different from that of Nozick— Chamberlain's success, like that of many African American athletes in America, helps erode real and perceived racist barriers to equality of opportunity in this society.

The summary Rawlsian rule of distributive justice, then, comes down to this elegantly simple statement:

All social primary goods—liberty and opportunity, income and wealth, and the bases of self-respect—are to be distributed equally unless unequal distribution of any or all of those goods is to be to the advantage of the least favored.[18]

Only superficially can Rawls's view of justice be accused of being a foil for either capitalist or socialist ideologies. Its roots and historical–cultural context closely resemble those of Nozick; both philosophers believe deeply in individual liberty as essential to the just society; both construct society through a series of contracts between and among individuals. Each depends, finally, upon basic assumptions about human nature and the functions of human reason. Both promote a rather abstract idealism. Rawls is clearly more at home with the idealism than is Nozick. Neither philosopher would be likely to accept the view of some economists that "justice is irrelevant to the functioning of the market" in the real-life business world. Both Nozick and Rawls seek to promote an ethical principle or philosophical criterion for judging justice in society, including its economy. Rawls, in particular, means to bring to bear a robust standard for measuring social justice in all sorts of societies. His standard reigns idealistically from above, so to speak, whereas Nozick's procedural test of justice reigns historically from behind. Both provide some leverage for criticizing the current state of human benefits as distributed in any society.

Of the two, Rawls furnishes the most leverage for assessing economic justice in the United States and other countries like France or Mexico where inequalities of wealth and the working of markets leave large numbers of the population with a very small share of the total national income. In the history of American capitalism, some business managers have portrayed their roles in national or world society as justified by something like the Rawls theory.

When in 1914 Henry Ford announced that in the future his workers would receive $5 for an eight-hour day—the average wage in manufacturing was less than $2 a day[19]—he undoubtedly believed that he was increasing his personal wealth, but in the process was serving the people least favored in American society. The vast migrations of black people from the subsistence farms of the South to the industrial cities of the North during and after World War I provide indirect evidence that the capitalism Ford typified provided a measure of Rawlsian justice. "With capital we create jobs" is still standard rhetoric in many a Chamber of Commerce gathering.

Not all the developments of capitalism accord as well with Rawlsian notions. Modern American investment dynamics may stand up very poorly to Rawls's plumb line. In the 1980s corporate investments all too often did not create jobs for the least favored people of this or any other society. Minority youth, particularly black teenagers with poor schooling, the mentally handicapped unfit for routine work, and a host of other disadvantaged people found little opportunity in the American labor market. Managers of multinational firms adopted Rawls's rule eagerly as they capitalized jobs in poor countries of Asia, but they did so only with that twinge of patriotic conscience that comes with the thought, "Is it just (fair) to displace American workers to hire Koreans or Malaysians?"

Given his reverence for free contract, Nozick's theory hardly admits us even to pose the question. As we have seen in earlier chapters, most

American business managers do not have an easy conscience about the destruction of American jobs in the movements of capital around the world by multinational corporations. Many have a profoundly *divided* conscience in the matter: Would not an ethic of justice for humanity sanction jobs for Asians as much as jobs for Americans? So in the midst of seeming to do justice to one group among the varied constituencies—stockholders, current American workers, and poor people around the world—managers may find themselves treating another unjustly. Even the most careful philosophical definition of justice seems inadequate for the dilemmas that the very definition helps to create. When admirably simple, do such definitions provide illumination for walking down these new managerial paths? Is there a way to live, act, decide, change, learn, and confront these complexities without falling pray to simple rules or to paralysis?

Another line of thinking, another context in which to inquire after justice and other values in contemporary human society can be urged as the boundary for asking such questions and for seeking some answers to them. We find such another approach in the writing of another philosopher, Alasdair MacIntyre, discussed in the next chapter.

NOTES

1. James G. March and Michael D. Cohen, *Leadership and Ambiguity:* The American College President, 2nd ed. (Boston: Harvard Business School Press, 1986), pp. 222, 228.

2. See this volume, Chapter 5.

3. Ralph S. Robinson and Donald W. Shriver, Jr., "The Case of the Constant Customer," *Harvard Business Review* 46(4) (July–August, 1968):150–58.

4. March and Cohen, *Leadership and Ambiguity,* especially Chapter 9, "Leadership in an Organized Anarchy," pp. 195–229. Although most business organizations may not be as anarchic as most universities, ambiguity afflicts their leaders as regularly. March and Cohen identify four areas in which it does so: organizational purpose, the nature of the leader's power, the nature of his or her experience as leader, and the meaning of success. What they see as the use of ambiguity to leaders can be gleaned from the following two quotations, along with the two that open this chapter:

> It seems to us that a bad man with good intentions may be a man experimenting with the possibility of becoming good. Somehow it seems more sensible to encourage the experimentation than to insult it. p. 227.
>
> . . . what we need . . . is playfulness . . . the deliberate, temporary relaxation of rules in order to explore the possibilities of alternative rules. p. 225.

5. The reader might pause to reflect on how he or she would have written a speech on this subject.

6. Louis B. Lundborg, "The Lessons of Isla Vista," (pamphlet) (San Francisco: The Bank of America, 1970), pp. 13–14.

7. Donald W. Shriver, Jr., ed., *The Unsilent South* (Richmond, Virginia: John Knox Press, 1965), p. 30.

8. Robert Nozick, *Anarchy, State, and Utopia* (New York: Basic Books, 1974), p. 160.

9. Ibid., pp. 158–9.

10. Ibid., p. 155.

11. "RJR Buy-Out Proposal Raises Bond Yields," *Wall Street Journal,* October 25, 1988.

12. *Brief for Metropolitan Life Insurance Company and Jefferson-Pilot Life Insurance Company,* United States Court of Appeals for the Second Circuit, August 16, 1989, p. 4. When coupon payments are fixed, the value of a bond changes *inversely* with interest rates.

13. Allan Sloan, "The rape of the bondholder," *Forbes,* January 23, 1989, p. 67.

14. "A Game of Greed," *Time,* December 5, 1988, p. 66.

15. "A Bid Too Far, *Financial Times,* November 24, 1988, p. 28.

16. John Rawls, *A Theory of Justice* (Cambridge: Harvard University Press, 1971), p. 302.

17. Ibid., p. 302.

18. Ibid., p. 303.

19. Ford's wage increase was a kind of profit-sharing plan which was to distribute up to $30 million (in 1989 dollars, $371 million) among his employees. The $5 a day he offered in 1914 is equivalent to about $62, in 1989 dollars. The wage was newsworthy, for it was 153 percent higher than the average wage currently paid in major American manufacturing firms.

11

MacIntyre: The Story of Our Life as Justice and Other Virtues

The Marxian "cure"—revolution and the evisceration of private property—would have been to [Adam Smith] like throwing the baby away with the bathwater. It is indeed ironic that while Smith and Marx saw "true" human development in "production" and "action," and while both emphasized the importance of "humanity" and social and aesthetic fulfillment, the one espoused means (markets and property) which to the other implied total destruction of the ends.[1]

Unlike their alleged mentor Adam Smith, twentieth-century economists have disdained to meddle with subjects as abstract as human nature. If Karl Marx has done nothing else for capitalists, he should have alerted them to the perennial pertinence of the big questions of human existence: Who are we? Toward what ends does our history tend? Where are we going and where do we want to go?

Alasdair MacIntyre, professor of philosophy at Notre Dame University, knows only too well that his own academic field, along with that of the economists, has helped educate the current generation of business leaders to ignore these questions. It is time, he believes, that universities and their disciplines again turn to them. The return will be difficult, however. Careful attention to such questions ceased generations ago to attract the brightest minds in philosophy and social science. Now such questions are "embarrassing strangers" in business schools; no one knows how to entertain them. We shall not stumble into something like wholeness for our fragmented society and social selves, says MacIntyre, until we learn to do so.

MacIntyre points out that Western people have been living for over three centuries in societies whose famous moral theorists have isolated morals and human values to the privacy of individual selves. One side of Adam Smith's theory moved in this direction, and the theories of

Rawls and Nozick regarding justice vividly illustrate it. "Individuals are in both [Rawl's and Nozick's] accounts primary and society secondary, and the identification of individual interests is prior to, and independent of, the construction of any moral or social bonds between them."[2]

MacIntyre also points out a long-standing and widely recognized fact about life and thought in the United States. Rawls and Nozick are both Americans, and their focus on individual liberties illuminates our cultural preferences. Is there anything surprising then in MacIntyre's observation? A minor illustration from his book provides an initial warning of an alarm he means to sound in assessing the notions of justice by European and American philosophers. Charles II once met with a group of scientists of the Royal Society[3] and asked, "Why does a dead fish weigh more than the same fish when it is alive?" Various sophisticated answers were advanced by eminent scientists, and then the king said, "But it is not true. The two weigh the same."[4]

The story, of course, is a study in the relationship of wise men to a powerful man. It is hard to believe that no one knew enough to challenge the presupposition of the king's droll little game. More than one person probably knew the truth of the matter but lacked the courage to contradict royalty. This hesitancy might be understandable, but was it justified? For a few minutes, at least, truth suffered in the conversation; courage was absent. Is it necessary to talk about truth and courage as well as justice when describing a "good" society? Are truth and courage to take a subordinate place in human affairs when persons of power manifest themselves? Is self-interest the criterion of choice, justifying the subordination of truth and courage? In a word, should ethics be concentrated upon the virtue of justice to the neglect of other virtues equally important for a "truly human" person or society?

THE PLURAL VIRTUES OF A HUMAN LIFE

MacIntyre invites us to consider if there are *virtues* or fulfillments in human life, individually and socially enacted, that are too important ever legitimately to be eliminated or diminished in the very definition of "human." Are these virtues that people should seek to practice as consistently as possible, for good, right, and proper reasons? Is there, in short, such a thing as behavior that we should expect always of each other, behavior that we may call moral or truly human? In asking these questions MacIntyre interrogates the assumption in modern Western culture that, ideally, individuals decide for themselves what is good, right, and fitting behavior.

Our cultural celebration of individualism makes difficult any traditional social definition of morality, says MacIntyre, nowhere more than in the modern business bureaucracy. What binds corporate managers, their subordinates, and their various constituencies together as an organization? The ordinary answer has to do with efficient production of profitable goods and services, noted in earlier chapters. Does anything else bind the parties to one another, providing a sense of loyalty and obligation among themselves and to their unity through the organization? Corporate recruiters at business schools report that the new breed of young executive recruits came straight from the "me-generation" into corporate employment, where they openly declared their intent to stay in a position only so long as it will serve their financial self-interest.

"Company loyalty" as a virtue disappears from their sense of ethics. Conceivably, of course, such young employees might stay on a job for a long time out of self-interest, if they receive promotions and satisfactory pay. Obviously they would receive not only money-value but also prestige-value as they made their way up within the organization. But staying for either value, or because of both of them, is a long way from participating in an organization because of the virtues of truth, courage, and friendship, all old-line values that MacIntyre believes to be greatly needed in contemporary human relationships.

MacIntyre is the first to concede that Western—not to say world—societies seem incapable of agreeing on any definition of "normative humanity," a conclusion that tends to cut the nerve of philosophical attempt like his own to recover a trustworthy, socially workable theory of human excellence like that prepared long ago by Aristotle. But the attempt to raise old questions about the "truly human" shows a form of intellectual courage on his part that conforms to one of the virtues acclaimed in his theory. To read *After Virtue*, for example, is to arrive at no definition of human virtue likely to be applauded everywhere in Western societies. But someone must begin to build ethical bridges between old and new concepts of human self-understanding, between old and new conditions of human existence; and the importance of books like MacIntyre's (and our own) may be what they *begin* to do.

A return to human experience in business organization may be a more promising way to begin than even the most careful philosophical reflection on a single ethical value, like justice. For example, are financial rewards and prestige alone sufficient bases for the commitment that business managers might be expected to display in their profession? Rather than treating the question as rhetorical, the reader might consider some data about corporate culture presented recently in a study by sociologist Robert Jackall.

Through participant observation in the headquarters of several large American corporations, Jackall sought to understand the moral environment of executive office life in large American corporations.[5] The title of his prizewinning article describes its findings: "Moral Mazes." In the several corporations Jackall studied, newly appointed young managers find themselves confronted with a "maze of relationships" with their peers and superiors where the ruling question is, "What must each of us do to stand the best chance of getting promoted to the rank of CEO?"

Success in the executive office does have a wider meaning than personal promotion, Jackall found. It includes service to the interests of others in *their* promotability; and often this serving involves the muting of criticism of any peer or superior, holding back new ideas that clash with favorite ideas of this or that person in power and—most of all—collaborating in every way with the known views and policies of the current CEO. Service to company profit is also a sine qua non of promotability in the corporate hierarchy. Promotion also involves an array of particular personal qualities that Jackall found as weighty as profits in promotion policy. They included dressing in conventional business styles, public affability, a team spirit that muffles direct criticism of corporate policy, urbanity that "gives the appearance of knowledge even in its absence," and above all, patron power—a connection with someone in the hierarchy who encourages and facilitates one's corporate advancement.[6]

Corporate culture at the upper levels of management calls for flexibility and adaptability that quite contradict the individualism, the devotion to freedom, and the primacy of profit that once supposedly characterized the American business creed. Justice, courage, and truthtelling take a back seat, too, to the lures of promotion, prestige, and power. One of the top managers interviewed by Jackall spoke of the corporate culture as a game in which individual players submerge themselves in others' definitions of the rules.

> One thing you have to be able to do is to play the game, but you can't be disturbed by the game. What's the game? It's bringing troops home from Vietnam and declaring peace with honor. It's saying one thing and meaning another.
>
> It's characterizing the reality of a situation with *any* description that is necessary to make that situation more palatable to some group that matters. It means that you have to come up with a culturally accepted verbalization to explain why you are *not* doing what you are doing. [Or] you say that we had to do what we did because it was inevitable; or because the guys at the [regulatory] agencies were dumb; [you] say we won when we really lost; [you] say we saved money when we squandered it; [you] say something's safe when it's potentially dangerous. This is the game.[7]

A very political game it is. At the individual level, the name of the game might be the dodging of responsibilities: personal flexibility without personal boundary; adaptability without moral core; or responsiveness without an *agent* who responds. "What matters in the bureaucratic world is not what [one] stands for but whom [one] stands with in the labyrinths of [the] organization."[8] Thus described, how does such a management culture differ from that at work in the relationship of Charles II to the scientists in the Royal Society? The Jackall portrait of managerial culture may not fit that of every large firm, but it describes the hierarchical style of many.

Consider a well-known rule of thumb in large firms: "If you want to have integrity in your work, find a boss who shows integrity." The rule applies especially to the CEO. Over decades of influence on their companies, some CEOs make profound impressions on their corporate cultures. "The way we do things here" reflects the way they do things personally. But so large a role for the CEO merely underscores the prevalence of the hierarchical style and suggests that very large organizations may depend on very few top people to determine their moral character. Certain *dis*advantages accrue to this style. By expecting subordinates to agree with them, top managers discourage individual initiative and deter whistle blowing. Both activities challenge reigning management assumptions and illusions. Initiative can signal few facts, emerging realities, and potential opportunities that otherwise may be overlooked. Whistle blowers may warn of slack and inefficiencies, in addition to corruption.

Consider also the puzzling questions: Why did it take the managers of Detroit's automobile companies so long to comprehend the significance of the increasing sales of Japanese and other foreign cars in the United States? After the first communications of information to managers about the likely health dangers of asbestos to workers in the John Mansville Corporation in the early 1930s, why did forty years pass before managers acted on it?

One answer to both questions lies in the problems top-level managers faced with data that clashed with their interests in their hierarchical organizations. As numerous recent studies of innovative organizations show, a priority for change, adaptation to new environments, and the search for new ideas may best be realized in a nonhierarchical culture. In a nonhierarchical culture the opinions and perceptions of people anywhere in the organization are more likely to be taken seriously by managers than in many-layered, bureaucratic hierarchies.[9]

For organizations struggling to survive in a new environment, unorthodox whistle blowers can save managers from costly mistakes. In 1986 had NASA managers encouraged whistle blowers openly to question the safety of the O-rings for the rocket casing, the nation might well have been spared the *Challenger* spaceship disaster and a serious setback to the national space effort. In 1987 had President Reagan's White House staff received more encouragement for whistle blowing and its members been expected to raise questions about doubtful, risky policies, the Executive Office, its policies, and reputation would have been far stronger and probably more effective than it was in the remaining

two years of the President's term. Even in safe, predictable environments, hierarchy and bureaucracy can exact unaffordable costs.

MacIntyre examines these costs with some care and subjects bureaucracies of all sorts to a suspicion akin to Jackall's. People may try to
bring their whole selves to work, but the work environment may or
may not permit them to express themselves wholly. It costs the human
character, MacIntyre might say, to suppress its real virtues. To speak of
courage, truthfulness, justice, and friendship as virtues of human life
generally is to speak of characteristics that ought to have a place in any
human organization and in any wholly human self. These words of
virtue, however, sound rather out of place, abstract, and idealistic in
many Americans' working experience.

In America do we not expect people to observe a certain decent separation between their personal values and the specialized work they
perform on the job or in a profession? Young people pledging limited,
self-interested loyalty to their corporate jobs, as many appear to be doing
in recent years, make clear their acceptance of such a separation. A job
supports their ability to put into practice the other values of their lives
in settings as diverse as family, church, vacation, and friendship. But
the price of this segregation of values in one's life can be high for the
person, and separation of employees' personal lives from organizational life can exact a high cost for the corporation itself.

Consider David Edwards, an employee of Citicorp's International
Banking Group in its Paris offices. In 1977 he blew the whistle on company managers who were "parking" funds illegally overnight in the
Bermudas with the intent of evading income taxes levied by various
European governments. Eventually the Securities and Exchange Commission (SEC) investigated the charges against Citicorp. After years of
considering them, it terminated its investigation, choosing not to consider the merits of Edward's allegations on the shocking theory that,
since Citicorp managers had never suggested to stockholders that the
company proceeded with "honesty or integrity," they were under no
legal obligation to reveal violations of these principles.[10]

Equally shocking to Mr. Edwards must have been the experience of
finding that his own government, whose laws and traditions motivated
him in the first place to risk the displeasure of his superiors, did not
truly support his protest. Certainly the upper managers of Citicorp did
not support it. The president asked for his resignation, and he could
not find an equivalent job in another large bank because he had become identified throughout the tight-knit investment banking community as a "troublemaker." His only alternative was to go into business
for himself.

In government, business, universities, and churches in America, whistle
blowers live under a pall of organizational suspicion. It may be human
nature to resist those of our fellows who call us to account for some

deficiency, but such instances raise the philosophical question, "What *kind* of human being is a better or worse human being?"

The answer "Any kind" is both absurd and trivial. Yet our culture falls into absurdity and triviality when it individualizes responses to that philosophic question. "Any kind" encourages many people to believe that such relativism *is* the culturally preferred answer. We have a long search ahead of us collectively, writes MacIntyre, if we are to emerge intellectually, educationally, and institutionally from the cul de sac of so extreme a cultural relativism. Its culmination may have been reached in the late 1960s with the popular motto of some youthful protesters: "If it feels good, do it." Not that personal feelings are irrelevant to a realization of our humanity; but there may be feelings that person *ought* to have—and must have—to deserve the designation, *mature.*

TOWARD RECOVERY OF SOCIAL SELFHOOD
AND SOCIAL ETHICS

MacIntyre's analysis of the history that brought us to the consideration of the question is long, demanding, and eminently worth reading. We can only sketch his argument here, suggesting the major directions in which it points, as we did with the ideas of Nozick and Rawls. Mac-Intyre argues for a mandatory cultural quest for values and virtues that will rescue us from individualistic, relativistic moral theory. If Robert Jackall's description of corporate culture has even a grain of general truth in it, we must charge Western culture, not just the corporation, with being a source of this relativism. When Americans bring no more than a thin slice of their real selves to work each day, or if they possess no real self to bring, employees will lack inner or outer permission to be more than shadow-selves at work. From one point of view they are not even individuals, they are conformist nonentities. They are ill-equipped even to be troubled by their conforming behavior, for neither the bureaucracy nor the general culture tells them that there are *any* solid virtues that human beings are bound by their very human nature to practice. But the individual as *agent* disappears; organization as a matrix of mutual interchange of personal *responsibility* disappears, too.

The ironic result of this relativistic individualism, MacIntyre makes clear, is the weakening of those personal commitments that bind individuals together in any common enterprise whatsoever. In the past three centuries, we have come to divorce personal morality from the constraints of theology, philosophy, government, family, and the church. "Each moral agent now [speaks] unconstrained by the externalities of divine law, natural teleology, or hierarchical authority; *but why then should anyone else now listen to him or her?*"[11] If the "he" or "she" happens to be the CEO of the place where you work, you might listen closely enough to secure financially desirable promotions, but the good would be what MacIntyre calls an external. The job will provide one of the things that people need—money—to survive in the economy. Despite the importance of this provision, does any reader think that money alone is enough to keep one's life human? The old reply of Jesus of Nazareth to the Devil, "[Human beings] do not live by bread alone," was not merely the expression of a human ideal, it was intended as a statement of fact.

Are there *moral facts* in human life? MacIntyre knows that the phrase sounds odd to most modern readers. We have been raised, some of us in Philosophy 101, to separate fact and value into distinct compartments. We might have been taught Ronald Dworkin's view: "The central doctrine of modern liberalism is the thesis that questions about the *good life for man . . . are to be regarded from the public standpoint as systematically unsettleable.*"[12]

But separating facts and values may be a source of momentous social dilemmas. In Birmingham in the 1960s, U.S. Steel did in fact have an influence on that city's acceptance or rejection of a new set of community norms of justice for black citizens. Was the power of U.S. Steel value- and virtue-neutral in everything but economic value? Not at all; questions of what is good for a corporation, good for a city, and good for human beings intermingled here. The questions were at once practical and theoretical. When groups of people with clashing values have to agree to do something together, even when compromise is their form of agreement, there has to be consensus about what *must not* be compromised in their relationship.[13]

"Professional" Practice

In sum, MacIntyre proposes a much more complex, challenging exploration of the roots of human reflection on human nature than is proposed by such writers as John Rawls and Robert Nozick. As a way into his exploration, he asks the reader to consider what we ordinarily mean by human practices, for example, the practice of a profession. Only in the last half-century have business schools begun to suggest to managers that they, as well as ministers, priests, doctors, and lawyers belong to a profession. The phrase "professional manager" now falls easily from the lips of MBA graduates.

But what are the characteristics of any human activity that entitles it to a claim of professional practice? All the characteristics, replies MacIntyre, involve things judged practical by the practitioner and by society; those valued things are the context of the technical-skill element in the practice. However, he insists, mere technical skill does not make a practice or a profession.

Tic-tac-toe is not an example of a practice in this sense, nor is throwing a football with skill; but the game of football is, and so is chess. Bricklaying is not a practice; architecture is. Planting turnips is not a practice, farming is.[14]

At minimum the acquired practices that define a profession always involve an assent to the social judgment that they, among many possible others, are valuable for at least some group of people in the society. No one undertakes to be a doctor, painter, or engineer without believing that society deems the practice of some value. Furthermore, the learning of a practice also involves being taught by others who have mastered the practice.

A learner may have the ability to make a startling new contribution to the current practice—as did Isaac Newton and Rembrandt van Rijn— but each began the profession by learning through apprenticeship. Even Newton recognized this condition when acknowledging his debt to his predecessors. "If we have seen further than our forebears, it is because we have stood on the shoulders of giants." Copernicus prepared the way for Newton, Newton prepared the way for Einstein, which is to say that every human practice, every profession has a history. One may not understand that history if one remains ignorant of it; but even in ignorance one depends upon it.

Equally necessary in learning practice is mastery of certain methods, sometimes mislabeled "skills" or "techniques." They may be so basic to practice that they have an internal, inherent relationship to it. The medical doctors who supervised the Nazi physical experiments on their concentration camp victims exercised much scientific knowledge and skill, but they lost their right to be called *professional doctors.* They lost the most internal purposes of their profession, at least to do no injury and at best to assist healing.

A business manager who purposely drives a corporation into bankruptcy to prevent a despised member of his family from taking over as head of the enterprise might be said to be no true manager by the same measure. The legal right so to behave is beside the point of the analysis here. True professionals know what is worth doing and why; they know who must be partners of the practice, and what are the rounded set of measures by which anyone can judge whether the practice is competent, excellent, and humanly adequate.

One of the marks of a profession is that its members are capable on occasion of identifying some feature of their own behavior that fails to measure up to their own standards. As MacIntyre writes:

. . . the standards are not themselves immune from criticism, but nonetheless we cannot be initiated into a practice without accepting the authority of the best standards realized so far. If, on starting to listen to music, I do not accept my own incapacity to judge correctly, I will never learn to hear, let

alone to appreciate, Bartok's last quartets. If, on starting to play baseball, I
do not accept that others know better than I when to throw a fast ball and
when not, I will never learn to appreciate good pitching let alone to pitch.
In the realm of practices the authority of both goods and standards operates
in such a way as to rule out all subjectivist and emotivist analyses of judgment.[15]

Human practices are not limited to those that one day will be associated with one's profession in the sense of job specialty. They also include virtues that we may call justice, courage, honesty, and loyalty,
which have suitable expression in many realms of human life. Justice
has something to do with "recognizing what is due to whom," says
MacIntyre, as when a seller must decide what information is due a customer regarding a product. Courage has something to do with "being
prepared to take whatever self-endangering risks are demanded" along
the way of producing and selling a new complex product—such as risk-
ing the loss of a car sale by telling the customer that a particular model
has often had carburetor trouble. Honesty has something to do with
"listening carefully to what we are told about our own [or a product's]
inadequacies," with taking steps to persuade the manufacturer, for example, to face up to the carburetor problem. And loyalty has something to do with knowing *what* and *who* most deserve loyalty, as when
loyalty to a customer's interest overrides the loyalty of a car dealer to a
manufacturer.[16]

But any listing of human virtues drives one back to a fundamental
question: How do we know what various virtues, standards, and rules
are the right, true, and proper ones for any practice whatsoever? Pressed
to answer, says MacIntyre, many of us have to repair to *history*.

To enter into a practice is to enter into a relationship not only with contemporary practitioners, but also with those who have preceded us in the practice, particularly those whose achievements extended the reach of the practice to its present point.[17]

Someone has said, "Explanation is where the mind is at rest." Recourse
to history is the way many of us explain our professions, practices, lifestyles, and life purposes to outsiders.

Organizations have the same recourse. Any new employee of GM,
GE, or AT & T who wants to know what these corporations mean to
older employees will have to become acquainted, not only with the corporate leaders of 1990, Roger Smith, Jack Welch and Robert E. Allen;
but also with earlier CEOs, with the names of "Engine Charlie" Wilson,
"Electric Charlie" Wilson, and Walter Gifford. They will soon find
themselves getting acquainted with those managers' predecessors, Alfred
Sloan, Gerard Swope, and Theodore Vail. Should any graduates of a

reputable business school be entitled to call themselves professional managers if they are ignorant of what Andrew Carnegie, John D. Rockefeller, Sr., and Cyrus McCormick did to shape (for good and evil) the aims and purposes of large business corporations in the nineteenth century? Who can explain business as a profession without locating its modern reality in a history that these leaders shaped?

LEARNING TO TELL OUR STORY

As human history goes, a century is a short space of time; but for people who see their success or failure only in the setting of the next quarterly report, it is a long time indeed. The imagination of some business leaders may be so fixed on the immediate future that they fall easily into agreement with Henry Ford that "history is bunk." But this famous declaration was the real bunk. Every day of his active life as producer of automobiles, Ford depended upon a host of inventions, engineering theories, and human skills that made the modern auto factory possible. One of the sober lessons that modern evolutionary science teaches us is the awesomely slow, complex, chancy, and ancient processes by which this planet finally developed its hospitality to life in any of its forms. We do not know just who, some thousands of years ago, invented the *wheel,* but the individual or collective genius who called it forth prepared the way for the gigantic Ford River Rouge plant. Realism and a decent sense of indebtedness should incline all of us in the modern world toward gratitude for those predecessors of ours who made possible both the natural and the humanly devised environments of our lives.

Knowledge of our debts to the recent and remote past is not only an ideal obligation for any modern person; it is also a matter of being *at least* as competent as our ancestors before we presume to improve on their accomplishments. Newton was being realistic when he recognized Copernicus as a giant; Einstein never downgraded Newton's genius, and modern critics of the American steel industry should never downgrade the genius of Andrew Carnegie and his like. But to appreciate the excellence of the past may be to appreciate its deficiencies, too.

Our ancestors may not have lived up to their own standards. In our time we may have discovered some new, better standards, a conclusion to which no one of us will leap too quickly. Progress is one of the slippery, debatable, misleading concepts of the past few hundred years of our history.[18] As Holmes Rolston, III, puts it in his recent book, *Science and Religion:*

> It is a mistake to say that the three million hunter-gatherer years [of prehistoric humanity] were bad years, and that only the last two hundred years,

since the Industrial Revolution, have been good ones. The hunter-gatherers generally enjoyed good health, we are told, and rarely worked more than eighteen hours a week! They ate a wider variety of foods than we do, were less warlike, and were more independent of remote powers and markets. They had fewer unsatisfied desires. The preliterates were not pointless because they were not modern, just as we do not pity ourselves because we do not live five hundred years hence. After all, some of the most historic achievements were made in those (so-called) prehistoric years: brains, hands, language, culture, fire building, clothing, toolmaking, agriculture, conscience![19]

In short, to speak of what we value most in ourselves, our organizations, our nations, or any human project whatsoever, we have to revert to history. We have to tell a story. MacIntyre is sure that in the most ordinary of our daily relationships, we tell stories to explain something important to us. "Why did you move from New York to this job in St. Louis?" "Because . . . ," and there follows a story. If there follows only a crisp explanation, "The higher pay," we may understand that answer but may wonder if higher pay alone justifies the wrenching impact of job transfers on all the parties involved. We may wonder if the explainer is really hiding from us the story that makes sense of this job-change chapter in his or her life.

It is thus with other events and other communities of our lives. Numerous Americans recount to each other on any excuse in conversation something about the arrival of their families on these shores. Life in the "old country" is context for defining life in the new. Many are eager to talk about what life in Brooklyn was like before they moved to Scarsdale or in Oregon before they moved to San Francisco. On a much larger communal scale, we tell stories about the cities, the states, the nations, the churches, the clubs, and the voluntary associations of our lives when we want to explain these groups to each other. There is no way to make the name United States of America shine with importance in any conversation or public speech without reference to some updated version of Lincoln's way of beginning the Gettysburg Address, "Fourscore and seven years ago, our fathers brought forth on this continent a new nation . . ." Alex Haley made cultural history by expanding the definition of these American "fathers" to include the fathers and mothers of African Americans, who began arriving on these shores in chains shortly after the earliest white arrivals. To effect this expansion, Haley told a story, *Roots*.[20]

One striking appeal of this reasoning by reminiscence is that it asks us to reflect on what we think good, right, and valuable in our lives by letting our memories and imaginations roam in the realm of narrative rather than asking us to fit our thinking about values in the straitjacket of logical analysis and rules. The philosophical tradition in which Rawls and Nozick write expects a tight, logical set of rules to send us off in the right direction toward thinking about justice, freedom, and the like. But suppose that the most powerful, human, and truthful beginning to such thinking is the story we tell about how we got from "there" to "here" on our way to a possible future? Suppose this, and you have consented to the principle that MacIntyre advances as a central thesis: We are all essentially story-telling creatures. When we tell our stories, we know that we are not the only actors, or authors, in them.

> I can only answer the question, "What am I to do?" if I can answer the prior question, of what story or stories do I find myself a part? We enter human society, that is, with one or more imputed characters—roles into which we have been drafted—and we have to learn what they are in order to be able to understand how others respond to us and how our responses to them are apt to be construed. It is through hearing stories about wicked stepmothers, lost children, good misguided kings, wolves that suckle twin boys, youngest sons who receive no inheritance but must make their own way in the world and eldest sons who waste their inheritance on riotous living and go into exile to live with the swine, that children learn or mislearn both what a child and what a parent is, what the cast of characters may be in the drama into which they have been born and what the ways of the world are. Deprive children of stories and you leave them unscripted, anxious stutterers in their actions and their worlds. Hence there is no way to give us an understanding of any society, including our own, except through the stack of stories which constitute its initial dramatic resources.[21]

MacIntyre insists that a "real story" always has intelligibility—that is, its beginning, middle, and end fit together, making some sense of the whole. Stories, in contrast to random strings of events, hand on threads of continuity. Rolston makes the point concisely when he says:

> Meanings have a logical leaning toward relevance. One asks not merely, "What next?" as in a chronicle. One asks, "So what?" as in a plot. One wants not merely an account, but to know what in it counts.[22]

If, for example, we begin the next chapter of this book with a story from the recent struggle of American corporations with the injustice of racism in South Africa, we do so because we believe that the battle against racism in world society is one of the most important moral themes. One cannot rightly explain the importance of such an episode in the life of a single corporation, in fact, without some assumption about the "larger story" of twentieth century world history. Just as paragraphs form parts of chapters in a book, episodes form parts of even the shortest story and short stories are parts of larger ones. Every story we tell leaves us implicitly on the boundary of a larger story.

Our perennial problem is that we are likely not to know the nature of that large story. It may be so large as to escape the understanding of all humans whatsoever; and we are left tempted to believe, relativistically, that human beings are actors in their own disconnected episodes. What looks like no connection, however, may be a delusion. The hunter-gatherers of three million years ago really were preparing the way for us, their successors on planet Earth. There really is a story that connects us to them, as evolutionary scientists have begun to discover. But at best we are all left with fragmentary, open-ended glimpses into the "great plot" of human history. Meantime, as Rolston suggests:

> We can ask *when* history has meaning long before we ask about *the* meaning of history. There can be piecemeal episodes that have intrinsic value without necessary contribution further. The thrills of the hunt, the song, the dance, the poem, the rewards of springtime and harvest, of parenting and maturing, love and friendship, craftsmanship and leisure, the joy of laughter, the pangs of tragedy, the day spent at play, the life at a task—these are recurrent meanings in the routines of each generation. But they do not take place isolated from a distinctive narrative cultural flow. They are woven into and spiral out from and around it.[23]

The truth advanced here is two sided: If one concentrates on "cultivating one's own garden," living one's own neatly bounded episode, one still is acting in larger human stories. If one glimpses some larger story and willingly seeks to play a role in it, one has to find the proper boundaries or local requirements of the role. This is the common human situation: Living in a world too large to understand but required to act in it anyway. As MacIntyre would say, all the people in a story have character, identity, and accountability. No historian writes about people in some bygone era without assuming as much. Human beings are not automatons, mechanically determined. Perhaps animals are not so either. We, at any rate, must treat one another as actors, agents, and responding, responsible human beings. There are stories that can best be understood by acting in them; or as evolutionary scientists and philosophers like MacIntyre and Rolston would all agree, the human story stops being merely the story of other lives when, in connection with them, we are called on-stage in our own times and places.

> The story, which for long epochs moved without us, and which the sciences can help relate, now moves through us. Though it earlier moved through us over our heads, and indeed still does so, we have recently come to fuller awareness how its headings lie in our human decisions, made on the frontiers on which we now stand. The world is still being made. . . . Truth lies in the creation of what *ought to be,* beyond what *is.* . . . We are coagulating the possibilities this way and not that way. We are writing the text.[24]

This is an inspiring or a threatening truth, depending on how one reads the human story and one's personal story. Twentieth-century science, technology, economics, and politics make it easy for us to believe that human beings are more in charge of the future of this planet than any generation before. But what wise one among us is merely inspired by this easy belief? Personal and collective human responsibility these days has its awful side. Not only have we succeeded in inventing a weapon that could put an end to the human species itself, not only have we fought the most inclusive "world" wars in history, but our best personal and collective intentions have sometimes produced the most unanticipatable of consequences. Robert Oppenheimer, Director of the Manhattan Project, lived to regret his part in the construction of the nuclear bomb. The inventors and manufacturers of DDT, sure that they had found the way to destroy malaria and other enemies of human life, lived to perceive their deadly contribution to world environmental pollution. We come on the human scene and undertake roles that are both handed to us and invented by us; the outworkings of the plot leave our intentions only partly realized, sometimes outrageously foiled.

But that is the way with stories, as opposed to neat logical deductions, MacIntyre would say. The incidents, facts, persons, and elements of no human story fit neatly together. Tension, contradiction, conflict, and perplexing discontinuity afflict the biographies of all people. Such conflict is the stuff of fictitious stage plays, which seek to bring into focus the conflicts scattered across the years of many a human life. Whether we ask in the gloomy philosophical mood of Hamlet if suicide ("not to be") is better than living, or in the desperation of Willie Loman ask if the life of a failed salesman is worth continuing, we are wrestling with the connections of our little story to something larger that may or may not make it worthwhile, fulfilling of our humanity. These are the inner- and interpersonal debates of *every* human being; they belong to uncelebrated, ordinary, everyday actors.

The habit of work, for example, and the sense of responsibility that sends many people off to work each morning, has some deep kinship with the conflicts experienced by Hamlet and by Willie Loman. Various forms of suicide litter our human stories—the final response of actors to conflicts they see no way of living through. Sometimes men and women take this particular weapon "against a sea of troubles" because they find it no longer worth the effort to participate in interactions with their fellows. The suicides of 1929 among Wall Street investors included such folk. Mental hospitals and city streets have their own, other versions of the same withdrawal from action in the human drama.

Particularity and conflict are ordinary in our life-stories. Each of us is limited in our life story to particulars of birth, community, and time. Those particulars set the stage for much conflict in the drama of our lives.

> I inherit from the past of my family, my city, my tribe, my nation, a variety of debts, inheritances, rightful expectations and obligations. These constitute the given of my life, my moral starting point. . . . Without those moral par-

ticularities to begin from there would never be anywhere to begin; but it is in moving forward from such particularity that the search for the good, for the universal, consists. Yet particularity can never be simply left behind or obliterated. The notion of escaping from it into a realm of entirely universal maxims which belong to man as such, whether in its eighteenth-century Kantian form or in the presentation of some modern analytical moral philosophies, is an illusion and an illusion with painful consequences. When men and women identify what are in fact their partial and particular causes too easily and too completely with the cause of some universal principle, they usually behave worse than they would otherwise do.[25]

For MacIntyre a powerful living tradition is always the setup for an argument, as when the French revolutionaries set up an argument between three political values, hailed as liberty, equality, and fraternity.

When a tradition is in good order it is always partially constituted by an argument about the goods and pursuit of which gives to that tradition its particular point and purpose. So when an institution—a university, say, or a farm, or a hospital—is the bearer of a tradition of practice or practices, its common life will be partly, but in a centrally important way, constituted by a continuous argument as to what a university is and ought to be or what good farming is or what good medicine is. Traditions, when vital, embody continuities of conflict.[26]

And in such conflict is often the drama of the human story.

RECLAIMING THE CONTINUITIES AND CONFLICT OF AMERICAN HISTORY

All the major traditions mentioned in this study thus far embody some central continuities of conflict. Alexis de Tocqueville worried about the conflict between the individualistic and the associative sides of American life. He saw the newness of the society partly in terms of how intensely Americans seemed to put these two sides together. Nineteenth-century corporation builders, who assembled giant firms and whose successors now manage even larger conglomerates of cooperating members, really were dependent on a sense of individual responsibility to bring their workers to work every day, on time, and subject to the strict discipline of the machine, power tools, and assembly lines. They, too, worried about individualism in the American story and its tension with the cooperative, corporate requirements of factory production. Carnegie, Rockefeller, and the 1951 editors of *Fortune* worried about the same thing (see Chapter 8).

Civil rights leaders, especially Martin Luther King, proclaimed a form of liberty for American political life that presupposed liberty for all and therefore equality among all. The resulting debates of the 1960s showed how superficial were many claims for liberty on the lips of white people who had forgotten to engage themselves in the dialogue of liberty with equality in democratic history. The most thoughtful managers of multinational corporations in the 1980s were deeply perplexed by the real conflict between loyalty to the particular roots—often in the United States—of their national upbringing and the needs of poor people around the world for jobs. The conflict between jobs for us and jobs for them is profound and of a magnitude unprecedented in world economic history.

The local and the universal commitments of democracy itself are locked in this conflict also. One can understand the impulses of Detroit auto workers who, as a ritual act, take sledge hammers to a Japanese-made car. That very act is disturbing against the background of (1) free-market rhetoric growing out of the economic theory that auto managers swear by (though they do not practice it in asking for import quotas) and (2) egalitarian justice celebrated by religions as ancient as Judaism and Christianity and philosophers as modern as John Rawls.

Plainly this line of thought asks business managers to do their own thinking that, on their surface, has little to do with institutions of economics. Institutions such as the market "float," so to speak, in *culture*, in a tossing sea of beliefs, values, meanings, and stories that therefore provide a common environment for the diverse, vital elements of the lives of persons, activities of institutions, and the movement of whole civilizations. However indistinct or invisible this culture may be—like the depths of water beneath a ship's hull—it is nonetheless supporting the institution. He may express it differently, but this is MacIntyre's general view of the way in which certain general value commitments and storylines transcend the immediacies of human behavior and shape that behavior.

Among those recent scholars whose vision of human behavior in society parallels the one we have described here above is Robert M. Cover, a Yale professor of Law and Legal History, who has echoed MacIntyre's analysis in an eloquent way. In the essay *"Nomos and Narrative"* Cover says:

> The rules and principles of justice, the formal institutions of law, and the conventions of a social order are, indeed, important to [the world of right and wrong]; that ought to claim our attention. [But] no set of legal institutions or prescriptions exists apart from the narratives that locate it and give it meaning. For every constitution there is an epic, or each decalogue a scripture. Once understood in the context of the narratives that give it meaning, law becomes not just a system of rules to be observed, but a world in which we live.[27]

Along with MacIntyre, Cover observes that few human beings can "explain" their behavior to another without the aid of biographical stories that link one person's life to the lives of others.

Any person who lived an entirely idiosyncratic normative life would be quite mad. The role that you or I choose to play may be singular, but the fact that we can locate it in a common "script" renders it "sane"—a warrant that we share a *nomos* (law).

A combination of things that *are,* things that *ought* to be, and things that *can* be, compose the map of reality that each of us carries, and "narrative . . . integrates these domains."[28] And narrative integrates them not chiefly for our personal biographies, but for the communities in which we grow up. "The complexity of mutual understandings at work in the community is . . . revealed and transmitted in the narratives of the group."[29]

The democratic vision of a proper human society, Cover notes, requires not a mere single narrative in which all citizens find the meaning of their lives but a large, open-ended story always subject to new interpretation in light of new human experience. Totalitarian societies from the Nazi to the *1984* versions ask their citizens to accept a single, enforced, orthodox way of telling their social stories. Democrats, on the other hand, require of their society "the power of freedom of contract to create nomic insularity,"[30] that is, subgroups of people who agree to abide by some standard of behavior over which no other authority has definitional control. (Nozick, for example, makes this very freedom, in the economic realm, central to his definition of social justice.) A church, a corporation, and a school may all share this characteristic in a society like the United States, for as Tocqueville observed on every hand Americans are always "forming associations" outside of anyone's permission.

> Property and corporation law have always been bases for claims to creation of an insulated nomic reserve. The company town, mine, or plant often asserts a right to law creation and enforcement with respect to social relations.

Thus, between the new group, insulated in certain of its norms, and older groups "a wall begins to form, and its shape differs depending upon which side of the wall our narratives place us."[31] Furthermore, in the democratic vision of society, the right to have unique, insulated associational norms is often paralleled by an assumed right to try to convince outsiders ("across the wall") *that they too should behave in accord with the new norm.*

> People associate not only to transform themselves, but also to change the social world in which they live. Associations, then, are a sword as well as a shield.[32]

A democratic social order, then, expects its members to be diverse in their associations and expects its public debates to be structured in part around a diversity of associational stories. Hence arise "continuities of conflict" related to persisting but clashing group histories. In this light African Americans, Italian Americans, Anglo-Saxon Americans, and Mexican Americans will never have the same story to tell about what it means to be American. But all of their stories overlap: geography, contemporaneity, and the requirements of mutual survival ensure that.

Of all countries in the world, perhaps, the United States most expects this diversity of story to influence its public life, making more important and more dynamic the perpetual construction of a new "national" story out of the old stories that citizens bring with them in every public debate and confrontation. Martin Luther King, Jr. was an expert at seeking to tell the story of African Americans in relation to the premises of the United States Constitution that, in his view, had yet to find full meaning for his people. Making a bridge between the larger national story and the "up from slavery" narrative of his own ancestors and colleagues was basic to King's genius as an associational leader, as a citizen, and as a theologian. As Cover puts it:

> The principles that establish the nomian authority of a[n insular] community must . . . resonate within the community itself and within its sacred stories. But it is a great advantage to the community to have such principles resonate with the sacred stories of other communities that establish overlapping or conflicting normative worlds. Neither religious churches, however small and dedicated, nor utopian communities, however isolated, nor cadres of judges, however independent, can ever manage a total break from other groups with other understandings of law.[33]

One might say that the dividing line between democrats and autocrats—in religion, law, business, or family—falls between those who are sure that theirs is the only story worth listening to and those who feel an obligation to listen to stories not (yet) their own. This is a fundamental moral–religious surmise that slumbers under every notion of justice, equality, liberty, and responsibility discussed in any book written by democratic authors for democratic readers. "The spirit of liberty is the spirit that is not too sure it is right," said Judge Learned Hand,[34] and the corollary of this uncertainty is the spirit that listens to the stories of other people in the hope that *they* may contribute to one's knowledge of the right.

That is why even a book on business ethics must not pretend that it offers a *definitive* notion of the right, the just, the free, or the responsible society. Here the notion of responsibility resonates with the biblical command that human beings are to "love their neighbors as themselves." In Cover's words, "we ought to stop circumscribing the *nomos;* we ought to invite new worlds."[35]

Those "new worlds" may come from any one or any group of our

neighbors. This is one of the "moral facts" we have to recommend—
and not simply as an exhortation. The next chapter in the history of a
business firm, a church, or a government may come only with the help
of some person or group outside its ranks. It may come from poets,
novelists, historians, and theologians as well as from lawyers, economists, and legislators. It may come from "mere" consumers, common
workers, or ordinary community members and anyone else whose life
story may easily be dismissed or derogated as of little worth. More to
the situation of business managers: Help may come from those human
beings who *put one part of their life experience in touch with other parts to
make something like a dramatic story wherein small stories open up toward participation in larger ones.* Those who once thought of themselves as divorced from others' lives may thus find themselves together in a new,
inclusive drama.

TESTS OF A JUSTICE-TENDING NARRATIVE

As summary (or heuristic) questions, we ask ourselves: What happens
in a collective narrative that might signal to its participants that some
measure of justice is getting done? To ask this question is to inquire
about tags or abstract measures of justice but in the context of myth,
history, and collective story. It is to seek a virtue central to the ethical
dialogue of many human societies. In the next chapter we shall give
the name "responsibility" to the ethic we are describing, but justice will
always be a word on the lips of people who debate social rights and
wrongs. What happens, or what is failing to happen, when justice gets
done, or denied, in a society?

- A voice that was previously absent from the debate gets a hearing.
  The least favored speak up, someone listens, and the role of exit and
  absence in the story diminishes.
- A new or long-neglected social relationship arises, in the promise that
  a new conversation about mutual needs will now proceed and will not
  soon be broken off.
- An institutional rule, boundary, or routine falls under question, especially the question of whether the institution can become inclusive
  rather than exclusive in its benefits. Although limits of the system
  and its benefits remain, the question of constructing a new inclusiveness is now on the common agenda of the ongoing social conversation.
- Old contracts get reexamined historically, with a view not only to rectifying the injustices that were knowable at the time but to rectifying
  those that were knowable only in the light of their historical results.
  A certain respect for the past is given by those who currently benefit

or suffer from the past, as they understand relations among their mutual ancestors, who could not understand fully the implications of their lives and actions.

- A sense of *struggle* infuses most parties to the negotiations and conversations about justice. Those who seek justice articulate their suffering, and those accused of being unjust "suffer" that accusation in the hope of some fresh understanding.
- A new structure of power to ward off a recurrence of the old suffering begins to be put into place: Some balance of social power shifts, enforceable laws get invented, new forms of organization spring up, and (again) broader forms of inclusiveness are constructed. The habits of actors in the future story are now subject to certain pushes and nudges. They will find it easier to be just, harder to be unjust.
- But current participants in the debate and action concerning justice do not imagine that they have said the last word on the subject. They know and speak about justice yet to be done, and they end their particular time on the center stage of the struggle over justice conscious of times to come and grateful that others wait in the wings to undertake a search for justice in the future that the present actors are incapable of undertaking.

To speak of the nature of justice in this context is to combine a certain process with a certain content for defining the word, and to claim that the one without the other never tends toward genuine justice. Such a way of defining justice implies a kind of composure, an acceptance, on the part of the definer of his or her own limited powers of definition! Such thought about justice brings the thinker around to the consistency that acknowledges limit and openness in knowledge and the knower. The justice-tending narrative will get told best by those who, like Socrates, are still searchers in their telling and still part of the ongoing story.

Such storytellers, we believe, are practitioners of an ethic of responsibility. What such an ethic might look like in any human community, and especially in the world of the modern business corporation, is the summary concern of chapters 12 and 13.

We grant that language like this sounds strange in books aimed mostly at an audience of business managers. Again, we write as part of that company of contemporary human beings who wonder if, somewhere on our long journey from the past, we put ourselves in danger of losing our way into a livable human future.

We have a lot of remembering, revaluing, repenting, and redirecting to undergo. We sense our confusion because we are confused. But the most misdirected among us may be those who think they know the way. A great deal more consultation between previously isolated partners in this journey may be our only way out of the dangers in our confusion.

We may call it the way of responsibility and in a more careful, sober way—impossible in the opening explorations of this book—we may now ask ourselves, "What is responsibility in the context of the American business corporation for the years just ahead of us?"

NOTES

1. E. G. West, "Introduction," in Adam Smith, *The Wealth of Nations* (New Rochelle, N.Y.: Arlington House, 1969), p. xxxiii.

2. Alasdair MacIntyre, *After Virtue: A Study in Moral Theory*, 2nd Edition (Notre Dame, In.: University of Notre Dame Press, 1984), p. 250.

3. Founded in 1660, the year Charles returned to England, and incorporated in 1662. Its membership included leading scientists of the world.

4. MacIntyre, *After Virtue*, p. 92.

5. Robert Jackall, "Moral Mazes," *Harvard Business Review* (September–October 1983):118–30. See also Robert Jackall, *Moral Mazes: The World of Corporate Managers* (New York: Oxford University Press, 1988).

6. Ibid., pp. 122–24.

7. Ibid., p. 128.

8. Ibid., p. 130.

9. See the studies referred to by MacIntyre: Tom Burns and G. N. Stalker, *The Management of Innovation* (London: Tavistock Publications, 1968); and Herbert Kaufman, *Administrative Feedback* (Washington: Brookings Institution, 1973).

10. "S.E.C. Overruled Staff on Findings that Citicorp Hid Foreign Profits," *New York Times*, February 19, 1982. The SEC view is bound to shock business leaders whom we have described as holding strong convictions about the importance of obeying the law in this society.

11. MacIntyre, *After Virtue*, p. 64. Italics added.

12. Quoted by MacIntyre, Ibid., p. 119.

13. See T. V. Smith, *The Ethics of Compromise and the Art of Containment* (Boston: Starr King Press, 1956), p. 75.

14. MacIntyre, *After Virtue*, p. 187.

15. Ibid., p. 190.

16. Ibid., p. 191.

17. Ibid., p. 194.

18. Robert Nisbet, *History of the Idea of Progress* (New York: Basic Books, 1980).

19. Holmes Rolston, III, *Science and Religion: A Critical Survey* (New York: Random House, 1987), p. 281. These are the conclusions, Rolston notes, of Davydd J. Greenwood and William A. Stini, in *Nature, Culture, and Human History* (New York: Harper & Row, 1977), pp. 428, 436, 500.

20. Alex Haley, *Roots* (Garden City, N.Y.: Doubleday, 1976).

21. MacIntyre, *After Virtue*, p. 216.

22. Rolston, *Science and Religion*, p. 282.

23. Ibid., p. 280.

24. Ibid., pp. 337–38.

25. MacIntyre, *After Virtue*, pp. 220–21.

26. Ibid., p. 222.

27. Robert M. Cover, "The Supreme Court, 1982 Term. Forward: *Nomos and Narrative*," *Harvard Law Review* 97 (November 1983):4–5.

28. Ibid., p. 10. Italics added.

29. Ibid., p. 50, note 137.

30. Ibid., p. 30.

31. Ibid., pp. 30–31.

32. Ibid., p. 33.

33. Ibid., p. 33.

34. Learned Hand, "The Spirit of Liberty," *In the Spirit of Liberty: Papers and Addresses of Learned Hand*. Collected and with an Introduction and Notes, by Irving Dilliard. Third Edition, Enlarged. (New York: Alfred A. Knopf, 1960) p. 190.

35. Cover, *"Nomos* and Narrative," p. 68.

12

What Is Corporate Responsibility?

In the early 1980s the government of the Republic of South Africa invited a midwestern manufacturer to submit a bid on the operation of a plant whose output was vital, basic industrial machines. The value of the contract was $50 million. Senior managers and the company board of directors consulted among themselves, and, with little discussion, they decided not to bid. Asked by a group of business school students about their decision, the CEO replied: "We decided that it would not be in keeping with the character of our company to cooperate so directly in strengthening the government of South Africa." He added that after their decision, a West German producer bid for the operating contract and won it. Consequently the American firm will gradually lose its dominance in that product market, not only in South Africa but also in all of southern Africa.

Such a decision is only an incident in a much longer story of any firm. This company is a leading producer of industrial machinery; and its long-time, able CEO has stamped its "character" with the imprint of his own. Yet the decision to restrict rather than to expand business in South Africa had a much more complicated background than the moral principles of one executive. The firm has for years carried on its business there, with its managers only slowly coming to the conclusion that direct economic ties with a government founded on racial discrimination was incompatible with their business practice in the American setting. By the early 1980s the managers of many U.S.-based corporations, for mixed economic, political, and moral reasons, had begun to reach similar conclusions. Some had done so out of their dialogue with stockholders, church groups, and politicians, who argued that business had some responsibility for hastening the demise of apartheid. Earlier in the 1950s and even 1960s many corporate leaders scoffed at the notion of that kind of mixing of business and politics. Then in 1977, at the behest of the Reverend Leon Sullivan (pastor of Zion Baptist Church, Philadelphia, and member of GM's Board since 1972), General Motors took the extraordinary step of asking its South African subsidiaries to follow the "Sullivan Principles" for breaking down racially discriminating policies in the workplace. In 1987 even the Reverend Mr. Sullivan had grown impatient with the slow pace of change in South Africa and called for the departure of American companies from the country.

In their changing business relationships with South Africa, many American firms have moved from and altered their ethical storyline; they have moved from unconditional to conditional to minimal to no involvement with people responsible for legalized apartheid. The ethical story each is enacting is changing. It will change more. As we write, argument about the goods and evils of business involvement with South Africa continues among managers still buying from, selling to, or boycotting South Africa. Few managers claim to know exactly what their firms' policies ought to be, but many confess that over the decades their thinking on this complex matter has changed. What in the 1960s might have sounded like irrelevant moralism in the marketplace, in the 1980s sounds like the learning of new roles in a story larger than the one in which some originally intended to act.

The unpredictability and number of factors that have entered into business thinking here seem unprecedented to many managers, making them uncomfortably aware that their new role increasingly resembles that of the politician.[1] Seldom now do managers find or maintain their policies with much sense of simplicity, consistency, and coherence over time. Standing fast by the character of a company, producing benefits for a specific set of customers at a profit, attempting to influence social structures beyond the company's ability to effect decisive change, and coping with much uncertainty about the next step in the learning of new collective roles—all these challenges must be met as managers confront questions of doing business with South Africa. No wonder that some confess, "Business school never prepared us for this."

What actually prepares them is their own experience. Central to the thesis of this book is the claim that, in our time, *hospitality to multiplicity belongs in the new definition of good business practice.* As long as a competitive structured economy exists, business has to meet the market test; but the business world now consists of more than traditional markets. The firm must obey laws; attract employees; guarantee the health and safety of its workers, customers, and neighbors; maintain public and consumer confidence; serve its country; respond to international emergencies; and perhaps do a host of things for which its history has poorly prepared it. Its managers may, in MacIntyre's terms, have to invent new practices that recognize others' truths as defined by new constituency interests, principles, and purposes. Neither the first Ford nor the first Rockefeller would have been able—or willing—to grant legitimacy to such truths, or even to have conceived of them. Modern managers need to recognize that the multiple right and wrong claims, "It's wrong." "It's good for the company," "It's good for the country and bad for the company," "It's a mixture of good and bad, right and wrong" *all* have a place in policy deliberation worthy of being called *responsible*.[2] Many participants in public debate, such as that about political and business policy toward South Africa, have found themselves advocating a view that used the word "responsible" in one or another part of their conclusions. They are apt to say something like:

> We are uncertain about the mixture of good and bad results in anything we might do, and we are not clear about the balance between right and wrong, but on balance this . . . or that . . . seems to be the responsible thing to do.

In the intellectual game of moral theory, the terms "right" and "good" have dominated the playing field. Only in recent decades of the twentieth century has the word "responsibility" forcefully entered the rhetoric of those who have had to justify one course of action over another. We have claimed throughout this study that the modern business corporation has a social *responsibility*. It is time we looked carefully at the meaning of that word.

INGREDIENTS OF RESPONSIBLE ACTION

Response to Action upon Us

No modern philosopher has looked more carefully at the meaning of responsibility than the late theologian H. Richard Niebuhr. His posthumously published book *The Responsible Self*[3] begins with a description of the two styles of ethics—right principle and good purpose—which have set the terms of debate about ethics among most Western philosophers. His description applies not only to ethics in the narrow sense but also to the style of decision making generally in both our personal and organizational relationships. As a result of our upbringing or other education, some of us make our decisions by asking, "What are the rules, the laws, the principles that should guide this action?" Others of us like to ask "What is the purpose, the goal, the future good at which we should aim?" On the whole, American culture inclines most of us to prefer the second, good-purpose style of reasoning to the first; it comports best with our pragmatic "can-do" spirit. A strong argument for *breaking* some law or principle usually has to appeal either to a "higher law" or a "greater good"; and frequently such laws and goods compete among themselves for a priority in our decisions to act. It is hard exclusively to do one's duty or to do good. Consistency in the use of one style or another is not likely to characterize either individuals or organizations in our society.

For example, the spirit of the Puritan tradition still imbues some business managers. It sternly forbids lying, even to enhance company profit, though profit remains a justification for most company policies. Lying for good purposes can be raised to very high levels of moral rhetoric, of course, as it is for covert political operations in service to one's country—witness the testimony of Lt. Col. Oliver North in 1987.

Lying for good purposes in certain circumstances is virtuous, some would urge. One of those circumstances is war, as is any other life-threatening situation in which existence itself mounts to the priority-purpose of all possible action. Some religions give highest rank to acts that may sacrifice not only truth but life itself in the service of other life. But consistency may not be sustainable for long by those who accept such theory. Even the principle of survival may have to give way to the principle of truth or take second place to other human benefits or interests. Faced with these perplexities in the grip of circumstance calling for decision, most of us settle for different arguments in different circumstances, laying ourselves open to the charge of "relativism." A manager from an Atlanta corporation once commented, "Most of my business friends talk in public about managing their businesses 'by the bottom line' of profit and loss. But I have noticed that often they cut through some complex decision by 'doing what seems right.' "

Most of us cannot avoid thinking of both principles and purposes, rights and goods, lesser right and larger good, as we go about our decision making. But in our motivation may be something more basic to our humanity than our capacity to obey law and frame purposes, or even to balance the one against the other: our capacity to respond to what *happens* to us. Modern biological science fortifies ordinary human experience in that surmise. It suggests continual interaction of the living fetus and its environment from the moment of conception on! At first such interaction hardly includes what we mean by human "agency." All of us are shaped by forces outside ourselves before we realize that we exercise influence. We have questions asked us before we knowingly ask. Circumstances demand answers from us, and sometimes our present principles and purposes are insufficient for answering appropriately. Every action of ours is something of a reaction, and in this sense a receptive passivity always precedes our initiated activity. Biology, memory, and a little reflection make absurd the old idea of the "self-made" person. Modern social science, as Niebuhr points out, has fortified our sense of absurdity, even threatening us with a contrary sense that we are not actors at all, just reactors to circumstance. But most of us know that circumstances require an "answer" from us, says Niebuhr, and to keep on living is to keep on answering.

What is implicit in the idea of responsibility is the image of man-the-answerer, man engaged in dialogue, man acting in response to action upon him . . . To be engaged in dialogue, to answer questions addressed to us, to defend ourselves against attacks, to reply to injunctions, to meet challenges—this is common experience. And now we try to think of all of our actions as having this character of being responses, answers, to actions upon us. The faculty psychology of the past which saw in the self three or more facient powers, and the associational psychology which understood the mind to operate under laws of association, have been replaced by a psychology of interaction which has made familiar to us the idea that we react in reaction to stimuli. Biology and sociology as well as psychology have taught us to regard ourselves as beings in the midst of a field of natural and social forces, acted upon and reacting, attracted and repelling. We try also to understand history less by asking about the ideals towards which societies and their leaders directed their efforts or about the laws they were obeying and more by inquiring into the challenges in their natural and social environment to which the societies were responding. . . . The pattern of thought now is interactional, however much other great images must continue to be used to describe how we perceive and conceive, form associations, and carry on political, economic, educational, religious and other enterprises.[4]

The idea of response must be distinguished carefully from the idea of mere reaction. Response-ability in humans implies something originated, the act of an *agent* who, in reacting, contributes something new that cannot be reduced simply to external causes. Parents rightly marvel at a certain unpredictability in the behavior of even their youngest children. Presidents discover that appointed judges have minds of their own. In even ordinary conversations we continually say things we did not plan to say and have never said before.

Deep philosophical issues underlie these experiences. It is notoriously difficult to prove or disprove the reality of human agency. Are we free or determined? Reflection convinces most of us that we are both; and leaders of institutions such as business, government, and voluntary associations cannot afford to ignore *either* the shapers or the shaping of their own behavior.

Interpretation

What intervenes, then, to turn a human reaction into a human response? A second dimension of the human experience, says Niebuhr: An *interpretation* of "what is going on" in that outside world which against our intentions impinges upon us.

All actions that go on within the sphere of our bodies, from heartbeats to knee jerks, are doubtless also reactions, but they do not fall within the domain of self-actions if they are not accompanied and infused . . . with interpretation. Whatever else we may need to say about ourselves in defining ourselves, we shall need, apparently, always to say that we are characterized by awareness and that this awareness is more or less that of an intelligence which identifies, compares, analyzes, and relates events so that they come to us not as brute actions, but as understood and as having meaning. Hence though our eyelids may react to the light with pure reflex, the self responds to it as *light,* as something interpreted, understood, related. But, more complexly we interpret the things that force themselves upon us as parts of wholes, as related and as symbolic of larger meanings. And these large patterns of interpretation we employ seem to determine—though in no mechanical way— our responses to action upon us. We cannot understand international events, nor can we act upon each other as nations, without constantly interpreting the meaning of each other's actions. Russia and the United States confront each other not as those who are reflexively reacting to the manufacture of bombs and missiles, the granting of loans, and the making of speeches; but rather as two communities that are interpreting each other's actions doing so with the aid of ideas about what is in the other's mind.[5]

Niebuhr illustrates responsiveness in terms of labor–management relations:

When we think of the relations of managers and employees we do not simply ask about the ends each group is consciously pursuing nor about the self-legislated laws they are obeying but about the way they are responding to each other's actions in accordance with their interpretations. Thus actions of labor unions may be understood better when we inquire less about what ends they are seeking and more about what ends they believe the managers to be seeking in all managerial actions.

In short:

We respond to these events in accordance with our interpretations. Such interpretation . . . is not simply an affair of our conscious, and rational, mind but also of the deep memories that are buried within us, of feelings and intuitions that are only partly under our immediate control.[6]

"Deep memories buried within us" is a definition of personally or collectively appropriated *history.* MacIntyre's resort to story and one manager's resort to the long-developing "character" of his company have much in common with Niebuhr's theory. So too does Rafael Pagan's new interpretation of the story of Nestlé's constituencies. History and character provide some of the frameworks for interpreting what is happening to us, and these frameworks are very practical ingredients of responsible human behavior. The history of our species is replete with examples of how *false* interpretations can wreak havoc in the lives of persons and societies, from mushrooms falsely interpreted as edible to wars falsely interpreted as winnable—and to products falsely interpreted as marketable.

None of the theories of management described in this book can be "explained" outside of the historical contexts in which their proponents made sense of the theories. Without the mixture of Calvinist faith in the sovereign God and Locke's sovereignty of property, George F. Baer's self-confident industrial paternalism in 1902 would have made no sense to him or his supporters. The social Darwinism celebrated by C. R. Henderson depended upon the combination of new biological theory and the new industrial corporation of the late nineteenth century. Frederick W. Taylor's "Scientific Management" required him and managers to make comfortable analogies between human behavior and machines—a development only "natural" in the context of the same new industrialism. Elton Mayo's "human relations" in industry arose at the same time unions were acquiring new legal rights in this country, and each was a new response to, and rejection of, the perceived danger that machines would make their operators into their own image—the same history against which Karl Marx revolted a century before. Indeed, perhaps the most radical philosophical proposal of Marx himself was that humans see themselves as actors in a story begun generations ago and to be completed generations hence. It was this sense of history that allied the spirit of Marxism with the spirit of the Judeo–Christian religion. The great philosophical question of these history-minded proposals is not, "To know or not to know the ideal good," but, "To know or not to know what is happening to you."

In 1968 Lynn Townsend, president of Chrysler Motor Company, offhandedly said to one of the authors, "Americans will never opt for small cars like the ones the Japanese and Germans are making." It was a costly error, a misreading of consumers' preference that eventually pushed Chrysler to the edge of bankruptcy and seriously weakened the other big American automakers. If Detroit executives are to be blamed, the blame rests not on their inability to read the future but in their sometimes arrogant refusal to face the questions and ambiguities that were accumulating on the perimeters of their workplaces. Consumers' preferences, an electorate's vote, or natural catastrophe seldom speak clearly and distinctly. Always there intervenes our or other's interpretations of what is going on, and interpretations become shapers of response to facts. Many a conflict between the "new constituents" of business in the 1970s stemmed from mixtures of true and false interpretations on both sides. As we have seen, charged with "irresponsible" marketing of infant formula in Third World countries, the top managers of Nestlé S.A. at first dismissed their critics, with the result that public criticism spread around the world. In the process the critics popularized the slogan, "Nestlé kills babies"—interpretive overkill, some would say—and Nestlé successfully argued in a Swiss court that it was libel. Nestlé managers went further, however. Even as they moved to meet their critics' particular demands, they tried to persuade university teachers to accept their interpretation of the long-lasting boycott. It originated, they maintained, in a powerful, highly organized, Communist-influenced conspiracy around the world.[7] American leaders of the boycott movement had another interpretation: organization of citizens to combat particular social evils is a democratic habit in America certainly, if not in Switzerland. Eventually some Nestlé managers such as Pagan must have switched to just such a new interpretation.

Human beings will always have diverse, sometimes erroneous, and almost always only partially adequate interpretations of what each other's actions mean. Without interpretation, however, we have no guidance at all for responding to those actions; and although one's own interpretations of them has its rightful day in court, other claimants are there also; thus until some future jury reports its findings, no claimant would be wise to assume an exclusive jurisdiction over the truth, the justice, and the responsibility of the matter. Responsible people "ought to stop circumscribing the *nomos;* we ought to invite new worlds."[8]

Accountability

A third dimension of responsibility is likely to be implicit in any guiding interpretation: Niebuhr calls it *accountability*, a readiness for future responses.

> Our actions are responsible . . . as they are made in anticipation of answers to our answers. An agent's action is like a statement in a dialogue. Such a statement not only seeks to meet . . . the previous statement to which it is an answer, but is made in anticipation of a reply. It looks forward as well as backward; it anticipates objections, confirmations, and corrections. It is made as part of a total conversation that leads forward and is to have meaning as a whole. Thus a political action, in this sense, is responsible not only when it is responsive to a prior deed but when it is so made that the agent anticipates the reactions to his action. So considered, no action taken as an atomic unit is responsible. Responsibility lies in the agent who stays with his action, looks forward in a present deed to the continued interaction.[9]

As we have seen, at the beginning of the corporate social responsibility movement in the late 1960s, the inclination of some corporate leaders—after the pattern of Nestlé—was to treat their "new constituents" as temporary external annoyances to business efficiency. Only gradually did some come to acknowledge the new publics as permanent participants in the development of some business policies.[10] Even ten years after the Nestlé boycott had begun, some Vevey officials of the company spoke as though the "new constituents" would one day go away. Yet it seems unlikely that a broad world public will ever again be indifferent to the effects, unintended or not, of infant-formula marketing.[11] Prudent policy makers in that and many another industry know that they are *accountably* responsible now to groups of people determined to remain in the conversation for the indefinite future. This implies a willingness to maintain a relationship with constituents that can be called "solidarity."

Solidarity

If our critics or our allies mean to remain in the conversation with us for a long haul, will we do the same? As the fourth element of his notion of responsibility, Niebuhr calls this commitment to long-term relation to one's partners-in-dialogue the element of "social solidarity."

> Our action is responsible . . . when it is response to action upon us in a continuing discourse or interaction among beings forming a continuing society . . . it implies continuity in the community of agents to which response is being made.[12]

The community of agents may be as small as two friends thinking about the effects of their action now upon their hoped-for relations to each other twenty years from now, or as extensive as global humanity, wherein some of its members ponder its survival against the dangers of nuclear war and ecological disaster. To be sure, few responsible individuals or members of responsible organizations can ordinarily conceive, afford, or undertake action truly in solidarity within the whole of humanity. No one knows the storyline that connects all people in a single plot; our occasional glimpses of such a story are not likely to show us exactly how to play out our own role in it.

Responsibility always has both scope and limit. Given the limits to human wisdom and virtue, we cannot even act with certainty in ways that will protect a friendship or a marriage into the far future. Furthermore, we are all bound into a stronger relation with some human groups and more deeply involved in some human histories than in others. The Soviets and the Americans both use the rhetoric of world peace, but they are likely to prefer their own respective national survival to that of any other—an interpretation on whose partial truth the current system of nuclear deterrence squarely rests. One's own family, country, company, region, and social class may—now and again—be the "solidarity" that one prefers against all other solidarities. But we count persons or organized groups of human beings *responsible* only if they value ongoing relations with persons or groups close to *and* remote from their primary loyalties. Like dialogue, responsibility is by definition inherently social and indefinitely expandable.

Academic people like the two authors of this book should be the first to appropriate that view of responsibility for themselves. Academics do

not always enjoy great popularity in the private polls of business leaders, who frequently characterize academics not only as "people who
never met a payroll" but also as mountaintop thinkers who look down
from on high at people who do such ordinary things as build cars and
meet payrolls. Is a theory that looks for wisdom *in* the ordinary valleys
of human decision one around which these parties can all rally? An
attractive, affirmative answer to this question has come recently from
an educator—Carol Gilligan—whose version of responsibility-ethics
matches many features of Richard Niebuhr's. But hers is based on some
sources very different from his—a series of interviews almost exclusively with women. It will help illuminate the notion of responsibility
here if we rehearse some of Gilligan's conclusions from her research.

THE WEB OF RESPONSIBILITY:
TOWARD AN ETHIC OF CARING JUSTICE

The majority of business managers are men. Twenty years from now,
if the growing number of women in American business schools is pre-
dictive, the proportion of women in managerial positions is likely to
grow; and already one may speculate on the difference that rise in
proportion may make to the style and substance of human relations
inside and outside the American business corporation.

Gilligan's research suggests that the difference may be profound. Her
1982 book, *In a Different Voice*,[13] reports and explores evidence from
her interviews with women and some men around questions of moral
choice. Her conclusions highlight the differences of decision styles implied in the theories we have discussed here and in Chapter 10.

Men in our culture, says Gilligan, tend to prefer ways of reasoning
compatible with those of philosophers like Rawls and Nozick—paring
logic down to basic principles, connecting them in a hierarchy, offering
derivative rules of right, wrong, good, and bad. Women are more at
home, she implies, in the thinking of those like MacIntyre and Niebuhr—in multifactored contexts, in choices that seek to respect as many
human relationships as possible, in decisions less intended to protect
the rights of individuals and more directed at sustaining and enhancing
networks of care between individuals. In other words, women (in
American society, at least) seem inclined by their socialization and education toward the ethics of responsibility. In her interviews, women
actually use the word "responsible" more frequently than do men. Gilligan's descriptions of this ethic display a remarkable match with the
Niebuhr and MacIntyre proposals. Three of Niebuhr's four elements
of responsibility—the responsive nature of human life, accountability,
and social solidarity—appear repeatedly in quotations and summations
from Gilligan's interviews. Less prominent in her analysis is the theme

of interpretation, though she frequently notes that women prefer telling a story to explaining an ethical decision by calling on the authority of a moral principle like justice.

A moral choice, said one woman, arises when many people and circumstances call for respect and care, and not all can apparently be cared for. To be moral means:

> . . . taking the time and energy to consider everything. To decide carelessly or quickly or on the basis of one or two factors when you know that there are other things that are important and that will be affected, that's immoral. The moral way to make decisions is by considering as much as you possibly can, as much as you know.[14]

Another woman made the same point thus:

> The only way I know is to try to be as awake as possible, to try to know the range of what you feel, to try to consider all that's involved, to be as aware as you can be of what's going on, as conscious as you can of where you're walking. (Q: *Are there principles that guide you?*) The principle would have something to do with responsibility, responsibility and caring about yourself and others. But it's not that on the one hand you choose to be responsible and on the other hand you choose to be irresponsible. Both ways you can be responsible. That's why there's not just a principle that once you take hold of you settle. The principle put into practice here is still going to leave you with conflict.[15]

As some critics of Gilligan have charged, what it may mean to be "careful" as well as just in human relationships, she and her interviewees have trouble specifying in detail.[16] But this charge should equally be leveled at the moral theories of Nozick and Rawls. Nozick retreats utterly from prescribing particular moral goods for human social relations, putting all his confidence in contracts and other procedural forms of justice. Rawls's elegant theory of justice elevates liberty and equality to the height of social value hierarchy but leaves these principles curiously abstract and divorced from interpretation through historical illustration. Gilligan could rightly charge these philosophers of the liberal tradition with failure to allow more than a secondary role for the particularities of history, context, and social roles:

> The proclivities of women to . . . request or supply missing information about the nature of the people and the places where they live, shifts their judgement away from the hierarchical ordering of principles and the formal procedures of decision making.[17]

> In this conception the moral problem arises from conflicting responsibilities rather than from competing rights and requires for its resolution a mode of thinking that is contextual and narrative rather than formal and abstract.[18]

Often this "shift" means moving from a search for "the" good in difficult moral situations to a search for the lesser of evils. Such choice is
real choice, because it cannot be formulated in advance of having to
respond to the demands of a particular historical situation. As Mac-
Intyre might say, the way you act has much to do with the role you
play and the plot in which you have to act. Or as Niebuhr might say,
only in an ongoing dialogue does one learn to respond in fitting, appropriate ways. How do we know what we *will* say to others until we
hear what others say to us?

Opposed here are two images of humanhood whose scope and importance extend far beyond ethical theory in the narrow sense. Gilligan
illustrates the "different voice" of women by a famous incident in the
Bible:

> The blind willingness to sacrifice people to truth . . . has always been the
> danger of an ethics abstracted from life. This willingness linked Gandhi to
> the biblical Abraham, who prepared to sacrifice the life of his son in order
> to demonstrate the integrity and supremacy of his faith. Both men, in the
> limits of their fatherhood, stand in implicit contrast to the woman who comes
> before Solomon and verifies her motherhood by relinquishing truth in order
> to save the life of her child.

"Unwillingness to learn from *anybody anything* except what was approved by the 'inner voice,'" was Gandhi's great weakness, said his
psychiatrist-biographer Erik Erikson.[19] As we all know, an ethic of "the
inner voice" can be very individualistic and even dictatorial. Many of
the profoundest ethical questions in society *begin* when "my" inner voice
conflicts with "yours." In that conflict, dialogue can and should begin.

The characteristics of an ethic of responsibility, as Gilligan portrays
them, include elements seldom recognized publicly in American culture as having anything at all to do with "ethics": attention to facts,
empathy for others, care to make and sustain relations, regard for limits, search for unpredictable possibilities, dialogue between opposite
convictions, and the meaning of particular moments in an ongoing story.
From beginning to end, this ethic presupposes not individuality as the
essence of humanhood but social solidarity. As one of Gilligan's interviewees put it eloquently:

> By yourself, there is little sense to things. It is like the sound of one hand
> clapping, the sound of one man or one woman, there is something lacking.
> It is the collective that is important to me, and that collective is based on
> certain guiding principles, one of which is that everybody belongs to it and
> that you all come from it. You have to love someone else, because while you
> may not like them, you are inseparable from them. In a way, it is like loving
> your right hand. *They are part of you;* that other person is part of that giant
> collection of people that you are connected to.[20]

As Gilligan herself summarizes this relational image of humanhood:

> The truths of relationships . . . return in the rediscovery of connection, in
> the realization that self and other are interdependent and that life, however
> valuable in itself, can only be sustained by care in relationships.[21]

If women in American society seem more likely to prefer an ethic of
care-for-relationship, this does not mean that this ethic is necessarily
superior to an ethic of loyalty-to-principle or an ethic of striving-for-
good; but it may mean that men and women in this society have something to teach each other about a richer way of *being* human than
Americans—including managers—have ever known. Speaking about the
stress on liberty and equality as prime values in our culture, one of
Gilligan's interviewees, a man, commented:

> People have real emotional needs to be attached to something, and equality
> doesn't give you attachment. Equality fractures society and places on every
> person the burden of standing on his own two feet.[22]

The insight is as old as Tocqueville, whose chief worry about the future
of the American people was that they would think so highly of individual freedom to choose and the equal worth of all humans that *reasons*
for association and solidarity with each other would mostly disappear
from their ordinary thinking until some drastic national emergency—
such as war—forced Americans to unite around some absolute leader
for the sake of national survival. Democracy would thus die a quick
death. Only the power of voluntary associations and the restraints of
religion, Tocqueville believed, would protect American democracy from
the double danger of anarchy and absolutism slumbering in the radical
individualism of American culture. This fear of democracy is as ancient
as Plato, and Americans may be closer to the two dangers than we often
think. Our dominant cultural individualism accounts for the subjectivist
"ring" of the word "ethics" in most conversations on the subject in a
time when Wall Street scandals, for example, raise major questions about
the dogma that says that ethics is "after all a matter of personal integrity." It is that, but it is more: ethics is also exploration, dialogue, relation building, and collective struggle. Individual honesty was *not* the
only ethical issue at stake in those scandals. The responsibility of brokers to customers was also at issue, and the responsible supervision of
brokers vis-a-vis their managers. The sins of individuals contrived with
those of *organizations*.

In the end, Gilligan calls for a conversation between men and women
in our society that will enable each to appreciate the place of care-for-
principle and the principle of care. "This dialogue between fairness
and care not only provides a better understanding of relations between

the sexes but also gives rise to a more comprehensive portrayal of adult work and family relationships."[23] There are limits to both styles of moral reasoning. The very spirit of dialogue and responsiveness requires attentiveness to other people's abstract principles as well as one's own. But circumstances modify our understanding of principles, and stories help us to discern the humanizing relation of one principle to another.

LIMITED AND OPEN: RESPONSIBILITY AS A JOURNEY FROM THE PARTICULAR TO THE INCLUSIVE

No one can be sure about his or her capacity to include all the relationships, all the relevant principles, all the empathy that "ideal" responsibility might require. We know that there are limits to this capacity. Furthermore, we know that only through experience and exploration are we likely to discover these limits. Responsible persons, therefore, in acknowledging their limits, open themselves to new experience of the limits and new understanding of where old limits may no longer apply.

As MacIntyre reminds us, it is easy for future-oriented Americans to forget that every human being is born somewhere, joins society in a particular location, sees the world still from the standpoint of a particular community, and relates to that world with blinders—even illusions—that frustrate all our attempts to see ourselves as members of a single species, the human race. Even that most universal lens for interpreting human life—religion—takes its root and origin from particular times and places. Almost half of the world's peoples, for example, profess a religion related to the ancient figure of Abraham. The seventeenth-century French philosopher–scientist–theologian Blaise Pascal called attention to the difference between the "god of the philosopher and the learned"—a Power that presides over the universe—and "the God of Abraham, Isaac, and Jacob."[24] One does not have to be explicitly religious to see the momentous difference in this example between a consent to live and think within a particular framework and the ambition to burst through the shackles of the particular to the "universal." The contrast here is clearly between the intellectual program of a Nozick or a Rawls and the program of a MacIntyre, Niebuhr, or Gilligan. Thinkers—and professionals—impatient with the limits of an American, a European, a Christian, a Jewish, a Moslem, a twentieth-century, or a Third-World perspective on the meaning of human life will ask of books like this that they shed the husks of these localisms in the name of truth that everyone can recognize and practice. But what if the human pilgrimage is exactly that—a pilgrimage toward wisdom whose very pathways may not be evident without a lot of entry into blind alleys and exploration of utterly strange trails? What if we are so limited by our times and places that our most basic choice is whether we will live complacently in their cocoon or will see them as gateways to new times and places?

We dwell at some length on that choice, for at stake is the (perhaps abstract?) question of whether people can live at all in the realm of abstraction. With MacIntyre and Niebuhr, we think not. Their and our senses of calling to live in the limited, particular, but open contexts of our lives means that we consent to thinking about our personal and collective responsibilities as employees in an industrial firm, as upper middle class members of a capitalist society, as heirs of a particular religious faith, and as citizens of a country much buffeted these days with unprecedented interventions from other countries of this planet. Our attitude here resembles that of the political philosopher Michael Walzer when, at the beginning of his book *Spheres of Justice*, he parts company philosophically with his colleagues Rawls and Nozick on just this point.

> My argument is radically particularist. I don't claim to have achieved any great distance from the social world in which I live. One way to begin the philosophical enterprise—perhaps the original way—is to walk out of the cave, leave the city, climb the mountain, fashion for oneself (what can never be fashioned for ordinary men and women) an objective and universal standpoint. Then one describes the terrain of everyday life from far away, so that it loses its particular contours and takes on a general shape. But I mean to stand in the cave, in the city, on the ground. Another way of doing philosophy is to interpret to one's fellow citizens the world of meanings that we share. . . . If [the society we seek] isn't already here—hidden, as it were, in our concepts and categories—we will never know it concretely or realize it in fact.[25]

The first question such a declaration suggests is not the one that Rawls asks: "What would rational individuals choose under universalizing conditions of such-and-such a sort?" Rather, says Walzer, the first question has to be:

> What would individuals like us choose, who are situated as we are, who share a culture and are determined to go on sharing it? And this is a question that is readily transformed into, what choices have we already made in the course of our common life? What understandings do we (really) share?[26]

It is no easy question. But it may be a very practical question for any group of people to ponder together as preparation for making any collective decision. What understandings do we share about American history? About world history? About the evolution of the business corporation into its current state of mixed goods and evils? About the meaning of our new global neighborhood with its mixture of friends and enemies? About the challenge of the rest of the century to Americans, to business, to educators, to politicians, to the religions of the world?

Even to pose these questions is to tempt some of us to throw up our hands and to retreat into that convenient American-style individualism and skepticism that (with Nozick) expects persons and communities to have few common understandings about the "basic truths" of human social existence. According to this view, Walzer, MacIntyre, Gilligan, Niebuhr, and we are ignoring the actual course of the particularistic thoughtways of our own nations: Trends toward greater and greater diversity, less and less agreement on "real" values and meanings. Instead of a pilgrimage toward new, more inclusive, wiser communities of interpretation, are humans rather on a collision course with more fragmentation and anarchy of meeting?

RELIGION AS INTERPRETER

That is one way to interpret the data of modern life, one way to project the current chapters of the human story, one way to characterize the story itself. But some humans refuse to tell the story that way. They say that, at some point in the telling, we have to draw into the story standards of interpretation that are religious. Religion supplies many people with the ultimate framework of their interpretation of everything in their experience. They may not have yet understood the connection between the little stories of their lives and the great story of religion—like the story that three great world religions begin with an ancestor named Abraham. But one element of religious faith is that there *is* a connection, and the faithful are bound to look for it.

On a less elevated plane, it would seem simple blindness to Walzer's "society already here" to suppose that no common understandings permit this society to *be*. We may focus on the differences among individual economic preferences that make the market system work, but the very priority given to many preferences is an important common agreement among all in the system. "Agreement to disagree" is a very *important* agreement in a democratic society, especially if—in Niebuhr's terms—it is a real agreement, implying intentions on both sides to continue certain arguments in an ongoing conversation whose end cannot be predicted. "Traditions," says MacIntyre, "when vital, involve continuities of conflict."[27] Such conflict composes part of the "understandings really shared" in all the organizations of human society: families, towns, states, congresses, schools, universities, churches, *and* business corporations. Modern humans often experience "booming buzzing confusion" in the variety of beliefs, values, commitments, behaviors, and costs which all these organizations seem to embody. One's personal problem of "making sense of it all" is very severe, for each of us participates in many systems, a variety of relationships, constellations of communities, and a multitude of structures. Is there a consistent possibility of meaning, sense, or story in it all? Again, this may be a high-level religious question; but it is a very human question whose most cogent answer could well be "Let's see!" rather than the premature dogmatic "Of course not!" or the agnostic "Who knows?"

The "Let's see" response comes from those who, with science and certain forms of religion, believe that exploration is a more human response to ignorance than resignation. Exploration is the way of responsibility, the way of willingness to move with partners from one moment of dialogue to a next in the confidence, more than rational, that they and we will find down the road new things to say and do.

IN WHAT SENSE CAN CORPORATIONS BE RESPONSIBLE?

Whether or not they make much of the word in their theories, all of the philosophers to whom we have given attention here could be adapted to some version of "responsibility." As a retrospect and summary, before we try in this chapter to specify some particular directions in which we think the responsible managers of American corporations should consider going in the next decade, we may well ask what sort of responsibility seems inherent in the theories most prominent in this and the previous chapter.

If Nozick is right about the impossibility of agreement in a liberal society among individuals on the good human life, then groups and organizations within society are conveniences for achieving this or that good as defined by contracting individuals and interest constituencies. An organization may be responsive to individuals by being responsive in its procedures for hearing and adjudicating their claims. It may owe every individual his or her day in court, which in the business corporation may be the making and the revising of contracts into which an individual can freely enter. While this is a narrow concept of corporate responsibility, it is not necessarily a formula for irresponsibility.

If Rawls is right about the indispensability of rules of justice in all human relationships, then managers in organizations will give time and energy to the formulation of those rules and to securing agreement with them among the members. If they agree with a Rawlsian notion of justice-for-the-poorest, and are critical of the merely procedural justice of Nozick, they will have at least one ethical reason for questioning the primacy of personal ambition, profit, and growth as the standards for judging whether a manager—or the firm—has "succeeded." A Rawlsian judgment of success in management has a strong focus on the consequences for the weakest among those affected.

If MacIntyre is right in his critiques of both Nozick and Rawls, the human worth of business in modern society cannot be judged simply by a new focus on justice in the workplace or in society at large. Human beings do not live by justice alone, either procedural or distributive. They live and thrive by other virtues like truth, love, and courage. They live in communities with histories. Their work is humanly fulfilling to the extent that it cultivates and expresses just these virtues and relationships in practice. Business corporations, like other parts of society, must embody practices that are humanistic in the most fundamental, value-laden sense of the word. All institutions, business included, are responsible not only for the quality of some material product but also for the quality of human relationships among its workers, suppliers, stockholders, and customers. Business managers, like other leaders of institutions, must beware of the tyranny of economic efficiency—the reduction of the purpose of economic enterprise to the continually expanding production of yet more economic enterprise.

If Niebuhr and Gilligan are right in implicitly sharing so much of MacIntyre's critique of liberal individualistic society, we Americans, including those of us who work in business corporations, must give up some of our illusions about the "self-made" individual of our national mythology and learn—perhaps from the experience of women—to critique, explore, and rejoice in the relationships that make us what we are. We must acknowledge our debts to our life companions from birth to death, our ability to be what we are because of what others have been to and for us. We must expect that our achievements will be roughly proportional not to our success as self-interested competitors but rather to our success as responsive collaborators. We will have to give up our narrow concentration on a specialized job within large bureaucracies and spend time inquiring with others, "How do we know that the roles we play here add up to something important in some larger scheme of things? What is our common story, where might we be heading, where would we like it to head, and what new roles should we undertake?"

Embarrassing as our culture makes the posing of such general questions, posing them may be essential to the living of a responsible human life. University teachers note that as business managers in highly technical fields of industry grow older, get promoted, and acquire new responsibilities, their requests for continuing education change. Five years after graduation from college or professional school, they want courses that will update them in the technical developments of their fields. Ten to fifteen years out, they have become middle managers and they request courses in "human relations." Then at age 50 or so, when some have become vice-presidents and chief executive officers, they express hunger for more education in the humanities—philosophy, history, and religion.

At stake here is the momentous, difficult, controversial issue of what modern human beings, including Americans, are learning about both the conflict and complementarity of personal and social meanings in our lives. Facing this issue seems very critical to the United States, for, as Tocqueville observed a century and a half ago, ours is the country that has most celebrated the benefits of maximum freedom for individuals to make their own way. As we have often noted here, Tocqueville believed that only the restraints of religion and the experience of the voluntary association kept American society from dissolving into an anarchy of individual self-interests. He was sure, as MacIntyre and Niebuhr are sure, that

> Feelings and ideas are renewed, the heart enlarged, and the understanding developed only by the reciprocal action of men one upon another.[28]

He would have joined these later thinkers in their alarm at a twentieth century American culture, rampant in many business corporations, whose members expect people to work for money but for no other agreed-upon human purpose. In modern America the vacuum of consensus on the plot of the human drama applicable inside the workplace seems to match the vacuum outside; so in one sense, corporations are no more to be blamed for ethical anarchy than is Robert Nozick, who gives anarchy a sophisticated and typically American philosophical expression in one intellectual interpretation of the facts of American life. Sensitive leaders of every institution in this society, however, may well wonder what their successors will do if they inherit even less philosophical coherence than did previous leaders.

Will the only justification available be appeals to economic efficiency or personal aggrandizement? Does ambition to win promotion (and to gain the CEO's position) supply enough meaningful consensus to keep the middle managers of corporations doing their best work for decades? Or does that ambition, eventually thwarted for many, curdle into alienation and resignation—or some form of early retirement—from strenuous effort to serve the corporation and its primary constituencies?

A corporation that fits Robert Jackall's description[29] may be programming the majority of its managers into at least *psychological* early retirement, no matter how many years longer they come to work. In MacIntyre's words, they will have lost a sense that their work has any intrinsic value. They will have given up on their work as a practice whose services are inherently valuable to themselves, their associates, and their external public. They will thus, ironically and tragically, become psychologically what many managers have falsely assumed is the "wageworker mentality"—putting in time to get a paycheck.

Wage workers have never been different from managers in this mentality. They too have never preferred to work for wages alone, and here Americans are having to learn painful lessons from the workplace psychology of the Japanese. While Americans are not apt to borrow specific psychological techniques from the Japanese, they have come to realize there are effective, productive alternatives to our traditional industrial mindset. It is not likely, either, that adding a power incentive to an income incentive (through unionization, for example) will meet the humanistic requirements for organizational loyalty among either managers or workers. Even the power to demand justice is not enough to insure humanly satisfied workers, many of whom are moral and religious enough to ask "big" questions like, "Does my work get appreciated by my associates? Are we producing something that is really *good* for our customers?"

Public exhortation that managers show "responsibility" to the interests and the life-stories of constituents external to the corporation is likely to fall on deaf ears if such responsibility has no place or procedure *inside* corporate culture. In their study *Middletown* in the depression-beset 1930s, the Lynds asked about the townspeople, "Why do they work so hard?" The typical answer, they found, was: "Why, to keep up with the cost of living." Is that enough to make any human being continue to work hard?[30]

The challenge here is awesome. We have seen that a real *philosophical* crisis brews inside *and* outside the American business corporation. Can the virtues of productive practice and a wide-ranging sense of personal–social responsibility be achieved in the country as a whole? The prescription that we mean to recommend is neither that business shrug off the crisis of relativism as irrelevant to its calling nor that it shoulder the whole task of solving it. Its leaders must rather ready themselves for participating in the finding of "solutions" that can only be found in dialogue with large new sources of wisdom whose possessors, in turn, can claim only a part of that wisdom.

Neither problems nor solutions here admit of clear "management planning." Like new inventions, new ways of work and new understanding of working together may have to emerge from very long periods of investigation. Like most research the emergence will be a confused, complex process involving many initially unpromising conversations. The pragmatic, problem-identifying, and problem-solving mind of our culture encounters this prescription with impatience. A truer pragmatism would counsel patience. New problems, when they rise above the horizon, are apt to be clouded. Further, there are problems that one will never see unless one sees them first through the eyes of others.

It was the genius of Adam Smith to look at commercial English society through the eyes of the small businessman, the genius of Karl Marx to look at industrial England through the eyes of its wageworkers, and the genius of Carol Gilligan to ask how our society looks through the eyes of women. At base, the corporate responsibility movement asks business to look at itself and its social effects through the eyes of those not represented in traditional corporate board rooms or the usual managerial offices. From such perspective, managers and others involved in the business enterprise can learn from each other, just as in any imaginable conversation between them Nozick, Rawls, MacIntyre, Gilligan, and Niebuhr might learn from each other.

To offer this general prescription, and to specify it further as we mean to do in the last chapter, is rather presumptuous and perhaps unavoidably personal on the part of the two of us. In the context of a Nozickian, let-many-values-compete culture, are we presumptuous in seeming to recommend *our* values as worthy of incorporation into business organizations? Since both authors have had our years of being managers of church and academic organizations, we know how far those organizations, and we ourselves, are from practicing some of these very proposals. But to admit this merely confesses that we all inhabit much the same culture and that we all face similar uncertainties about new ways of responding to our fellow humans inside and outside our particular institutional niches.

Claims like these are unavoidably personal in another sense: They have to do with the personal character that managers and others bring to their work and not only with abstract rules of ethics and professional practice. Personal integrity *does* have its place in the constellation of responsible business practice. A certain kind of person, not simply a certain series of activities, deserves the name "good manager." A stream of philosophers from Aristotle to Saint Paul to Alasdair MacIntyre have all said as much. The burden of much written here is that the upper reaches of American *corporate* structures have few of the personal *character* structures necessary for the humanization of the American world of work. Joanne B. Ciulla, in her 1984 exploration of the MacIntyre theses, offers this illustration:

> Several years ago psychoanalyst Michael Maccoby did a study of 500 corporation managers. What he discovered in his study was that the kind of personal traits a manager needed to reach the top of a corporation were not very attractive human traits. In his book *The Gamesman,* Maccoby paints a vivid picture of how, if one is to become a corporate success, a person must learn to disassociate his emotions from his daily activities (the study was only of men). The book was very popular. Many people thought that it was a "how to get to the top" book. Yet, Maccoby's conclusions contained a powerful indictment of the large American corporations that fostered the development of ruthless and amoral personalities in their managers.[31]

Writing in 1976, Maccoby concluded that by schooling and by experience, managers of large American corporations are trained to take a "detached," rational-objective view of the corporation's internal and external problems. He contrasts this with the capacity to take into account the "social/human effects" of what they do. On the whole, he says,

> . . . the managers made no effort to learn the social/human effects of their actions in the United States today and the rest of the world. So long as they remain unaware and unrelated, they avoid having to accept responsibility.[32]

In the years since Maccoby wrote, one could adduce evidence that growing numbers of American managers are learning to take into account the social-human effects of their actions on many levels of our new global society. We began this chapter with one such bit of evidence in one major corporation, that corporation's decision to turn down a profitable contract in South Africa. The story is one portrait of corporate social responsibility. A group of managers listened to people inside and outside of South Africa; brought to the decision a framework of interpretation that both decried racism and discerned its economic-political supports; remained in dialogue with others for clues about next steps toward social justice in South Africa; and sensed an obligatory solidarity with many others struggling for change in that country, including those who criticized and those who applauded this particular decision. They could see themselves as responsible in deciding not to build a heavy equipment plant for the South African government, because such a decision best fitted this mix of circumstance, corporate history, national deliberation, and hope for change. They accepted this responsibility, putting their corporate feet onto a new, unexplored trail that probably has new surprises and new learning in store for them in the world of the 1990s.

The opportunities that may be in store for American corporations to step out responsibly toward that new world is the focus of our next and final chapter.

NOTES

1. See Chapter 4.

2. On the mixture of such arguments over the morality of American business involvement in South Africa, see Donald W. Shriver, Jr., "South Africa and Symbol's Strength," *Wall Street Journal*, July 8, 1985.

3. H. Richard Niebuhr, *The Responsible Self: An Essay in Christian Moral Philosophy*, Introduction by James M. Gustafson (New York: Harper & Row, 1963).

4. Niebuhr, *The Responsible Self*, pp. 56–67.

5. Ibid., pp. 61–62.

6. Ibid., pp. 62–63.

7. This was the gist of a speech by a retired Nestlé vice-president to a gathering of faculty, Graduate School of Business, Columbia University, Spring Term, 1985.

8. Robert Cover, *"Nomos* and Narrative," *Harvard Law Review* 97 (November 1983):68.

9. Niebhur, *The Responsible Self,* p. 64.

10. See *Social Responsibilities of Business Corporations,* A Statement on National Policy by the Research and Policy Committee of the Committee for Economic Development, June, 1971. "The modern professional manager also regards himself . . . as a trustee balancing the interests of many diverse participants and constituents in the enterprise . . ."

11. The reader should note that a number of American producers of infant formula, American Home Products and Bristol-Myers among them, quickly, though not necessarily happily, came to terms with their critics. The managers involved were ready to admit the critics as legitimate, though troublesome, corporate constituencies.

12. Niebuhr, *The Responsible Self,* The same, p. 65.

13. Carol Gilligan, *In a Different Voice: Psychological Theory and Women's Development* (Cambridge: Harvard University Press, 1982).

14. Ibid., p. 65.

15. Ibid., pp. 99–100.

16. This is Jonathan King's passing criticism of Gilligan in his "Ethical Encounters of the Second Kind," *Journal of Business Ethics* 5 (January 2: 1986).

17. Gilligan, *In a Different Voice,* pp. 100–101.

18. Ibid., p. 19.

19. Ibid., pp. 104–105. The quotation from Erik Erikson, *Gandhi's Truth* (New York: Norton and Company, 1969), p. 236.

20. Gilligan, *In a Different Voice,* p. 160.

21. Ibid., p. 127.

22. Ibid., p. 167.

23. Gilligan, *In a Different Voice,* p. 174.

24. Blaise Pascal, "Memorial." *Encyclopedia of Philosophy* (New York: Macmillan Co. and The Free Press, 1967), Vol. 6, p. 52.

25. Michael Walzer, *Spheres of Justice: A Defense of Pluralism and Equality* (New York: Basic Books. 1983), Preface, p. xiv.

26. Ibid., p. 5.

27. MacIntyre, *After Virtue,* p. 222.

28. Quoted by Jonathan King, "Ethical Encounters of the Second Kind," p. 8.

29. See Chapter 11.

30. Quoted by King, "Ethical Encounters of the Second Kind," p. 6.

31. Joanne B. Ciulla, "Work and Virtue," (Ph.Diss., Temple University, 1984), p. 44.

32. Michael Maccoby, *The Gamesman* (New York: Simon & Schuster, 1976), p. 210.

13

Beyond the Market Ethic

Peter Drucker has argued that the autonomous managers of the Galbraithian business corporation needed qualities far more fundamental than alertness to market data, trust in expert staff work, and unrelenting pursuit of efficiency—they need *legitimacy*.

> Management has power. Indeed, to do its job, it *has* to have power. But power does not last, regardless of its performance, its knowledge, and its good intentions, unless it is grounded in some sanction outside and beyond itself, some "grounds of legitimacy," whether "divine institution," or "election," or "the consent of the governed." Otherwise, power is not "legitimate." It may be well-meaning, it may perform well, it may even test out as "highly popular" in surveys and polls. Yet illegitimate power always succumbs to the first challenger. It may have no enemies, but no one believes in it, either, and no one owes it allegiance.[1]

Drucker recommended that, to gain the necessary legitimacy, managers set about restoring and supporting strong boards of directors whose members would most likely be independent, not representatives of specific constituencies, even shareholders. He recognized that for such a board simply to replace the autonomous managers would accomplish little, merely the replacement of one set of "philosopher kings" by another. He urged another change in corporate governance, pertinent to the thesis of this study:

> . . . to mobilize new constituents, to bring in other "interests" to balance the former owners now become speculators, and to create new bonds of allegiance [between the constituencies and management]. . . . Thirty years ago it was popular in American big business to talk of management as being the "trustee for the best-balanced interests of stockholders, employees, plant community, customers and suppliers alike." . . . But even where there was more to this assertion than self-interest, nothing has been done as a rule to convert the phrase into reality. Few attempts have been made to institution-

307

alize the relationship of these supposed "constituencies" to the enterprise and its management. Will this now have to be undertaken in earnest to safeguard both enterprise and management? And what form should such institutional relationships take?[2]

We ask the same question of our readers; to the first part we answer, yes. We believe that the relationship of constituencies and managers needs to be regularized, recognized, and made formal. The reader may want to speculate concerning the form the relationship might take, examining the pluses and minuses of various arrangements. Different businesses will probably devise and need quite different forms. Experimentation is needed, for a priori, it is not likely that we can specify those forms that will best serve not only managers and constituencies well, but also the economy and the general welfare.

BEYOND SUCCESS: SOME FRONTIERS OF CORPORATE RESPONSIBILITY

Appeals to experience and experiment are the preference of American business culture: The entrepreneurial spirit assumes as much. Experience can mislead, however, and experiments can fail. Culture itself, as an accumulation of a society's experience over many generations, is a sort of data bank of failed and successful human experiments, making possible the shortcuts to wisdom of later generations who can learn from the experience of others and thus be delivered from having to endure the same experience.

In a society as widely subject to change as ours, though, every generation lives on the frontier between what its predecessors learned and what it can learn for transmission to its successors. However romantic Americans can be about frontiers, the grim facts of our own "western expansion" remind us that a frontier is mostly an uncomfortable, difficult, risky, threatening place in which to live. Those who demand from books like this a certain set of solid, specific guidelines for living there have good reason for their demand.

Reduced to utmost simplicity, the message here is merely that frontier crossing in the contemporary world requires a vast and multiplying number of partners. That requirement obtained more pervasively in the crossing of the American frontier than our myths suggest. The role of Native Americans in the survival of the Plymouth colony in 1620 was always a more instructive memory than the illusion that the typical "conqueror" of the wilderness was a lonely trapper or an individualistic cowboy. The frontiers of the modern global economy invite new worldwide adventures of collaboration.

We end our study with two sets of guidelines. One set identifies significant issues, relatively new in the experience of modern corporate managers, that impress us as major concrete issues on the frontier between the old and the new relationships of corporations to their constituencies. The other identifies major unresolved problems in this ongoing dialogue.

From Workers as Employees to Workers as Owners

A truism of the corporate responsibility movement is that customers, workers, and the public all have a stake in what corporations do and do not do. The legal power of management to turn a listening or a deaf ear to the claims of these constituents is still very large in our society; and nothing sounds more radical to most capitalist stockholders and their elected managers than the proposal that employees belong in the ranks of the legal owners of the firms for which they work. This is a far cry from state ownership of the forces of production in a Marxist-Leninist society, but it challenges the neat conventional division of a capitalist economy between owners and workers.

Why should stockholders be the only legal owners of a productive enterprise? The question is receiving more and more serious consideration among American managers beset by world competition. Some ask themselves if ownership is the missing ingredient in the formula for restoring motivation for efficiency among workers in their factories and offices.[3]

Take, for example, the instructive negative and positive chapters in the recent history of McLouth Steel Products of Trenton, Michigan. Part of the crisis of the 1970s and 1980s in the American steel industry, it filed for bankruptcy in 1981. Shortly thereafter, the director of the local United Steel Workers union proposed a worker-buyout (ESOP—Employee Stock Option Plan) of the company. Neither the company's creditors nor the international union leaders thought the plan worth pursuing. Instead Cyrus Tang, a Chicago scrap metal dealer, bought the failed company, and then negotiated pay cuts of one-third with the employees.

Tang did not seek his workers' loyalty. He continued the old managerial practice of laying them off when sales lagged. Morale dropped and productivity with it, generating operating losses. Even when sales regained strength, the workers refused to offer full cooperation. They would not perform well enough to help McLouth operate profitably.[4]

Five years after bankruptcy, old-style confrontation between managers and unions once again was failing. In 1986 a group of creditors reconsidered their objection to the ESOP proposal and decided to try it. The rest is a story of corporate revival. A new board was instituted, composed of labor, management, and creditor representatives. Cyrus

Tang's share of the company stock was reduced from 65 percent to 10 percent, and workers gained 85 percent in return for a 10 percent pay cut and 10 percent reduction in the work force. Within six months, production rose by 50 percent, producing a cash flow sufficient for major capital improvements, and by June, 1988, a new CEO was able to testify: "This presents a real opportunity to show what can be done with an employee-owned company."[5] The McLouth story illustrated a new reality at the place of work, recognized by labor scholars but often ignored by managers: Union leaders and members increasingly reject the notion that only managers manage and workers can but grieve unilateral decisions and practices.[6] Employee ownership and participation in managing is one of the new frontiers of American business history. Not all of the current experiments are or will be economic success stories. The proposal addresses only one side of the crisis of global competition, and it lays workers open to *other* constituency pressures that once were managers' concern alone. But the increase of worker-owner is an experiment that even its critics will be wise to observe carefully.

From Fighting Consumers
to Facilitating Their Voice

One of the great ironies of recent corporate history in capitalist America is the contempt with which the word "consumerism" issued from the lips of many a manager in the 1970s. Only a real suspicion of market-oriented capitalist economics could produce such an expression of feeling, as it did among General Motors' executives who blithely assumed that if Ralph Nader knew little about the technology of gears and metallurgy, he could know little about the meaning of accident statistics related to the Corvair. In recent years many producers and sellers have taken elaborate measures to solicit the opinions of those who buy their products. Feedback from customers is likely to be a norm of good management in companies that should never have acclaimed market economics without diligently observing that norm in the first place.

If consumers are the only real authorities on some dimensions of "good management," should a well-managed company vigorously solicit and publicize negative as well as positive reactions to its products? Advertising emphasizes the positive, but a negative consumer response may be the place where product-improvement frontiers now exist; thus it was not necessarily an antimangement move for Nader, in 1976, to propose a model bill to establish citizen utility boards (CUBs) to help consumers monitor, oversee, and evaluate company policies *and* rulings of utility regulatory agencies. (What is a more nagging problem of regulatory commissions than their record for serving the interests of the industries they regulate more consistently than the interests of the public they are supposed to serve?)

In 1979 the state of Wisconsin established a Nader-inspired board, giving CUBs right of access to the billing envelopes of utility companies. Other states followed, including New York, Illinois, Oregon, and California. Hundreds of thousands of customers began to see their views included in their monthly billing envelopes, sometimes accompanied by management replies. Far from being uniformly correct or trustworthy, some of the consumer complaints were misinformed or less than valid. Many a utility manager wondered if it was really their job to publicize such error or to be compelled to respond to it. Incensed at being compelled to respond to invalid and inaccurate complaints, the Pacific Gas and Electric Company in 1986 took the issue to the Supreme Court and won the case, with the claim that the company's First Amendment right of free speech—including its right not to speak—had been violated.

In a dissenting opinion, Justice William Rehnquist elucidated an issue that deserves continuing exploration: A corporation does not have the same right of silence that individuals have. To the contrary, he said, giving consumers a voice inside the communication system of a firm will "facilitate and enlarge public discussion; it therefore furthers rather than abridges First Amendment values" in our society as a whole.[7] Our purpose in mentioning this case is not to argue its legal merits but to underscore an issue of change that responsible contemporary management might consider. Consumers are *internal* to a company's network of relations. They are already among the potential *allies* of the company, certainly more so than many a stock trader, investment bank, or stockholder interested in the short-term gains to be made by trading in the firm's stocks and bonds rather than the practical uses of electric power. We wonder if utility managers have not responded to the innovative CUBs with an unthinking hostility rather than with an appreciation of how vulnerable they can quickly become to some *real* enemies and how valuable consumer organizations might be as their real allies.

From Obstruction to Partnership
in Environmental Preservation

The epithets "nut," "freak," and "kook" have appeared frequently in the private and even public rhetoric of managers faced with environmentalist critics in recent years. Environmental values are uniquely new, intrusive concerns for many of these managers. Industrial economies have proceeded in much ignorance of environmental consequences, and our societies are still far from possessing the knowledge necessary for remedying or preventing the damages to the planet of which human beings are obviously now capable. We are only beginning to discover this knowledge, and we are very far indeed from a collective commitment to the economic, social, and political costs of remedial and preventive action.

Environmentalists have met with some success in teaching the American public that we neglect our dependence upon the air, water, and the rest of our biosphere at our peril. The problem of irreparable global impairment is real, pervasive, and long-term. Its solution will require at minimum some new partnerships, both between the human species and its environment and among members of the species itself.

The survival of a certain butterfly in California—the Bay checkerspot *(Euphydryas editha bayensis)*—may at first suggest a trivial evocation of the environmental issue. However, a pair of corporate responses to a threat to this insect illustrate the difference between managers willing to explore the frontier between the value of our environment to us and its value to many another specie.

United Technologies (UT), a large defense contractor, began the development of a testing range for its missiles on 5,200 acres adjacent to Kirby Canyon near San Jose, California. A professor of biology at Stanford discovered that a part of the range was a critical habitat for the Bay checkerspot butterfly, and in 1981 with the aid of other concerned citizens, he petitioned the U.S. Fish and Wildlife Service (FWS) to list it as an endangered specie. UT managers began an aggressive campaign against the listing, arguing that "the unnecessary protection of the butterfly" would endanger national defense. They persuaded the Undersecretaries of the Navy and Air Force to warn the Interior Department that listing of the butterfly could slow delivery of Minuteman and Tomahawk missile propulsion systems. They hired a San Francisco law firm to raise technical challenges to the listing, and in the meantime, installed a water line through the butterfly's habitat.

In reviewing UT's strategy, Dr. Dennis Murphy, associate professor of biology at Stanford University, concluded, "[They have] spent far more money fighting this listing than it would ever have cost them to accommodate the butterfly."[8] He noted that only a portion of the total range would ever be restricted for the insect's habitat and that those areas already served as buffers to keep prying eyes away from the testing positions.

Another company, Waste Management Inc., in 1984 sought approval to use an 800 acre plot adjoining the missile range site as a $15 million garbage disposal site. Its managers learned about the butterfly's habitat problem and the citizens' concern for it in 1986, more than a year before the FWS had listed the butterfly. They then joined with the City of San Jose and the FWS (using the biologists as expert guides) in agreement on a plan that was to minimize alteration of the critical butterfly habitat.[9] The company then funded research biologists in enforcing rigorous conservation procedures and agreed to set aside 250 acres as a permanent preserve and to furnish over $1 million over ten years for further and continuing research. The purpose was to learn more about the insect's life cycle and requirements, enabling more effective protection. The company also agreed to reseed the landfill slopes to enlarge the habitat area for new colonies.

Phillip B. Rooney, president of Waste Management, told a reporter that the company had considered fighting the environmentalists, but, reflecting on their own negative experience with such conflicts, "we chose preservation as the way to go." Waste Management then erected a large sign at the entrance of the landfill, warning, "Butterfly Crossing!" and made the checkerspot the landfill mascot, starring it in television commercials aimed at showing the company was, indeed, a convert to environmentalism.

To be sure, the managers of Waste Management are subject to the skepticism of an officer of the Sierra Club who commented that a lot of companies are "long on taking credit, after certain societal prodding, for addressing problems they themselves are causing." [10] The presence of mixed motives in these cases, however, is no strong criticism of any collective, cooperative action taken on any side. Single-minded devotion to some alleged overarching human purpose, e.g., national defense, is not necessarily noble. Far more noble and normatively *human* might be those decisions that mix a variety of values, interests, and high ethical claims in as rich and consistent a way as possible. Like partnerships in nature, multivalued choices in society may be the most genuinely human choices of all. By that norm of humanity, the behavior of Waste Management rates a higher ethical grade by far than that of United Technologies.

Furthermore, constituencies affected by some action may begin their relationship with each other in great conflict, but they can learn to cut down its degree even if they cannot eliminate it. They have both the right and obligation, on occasion, to affirm each other's values even while *ranking* them differently. Conceivably, Waste Management's president and the president of the Sierra Club both have an interest in both butterflies and waste disposal. One would suppose that their respective responsibilities might require them to rank these two concerns in opposite priority order. What they are *not* required to do, as UT's managers apparently thought, is to downplay and scoff at values that are primary for certain constituents. Even for business managers, then, there is nothing necessarily absurd, unbusinesslike, or unprofessional about valuing butterflies and the people who value them. Morality and professionalism, not to say citizenship, may require managers both to value these others and to act as often as possible on their behalf.

From Routine Responsibility to Emergency Ethics

We have stressed in these stories the virtues of patience, inquiry, learning from many points of view, and the search for responsible, multivalued choices. Such response to new constituency pressures takes time. In the case of Nestlé it took nearly a decade; in the case of the manufacturing and marketing of asbestos products it took the managers of Manville about forty years; in the case of global response to the pollution threat of the Earth's environment it may take even longer.

But what of those events that come crashing into the professional lives of managers, that demand almost immediate response if the organization, its interests, and its commitments are to be salvaged? What about events like a nuclear power plant explosion, a highjacked airplane, the escape of lethal gas from a Union Carbide plant in Bhopal, and threats of murder against the staff of the publisher of a book that some constituency wants banned or burned? In some ways such events illustrate ethical frontiers of corporate responsibility only in a vague sense, so unpredictably do they erupt in the moment without warning. But the interconnections of modern social life are so many and fragile that their incidence is more frequent than ever. Among the problems confronting business managers, these may be among the most difficult.

In a decision that is likely to be remembered as a classic positive example of ethics in business, James E. Burke, chairman of Johnson & Johnson, on October 5, 1982, ordered the recall of 31 million bottles of Tylenol capsules, at an immediate pretax cost of almost $100 million. Six people in the Chicago area had died after taking Tylenol capsules that had been injected with cyanide. The news had broken unexpectedly the previous Thursday when news reporters called Johnson & Johnson headquarters to say that the Cook County Medical Examiner's Office had found three people killed as a result of ingesting poison contained in Tylenol capsules. Quickly Burke assembled a management team to manage the crisis, first to discover what had happened, and second to determine what should be done. He opened the doors of the company to news reporters, telling them any and all information his team had discovered; in return he found that "the company was getting some of its most accurate and up-to-date information about what was going on around the country from the reporters calling in for comment."[11]

After meeting with officials of the FBI and the Food and Drug Administration in Washington, and against their advice, Burke decided that the capsules had to be recalled. He was acting against the advice of his own vice-president for marketing, too. The managers of the subsidiary company producing Tylenol, McNeil Consumer Products, opposed the decision. Said Burke, "Often our society rails against bigness, but this has been an example of where size helps. If Tylenol had been a separate company, the decisions would have been much tougher. As it was, it was hard to convince the McNeil [wholly owned] people that we didn't care what it cost to fix the problem."[12]

The decision was so immediate and so sweeping, and so clearly involved a clear moral issue—protecting the lives of consumers against death—that most people readily recognized the courage and high sense of responsibility that must have motivated Burke. Particularly important for the ethical stance taken in this book was the unpredictability of the *results* of the recall for the company. As events turned out over the next two years, no more tampered bottles of Tylenol were discovered and no other people taking Tylenol died of cyanide poisoning. At the moment of decision making, however, Burke could not offer a rational, empirical justification; he chose life over profit. To act in the face of ignorance and on the basis of speculation is a profound test of responsibility, resembling as it does the search with a broken compass for a coastline in a fog.

What does the responsible decision maker do when the aids to responsibility are absent, few, or in dire conflict? One answer from the realm of pragmatism has been, "Do the best you can." Another, from the realm of religion, has been "Sin bravely." But both answers obscure the critical role, precisely in emergency situations, of knowing what *the* primary one or two human standards of conduct have to be when only one or two can be followed. One might speculate that James Burke was betting on a future that would see the rehabilitation of the reputation of Johnson & Johnson and Tylenol rehabilitated more quickly because of his prompt response to the emergency. One could assume, skeptically, that the decision did not involve sacrifice on his part or that of his company; it was simply a very shrewd act of prudence.[13] Yet one is reminded of that Atlanta executive when he observed that his associates in business were always "talking bottom-line" about their decisions but who characteristically "did the right thing."

The moral calculus with which Burke and his managerial colleagues wrestled came down to a possible conflict between the claim of the bottom line and the claim of the life of some unknown and unknowable customers. Care for the life of so much as one such customer constitutes a rare form of regard for the principle that is probably the oldest in historic social ethics: "Thou shalt do no murder." Collaborating even passively with murder *at that time* was the course that Johnson & Johnson avoided. One has to be cynical indeed about everything that business leaders do in order to discount the social responsibility in the recall action.

The reader should note, however, that the company did *not* immediately stop the production and sale of *all* capsuled Tylenol, even though it was far more vulnerable to tampering than the caplet form. It took another poisoning and another death, a little over three years later, to convince them that all capsule drugs had to be removed from over-the-counter sales.[14] Burke and his company crossed one frontier in 1982; they crossed another in 1985. In the 1990s drug companies will undergo a lot of learning from such events—if their managers want the public to judge them responsible.

FRONTIER ISSUES: SOME DANGERS AND PROBLEMS

The recommendations of this book promise no rosy, romantic course for the work of business managers. If anything, the perspectives promoted here increase managers professional difficulties. But coping with them, we believe, increases the chances that they will acquire the rudiments of real business professionalism: An integration of theory and practice, principles of truth and justice, and the skills to serve people without merely using them. That professionalism will match MacIntyre's definition. It will also match the ethic of responsibility described by Richard Niebuhr and Carol Gilligan.

Our recommendations entail certain dangers and quandaries, however, many of which will have occurred to the reader. Below we examine those we believe to be especially critical.

Constituency Co-option

Any constituency that cooperates with corporate managers may find that its members and leaders have become a part of the system about which they protested originally. Some would say that Dr. Dennis Murphy and his associates who worked with Waste Management to save the Bay checkerspot is an example of such co-option. Others will note that dissident stockholders who sponsor activist resolutions at annual meetings, such as those affiliated with the Interchurch Center On Corporate Responsibility (ICCR), have much at stake in wielding the power of their large holdings. They are fiduciary trustees of pension funds, upon whose performance the livelihood of many employees depends. These shareholders are thus hardly in a position to be outside the system, whatever policies they seek or changes in corporate governance they propose. Simply by holding and voting their shares, they are part of the business system that collects the savings of hundreds of thousands of people and invests them in productive, capitalistic endeavors. Where shareholders' fiduciary responsibilities conflict with their advocacy of and search for justice, they will almost surely have to choose advocacy, for that is imposed by law.

In the long run, to avoid such conflicts between their legal and moral responsibilities, constituencies need to be able to argue cogently and convincingly that the two responsibilities should coincide. For example, they may have to find a coincidence of a "fair return on investment" and "justice" in preserving an endangered species or helping the poor. The search may require them to moderate their own demands if they are to protect a business return. Unfortunately there has been too little research and investigation of the returns—or lack of them—in socially responsible investing of any kind. In the future, constituencies will need to be much better prepared than they are at present for convincing both courts and their members that the costs of their advocated policies are well balanced by the gains, "properly" and legally defined. Should constituencies fail to promote such research to prepare themselves for defending their moral stands, their arguments and policies will inevitably be measured against the performance standards of economics or other current business practices. Co-option can be avoided, therefore, only if constituencies develop their own social accounting and gain recognition for its applicability, usefulness, and value to society at large. They can thus maintain, with their own constituencies, and in courts of law, that self-interest, like sound business policy, justice, and humane society, has more than a financial dimension.

Enforcement of Agreements with Managers

In the short run, constituencies can face the difficult problem of enforcing the pacts and agreements they reach with corporate managers. Constituencies usually operate with small budgets compared to those of business firms, and their staffs are far more limited. Compared to the size of the operations they seek to monitor they are puny. The settlements they reach are often not legally binding, and if they rely upon a monitoring firm, it is likely to owe its funding (and thus its conclusions) to the firm's own managers rather than to the constituencies.[15] Insofar as enforcement is a problem for constituencies, they will probably push for legal recognition of their efforts and their outcomes. Just as Congress made collective bargaining agreements legally binding,[16] so may constituencies seek to base their compacts and settlements upon the force of law. A future continuing issue between constituencies and managers may be the legal standing of constituencies and the agreements they make.

"Irrational" Constituency Demands

Various groups and interests bombard business managers with requests for help and demands for actions, many of which may appear to be preposterous or irrelevant to any sensible definition of company responsibility. However absurd mangers may find some of the demands made upon them or charges leveled against them, they must screen the charges and examine the interests carefully.

When managers of Procter & Gamble (P & G) discovered in the early 1980s that various Christian groups accused them of promoting Satanism, they could hardly believe that it was a serious matter. For half a century the company had used a logo of a stylized Man-in-the-Moon in a circle with a constellation of thirteen stars. People with a keen eye for the occult and a crafty regard for the subtle ways of the devil found in the beard of the Man-in-the-Moon some curls that, when reflected in a mirror, revealed the ominous biblical symbol of Satan,[17] 666! It was obvious to these people that P & G was secretively supporting Satanism. Rumors abounded, some charging that the company had admitted it gave 10 percent of its profits to the church of Satan. Others inferred that the logo meant that P & G was connected with the religious movement of The Reverend Sun Myung Moon.[18]

Since many people believed the rumors and began to boycott P & G products, the managers decided they had to act. First they offered to explain to anyone who called their toll-free telephone information number that the rumors were false and the charges had no substance. They received thousands of calls each month for more than three years. At the same time the managers brought libel suit against several people in various communities across the country whom they accused of spreading false rumors about the company.[19] They won the suits, but they could not quiet the charges. When the rumors spread to Pennsylvania, New Jersey, and New York, and the calls kept coming, the managers decided that the best response was to eliminate the logo from the company's packaging which they did in the spring of 1985. A few years later the company reinstated the logo on four products to protect its place as a trademark. Within months it again had to contend with accusations of sponsoring Satanism and began to suffer boycotts.[20]

P & G's experience with its logo warns managers that they can be subject to apparently irrelevant charges and suffer from irrationally motivated boycotts. There is, however, an irony here. Through advertising, business managers themselves have taught American consumers to admire and desire irrational product attributes. One thinks of De Beers' claim that "diamonds are forever" or Chrysler's assurance to automobile customers that "The Pride's Inside," or General Motors' claim that Chevys are "The Heartbeat of America." Who is to say that such qualities are not mythical? And whose myth is necessarily superior?

Constituencies may well follow suit as the Satanism incidence suggests. They can construe their own product attributes and act on their evaluation of them. We see no way of limiting either constituencies or managers in making foolish or irrelevant charges about products. The American public, however, has its sophistication. It has been exposed to much advertising and knows the value of many a claim whose validity must be tested by other authorities than advertising. Behind the "market signal" of what it actually purchases lie norms and stories rich with the potential to influence the public's purchases of the future.

Constituency Infringement on Management

Some managers fear that the rise of the many new constituencies will force them to give over the making and carrying out of corporate policy, thereby impairing managers' ability to oversee the efficient production of goods and services. Even if they can sort out and handily deal with the irrational and irrelevant demands made upon them, they may be pushed and pulled in so many directions by competing constituencies with conflicting demands that the managerial realm will be impossibly constricted.

We do not believe that this fear is well founded, for it assumes that managers' functional role is to make and direct policy. Managers play an important, even vital role in policy making; many constituencies, beginning with stockholders and employees, also contribute to it. If those constituencies can contribute to policy making because they have both legal and vested interests in the outcomes of policy, other groups are pressing hard to play a role as well. Managers are not, and seldom have been, the sole source of policy in organizations. Their unique, inescapable functional role is not in policy making, but in coordinating all the various deals made with all the constituencies, to make sure that the inflow of resources (revenues) matches the outflow of resources (total costs). Managers are primarily and basically those who have to balance the different transactions, making sure that no one of them makes such drafts upon resources that the organization is endangered.

The more constituencies and groups with whom managers must deal, the more complex becomes the coordination and the more complex the accounting procedures. But difficulties are what managers are paid to surmount. Their special expertise should be, and usually is, in solving coordination problems of their organizations. That expertise will need to be even more developed in the future than it has been in the past; they must learn to incorporate the wider storylines of constituencies who have declared themselves—or surprisingly discover themselves—to be integral members of the corporate community.

As we have suggested earlier, managerial activity increasingly is becoming more political and less economic (as economists have mistakenly defined that word in recent decades). They will find it worth their effort to recognize explicitly that they are involved in social systems where such values as justice and compassion rate as highly as efficiency and productivity. They may be—or are supposed to be—the experts on efficiency and productivity, but they need the help of their constituents, who will join with them in defining justice and compassion. And managers may even find that their constituencies can assist them in winning more and better efficiency than they have recognized.

MANAGING IN THE NINETIES

Business managers need their constituencies not just as suppliers of raw materials, parts, capital, employees, and consumers, but also as members of the large community that makes possible productive endeavor. If managers desire a legitimate role in that community and a widespread acceptance of their activities, they will have to conduct themselves in accordance with the expanding and changing social definitions of justice, truth, and compassion. Conventional definitions of business and its managers are too confining to describe adequately the roles and contributions of constituencies. Constituencies, too, are a part of the dynamic business system—part of the impressive services industry—led by modern entrepreneurs who have created new markets for goods and services, won popular support for their "products," and demanded a regular place in business decision making. Managers will have to widen their storyline to include the storylines of their constituents, and in doing so may realize more efficiency and acceptance than they have realized in the past.

Constituencies well understand the power and influence of business managers, but are having to learn the economic costs and limits of their own pursuit of equitable treatment. Business firms are not bottomless purses from which funds and resources can be drawn to assure the best and most of life's desires. They must be operated with due regard for efficient production. In a poor world, wasted resources and low productivity are heavy burdens indeed. But efficient production, if unaccompanied by a full and wide participation of all who can contribute, and fair distribution of goods and services, will not satisfy people who must voluntarily cooperate to sustain major institutions like business corporations.

Both parties need each other as checks on the partial visions of each, if some larger vision is to guide us beyond success.

NOTES

1. Peter Drucker, "Corporate Takeovers—What Is To Be Done?" *The Public Interest* (Winter 1986):20. [Reprinted with permission (emphasis in the original).]

2. Ibid., p. 22.

3. See William Foote Whyte and Joseph R. Blasi, "Worker Ownership, Participation and Control: Toward a Theoretical Model," *Policy Sciences* 14 (1982):37–164.; Joseph R. Blasi, *ESOPs* (Cambridge: Harvard University, Belknap Press, 1987); Corey Rosen and Michael Quarrey, "How Well is Employee Ownership Working?" *Harvard Business Review* (September–October 1987):126–32; and

William Foote Whyte, *Making Mondragon: The Growth and Dynamics of the Worker Cooperative Complex* (Ithaca: Cornell University Press, 1988).

4. James R. Norman, "A Hardheaded Takeover by McLouth's Hardhats," *Business Week,* June 6, 1988, p. 90.

5. Nina Andrews, "A Steelworker's Son Will Head McLouth," *New York Times,* June 15, 1988.

6. Aaron Bernstein, "Move Over Boone, Carl, and Irv—Here Comes Labor," *Business Week,* December 14, 1987, p. 124.

7. Justice Rehnquist wrote: "Extension of the individual freedom of conscience decisions to business corporations strains the rationale of [earlier] cases beyond the breaking point. To ascribe to such artificial entities as "intellect" or "mind" for freedom of conscience purposes is to confuse metaphor with reality." *Pacific Gas & Electric Co.,* v. *Public Utilities Commission of California et al.,* No. 84-1044, (slip opinion), decided February 25, 1986, p. 9 of Rehnquist dissent.

8. The story of the two companies is told by Ken Wells, "One Company Treats a Butterfly Gruffly, Another Is Friendlier," *Wall Street Journal,* April 23, 1987. A month later, the *Wall Street Journal* reported a similar story. The managers of the Florida Power & Light Co., like those of some other utility companies, have learned to work with environmentalists, winning their praise and support. They have so channeled the waste hot water from the Turkey Point nuclear plant, on Biscayne Bay, that the area has become a nature preserve for a variety of birds, sea animals, and swamp creatures. See Eric Morgenthaler, "A Florida Utility Wins Naturalists' Praise for Guarding Wildlife," *Wall Street Journal ,* May 7, 1987.

9. The terms of the plan, in part, read, "in order that the conservation and operation of the landfill will not appreciably diminish the likelihood of survival and recovery of the Bay Checkerspot Butterfly, and is intended to minimize alteration of critical habitat so that the landfill is not likely to jeopardize the continued existence of the Bay Checkerspot." The parties to the plan created a trust fund (designated The Kirby Canyon Habitat Conservation Trust Fund) into which Waste Management deposits $50,000 annually to finance habitat acquisition and mitigation or enhancement measures for conservation purposes. The trust was to be administered by a technical advisory committee of representatives of U.S. Fish and Wildlife Service, the City of San Jose, Waste Management of California, Inc., the California Department of Fish and Game, and a private citizen who is qualified as a lepidopterist. The technical advisory committee reviews and recommends modifications in the design, cost, and operation of the conservation agreement. See Dennis D. Murphy, "The Kirby Canyon Conservation Agreement: A Model for the Resolution of Land-use Conflicts Involving Threatened Invertebrates," *Environmental Conservation* 15 (Spring 1988):47–48.

10. Eric Morgenthaler, "A Florida Utility Wins Naturalists' Praise For Guarding Wildlife," *Wall Street Journal,* May 7, 1987.

11. Thomas Moore, "The Fight to Save Tylenol," *Fortune,* November 29, 1982, p. 46.

12. Ibid., p. 46.

13. Afterward Burke recognized the enormous stake he and his company had in keeping the trust of Tylenol customers. "People forget how we built up

such a big and important franchise. It was based on trust. People started taking Tylenol in hospitals or because their doctors recommended it. In other words, they were not well and in a highly emotional state." The *Fortune* writer, however, pointed out that "The contrary view, it can be argued, is that those same people who originally bought Tylenol because they didn't want to take a chance on aspirin's side effects are the last people who would want to take a chance now with the emotionally charged brand name. Ibid., p. 49.

14. As a consequence, on February 6, 1986, a little over three years later, Diane Elsroth, Yonkers, New York, was killed when she took a cyanide-laced capsule. On February 17, 1986, three days after a second bottle of cyanide-laced Tylenol was found, the company announced that it would no longer sell over-the-counter capsule products. See Richard W. Stevenson, "Tylenol Sales In Comeback," *New York Times, July 5, 1986.*

15. For examples of the difficulties some constituencies have had see Linda Williams, "Minorities Find Pacts With Corporations Are Hard to Come By and Hard to Enforce." *Wall Street Journal.* August 23, 1980.

16. In 1947 congressional passage of the Labor-Management Relations (Taft-Hartley) Act allowed either party to a labor-management agreement to sue for enforcement of its terms and provisions. At the time management organizations strongly supported the action and labor unions opposed it.

17. Actually the number 666 is the mark of the beast, the Antichrist, as a reading of *Revelation* 13:18 will show.

18. Margaret Yao, "Procter & Gamble Co. Enlists Aid of Clergy To Blitz Satan Jinx," *Wall Street Journal,* June 25, 1982.

19. Sandra Salmans, "Satanism Tales Spur P. & G. Suit," *Wall Street Journal* July 2, 1982.

20. Sandra Salmans, "P.& G. Drops Logo; Cites Satan Rumors," *New York Times,* April 25,1985, and *Time,* May 6, 1985, p. 63 and Alecia Swasy. "P&G Once Again Has Devil of a Time With Rumors About Moon, Stars Logo;" *Wall Street Journal,* March 26, 1990.

INDEX

The United Library
Garrett-Evangelical/Seabury-Western Seminaries
2121 Sheridan Road
Evanston, IL 60201